VA Mycorrhiza

Editors

Conway Ll. Powell, Ph.D.
Scientist-in-Charge
Horticultural Research Station
Ministry of Agriculture and Fisheries
Pukekohe, New Zealand

D. Joseph Bagyaraj, Ph.D.
Associate Professor
Department of Agricultural Microbiology
University of Agricultural Sciences
GKVK Campus
Bangalore, India

CRC Press, Inc.
Boca Raton, Florida

Library of Congress Cataloging in Publication Data
Main entry under title:

VA mycorrhiza.

Includes bibliography and index.
1. Mycorrhiza. I. Powell, Conway. II. Bagyaraj, D. Joseph.
QK604.V25 1984   589.2′0452482   83-17468
ISBN 0-8493-5694-6

This book represents information obtained from authentic and highly regarded sources. Reprinted material is quoted with permission, and sources are indicated. A wide variety of references are listed. Every reasonable effort has been made to give reliable data and information, but the author and the publisher cannot assume responsibility for the validity of all materials or for the consequences of their use.

All rights reserved. This book, or any parts thereof, may not be reproduced in any form without written consent from the publisher.

Direct all inquiries to CRC Press, Inc., 2000 Corporate Blvd., N.W., Boca Raton, Florida, 33431.

© 1984 by CRC Press, Inc.
Second Printing, 1986

International Standard Book Number 0-8493-5694-6

Library of Congress Card Number 83-17468
Printed in the United States

THE EDITORS

Conway Ll. Powell, Ph.D., is currently Scientist-in-Charge of the Pukekohe Horticultural Research Station, Ministry of Agriculture & Fisheries, Pukekohe, New Zealand. He completed his B.Sc.(Hons) in Botany at the University of Otago (New Zealand) in 1970 and carried out his postgraduate doctoral research on VA mycorrhizae in rushes and sedges at the University of Otago, under Professor G.T.S. Baylis. He graduated with a doctorate in 1973 and was a postdoctoral research fellow at Rothamsted Experimental Station, England, working with Dr. Barbara Mosse until 1975. From 1974 to early 1983, Dr. Powell has been a scientist at Ruakura Soil and Plant Research Station (MAF), Hamilton, New Zealand, before taking charge of Pukekohe HRS.

His research interests have always been on the practical application of VA mycorrhizae in agriculture, earlier in pastoral farming and more recently in horticulture. He has authored and co-authored over 40 scientific papers, reviews, and book chapters. He has been a frequent assessor for NSF and BARD grant applications and referees papers for several international journals. In 1982-1983 he was FAO consultant to Tamil Nadu Agricultural University, Coimbatore, India, charged with initiating a relevant VA mycorrhiza research program under UNDP/FAO funding.

D. Joseph Bagyaraj, Ph.D., is Associate Professor in the Department of Agricultural Microbiology, University of Agricultural Sciences, Bangalore, India.

He graduated in 1961 from Mysore University, India with a B.Sc. degree in Agricultural Sciences. He obtained his M.Sc. (Agriculture) in Plant Pathology from Karnataka University, India in 1963 and his Ph.D. in Agricultural Microbiology in 1971 from the University of Agricultural Sciences. He was the recipient of the University Gold Medal for a grade point average of 4.00 during his Ph.D. program.

Dr. Bagyaraj has been on the faculty of the University of Agricultural Sciences since 1966 where he formerly was Assistant and later Associate Professor of Agricultural Microbiology. In 1975, he spent 6 months in New Zealand and Australia as an UNDP/FAO Fellow studying the taxonomy and ecology of rhizosphere microorganisms. He spent 1976 at the University of California, Riverside, as a Fulbright Fellow working on versicular-arbuscular (VA) mycorrhizae. From 1981-1983 he was at Ruakura Soil and Plant Research Station, Hamilton, New Zealand on the National Research Advisory Council Fellowship working on VA, eriocoid, orchidaceous, and ectomycorrhizae.

He is a Fellow of the Indian Phytopathological Society and a member of the Association of Microbiologists of India, the Indian Society of Soil Biology and Ecology, and the Indian Society of Soil Science. He has served on seveal committees of various professional societies and has been the principal investigator of many research projects. He has also trained many students in agricultural microbiology.

Dr. Bagyaraj has published more than 90 research papers on the ecology of rootzone microflora, biological nitrogen fixation, and mycorrhizae. His major research interest at present is mycorrhizal fungi and their role in crop production.

CONTRIBUTORS

L. K. Abbott, Ph.D.
Soil Science and Plant Nutrition
School of Agriculture
University of Western Australia
Nedlands, Australia

D. Joseph Bagyaraj, Ph.D.
Department of Agricultural Microbiology
University of Agricultural Sciences
GKVK Campus
Bangalore, India

Paola Bonfante-Fasolo, Ph.D.
Centro di Studio sulla Micologia
  del Terreno del CNR
Turin, Italy

Karen M. Cooper, Ph.D.
Division of Horticulture and Processing
Department of Scientific and
  Industrial Research
Auckland, New Zealand

I. R. Hall, Ph.D.
Invermay Agricultural Research Centre
Ministry of Agriculture & Fisheries
Mosgiel, New Zealand

Christine M. Hepper, Ph.D.
Department of Soil Microbiology
Rothamsted Experimental Station
Harpenden, England

Barbara Daniels Hetrick, Ph.D.
Department of Plant Pathology
Kansas State University
Manhattan, Kansas

John A. Menge, Ph.D.
Department of Plant Pathology
University of California
Riverside, California

Conway Ll. Powell, Ph.D.
Scientist-in-Charge
Horticultural Research Station
Ministry of Agriculture & Fisheries
Pukekohe, New Zealand

A. D. Robson, Ph.D.
Soil Science and Plant Nutrition
School of Agriculture
University of Western Australia
Nedlands, Australia

TABLE OF CONTENTS

Chapter 1
VA Mycorrhizae: Why All the Interest? 1
Conway Ll. Powell and D. Joseph Bagyaraj

Chapter 2
Anatomy and Morphology of VA Mycorrhizae 5
Paola Bonfante-Fasolo

Chapter 3
Ecology of VA Mycorrhizal Fungi 35
Barbara A. Daniels Hetrick

Chapter 4
Taxonomy of VA Mycorrhizal Fungi 57
I. R. Hall

Chapter 5
Isolation and Culture of VA Mycorrhizal (VAM) Fungi 95
Christine M. Hepper

Chapter 6
The Effect of Mycorrhizae on Plant Growth 113
L. K. Abbott and A. D. Robson

Chapter 7
Biological Interactions with VA Mycorrhizal Fungi 131
D. Joseph Bagyaraj

Chapter 8
Physiology of VA Mycorrhizal Associations 155
Karen M. Cooper

Chapter 9
Inoculum Production 187
John A. Menge

Chapter 10
Field Inoculation with VA Mycorrhizal Fungi 205
Conway Ll. Powell

Index 223

Chapter 1

VA MYCORRHIZAE: WHY ALL THE INTEREST?

Conway Ll. Powell and D. Joseph Bagyaraj

Frank[1] coined the term "mycorrhizae" to describe the symbiotic association of plant roots and fungi in 1885. Mycorrhiza literally means "fungus root", and by far the most common mycorrhizal association is the vesicular-arbuscular (VA) type, which produces fungal structures (vesicles and arbuscules) in the cortex region of the root. The VA mycorrhizal association is found in most plant families so far examined, although it may be rare or absent in families such as Cruciferae, Chenopodiaceae, Caryophyllaceae and Cyperaceae.[2] In addition to the widespread distribution of VA mycorrhizae (VAM) throughout the plant kingdom, the association is geographically ubiquitous and occurs in plants growing in arctic, temperate, and tropical regions. VAM occur over a broad ecological range, from aquatic to desert environments.[3] VAM are formed by nonseptate phycomycetous fungi belonging to the genera *Glomus, Gigaspora, Acaulospora,* and *Sclerocystis* in the family Endogonaceae. These fungi are obligate symbionts and have not been cultured on nutrient media. VA endophytes are not host-specific, although evidence is growing that certain endophytes may form preferential associations with certain host plants.[4]

Early work with VA mycorrhiza was not at all in proportion to their abundance and mostly dealt with their anatomy and occurrence. Until fairly recently VAM were virtually ignored by most soil and plant scientists. Ever since Baylis,[5] Gerdemann,[6] and Mosse[7] in the last two decades showed VAM could increase phosphorus uptake from soil by plants, researchers have been trying to extend and manipulate this "phosphorus-sparing" effect of mycorrhizae in agriculture.[8,9] The possible role of mycorrhizae in the biological control of root pathogens, biological nitrogen fixation, hormone production, and drought resistance attracted workers from different disciplines. There has been a rapid and continued increase in the numbers of scientists working and publishing on the biology of VAM; one only has to watch the ever-increasing number of VAM researchers attending successive North American conferences on mycorrhizae to witness this. As yet however, we as mycorrhiza workers are still unable to predict with confidence any sort of agronomically significant mycorrhiza growth response in field sown crops. It can even be argued (see Chapter 10) that most VAM workers are avoiding field inoculation experiments and are sticking to "academically safer" experiments in sterilized soil in pot trials.

Nevertheless, there has been great progress in the understanding of the VAM symbiosis in the last few years and it is our hope that in this book, the first devoted solely to the study of VAM, the contributors have been able to summarize the present state of knowledge of VAM and highlight the direction for future research. It is quite timely that one should review progress at this point as research into VAM seems to be leading in two main directions. On the one hand, there is continued emphasis on the measurement and prediction of plant growth responses (Chapter 6), inoculum production (Chapter 9), and field inoculation (Chapter 10). VA mycorrhizal inoculation should soon be accepted as a part of standard nursery practice in raising containerized seedlings or cuttings in sterilized mix, or unsterilized mix with low or inefficient strains of mycorrhizal fungi. This in itself is a very valuable "use" and return on our collective efforts in VAM research.

Additionally, there has been great progress in our basic understanding of the VAM symbiosis especially in the areas of anatomy (Chapter 2); taxonomy (Chapter 4); phosphorus uptake physiology, carbon use, water relations, and hormone production

(Chapter 8); ecology (Chapter 3); axenic culture (Chapter 5); and biological interactions (Chapter 7). As usual, our past discoveries often pose more questions than answers.

One of the major impediments to the use of, and research into, VAM fungi is our current inability to grow the organisms in pure culture. This is most strongly felt in physiology experiments where any mycorrhizal effect (on carbon use, P uptake, hormone production, etc.) can only be judged by subtracting the level of the compound in question in nonmycorrhizal plants from that in similar-sized mycorrhizal plants. It is also a problem that mycorrhizal inoculum for most experiments has to be raised in sterilized potting medium (of soil, sand, vermiculite, or whatever) with a suitable host plant. It would be much more convenient to produce inoculum in broth culture.

In our future field trials, we will have to show that inoculation with VAM fungi (whether in the nursery or the seed bed in the field) leads to predictable, persistent, and economically viable plant growth responses and/or reductions in phosphorus fertilizer use. A tall order perhaps? Yes, and this emphasizes that VAM inoculation must be judged, as with all other plant husbandry practices, on its biological significance and economic usefulness.

There is also a great need for continued basic research on the mechanisms of plant growth responses (and depressions), the extent of extramatrical fungal growth and consequent ion uptake and transfer, the cause of differences in fungal "efficiencies", and many more problems. It is vital therefore, in these times of shrinking science budgets, that the preoccupation that science must pay for itself immediately does not prevent our working on basic research. History has shown time and again how seemingly academic and unrelated research has been the springboard for great advances in applied research in later years.

Many of the best leads in mycorrhiza research, as in all research, can happen by accident or start off as a "way-out" idea. As an example, we had a colleague come to us at Ruakura Soil and Plant Research Station complaining that in a 600 m² plot of soil which he had previously sterilized with methyl bromide, he had very poor growth of an asparagus crop, while around the outside of the plot in unsterilized soil, the asparagus was growing very vigorously. It was in fact a classical mycorrhiza response which we were able to reproduce by inoculation of some of the stunted plants with mycorrhizal soil inoculum. The resultant 17-fold increase in plant dry matter production even at P fertilizer rates as high as 500 kg P/ha far exceeded any response we would have predicted. These "demonstration" plots have been more effective than any book or research paper at convincing those who saw them of the role and usefulness of VAM fungi in plant production.

In recent years the importance of VAM fungi for plant growth in most soils has been accepted and acknowledged by soil chemists, agronomists, horticulturists, and farmers. The following chapters in this book will describe in much greater detail these recent advances in VAM, and speculate on our prospects for manipulating the VAM symbiosis to our advantage in the future.

REFERENCES

1. Frank, A. B., Über die auf Wurzelsymbiose beruhende Eranahrung gewisser Baume durch unterirdische Pilze, *Ber. Dtsch. Bot. Ges.,* 3, 128, 1885.
2. Hirrel, M. C., Mehravaran, H., and Gerdemann, J. W., Vesicular-arbuscular mycorrhizae in the Chenopodiaceae and Cruciferae: do they occur? *Can. J. Bot.,* 56, 2813, 1978.
3. Mosse, B., Stribley, D. P., and Le Tacon, F., Ecology of mycorrhizae and mycorrhizal fungi, *Adv. Microb. Ecol.,* 5, 137, 1981.

4. Mosse, B., The role of mycorrhiza in legume nutrition on marginal soils, in *Exploiting the Legume-Rhizobium Symbiosis in Tropical Agriculture,* Vincent, J. M., Whitney, A. S., and Bose, J., Eds., University of Hawaii College of Tropical Agriculture Misc. Publ. No. 145, 1977, 175.
5. Baylis, G. T. S., Experiments on the ecological significance of phycomycetous mycorrhizas, *New Phytol.,* 66, 231, 1967.
6. Gerdemann, J. W., The effects of mycorrhiza on the growth of maize, *Mycologia,* 56, 342, 1964.
7. Mosse, B., Growth and chemical composition of mycorrhizal and non-mycorrhizal apples, *Nature (London),* 179, 922, 1957.
8. Hayman, D. S., Endomycorrhizas, in *Interactions Between Non-Pathogenic Soil Microorganisms and Plants,* Dommergues, Y. R. and Krupa, S. V., Eds., Elsevier, Amsterdam, 1978, 400.
9. Tinker, P. B., Mycorrhizas: the present position, in *Whither Soil Research,* Trans. 12th Int. Congr. Soil Sci., New Delhi, 1982, 150.

Chapter 2

ANATOMY AND MORPHOLOGY OF VA MYCORRHIZAE

Paola Bonfante-Fasolo

TABLE OF CONTENTS

I. Introduction 6

II. Fungal Infection in Angiosperms 6
    A. The Extramatrical Phase 8
    B. The Intraradical Phase 9
        1. Intracellular Hyphae in the Outer Cortical Layers of the Root .. 9
        2. Intercellular Hyphae 11
        3. Arbuscules 14
        4. Vesicles 17
    C. Host Response to Fungal Colonization 19
    D. Host-Fungus Relationships 20

III. Fungal Infection in Bryophytes 23

IV. Fungal Infection in Pteridophytes 24

V. Fungal Infection in Gymnosperms 26

VI. Conclusions 26

VII. Acknowledgments 29

References 29

I. INTRODUCTION

Nearly 100 years have passed since the most common endomycorrhiza in nature, the one now known as vesicular-arbuscular mycorrhiza (VAM), was first identified.

Traditionally, Frank[1] is considered the first author to have distinguished between ecto- and endotrophic mycorrhizae. Ectotrophic mycorrhizae (now usually called ectomycorrhizae) are characterized by the presence of fungal mantle or sheath covering the root surface, while endotrophic mycorrhizae (now usually called endomycorrhizae) do not possess such a fungal sheath and have extensive intracellular fungal penetrations. Frank[1] reached these conclusions early in 1887 and was able to formulate the distinction between ecto- and endotrophic mycorrhizae on the basis of his detailed studies on Cupuliferae for the former and on ericales and orchids for the latter.

However, it was not until the works of Schlicht (1889),[2] Janse (1896-97),[3] Dangeard (1900),[4] Petri (1903),[5] and Gallaud (1905)[6] that endomycorrhizal fungi found in other than ericales and orchids (i.e., those now called VAM) were described in detail. This early research was followed by many other important studies, including those of Peyronel (1924),[7] McLennan (1926),[8] and Butler (1939).[9] They established the following fundamental points which became the basis of our knowledge of VAM up to the 1970s, i.e., (1) The vast range of plants is colonized by VAM fungi; Schlicht[2] described 71 species of herbaceous angiosperms, Janse[3] studied tropical plants from Java (bryophytes, gymnosperms, woody dicotyledons, and pteridophytes), while Peyronel[7] listed about 150 mycorrhizal species from the Alpine areas. (2) During its complex life cycle in contact with the root, the fungal symbiont gives rise to different structures: extramatrical hyphae; intracellular hyphae; globose or oval terminal swellings called "vesicles"; intracellular structures with a form like small trees, or "arbuscules", as Gallaud[6] named them; and irregular clumps (when the fine branches of the arbuscules are no longer visible) termed "sporangioles" by Janse.[3] These different fungal structures illustrated in Figure 1 have different positions in the root.[8] (3) In nature the same root can be infected by two different endophytes at the same time. This hypothesis of a double infection was put forward by Peyronel,[7] who, on the basis of the endophyte's morphological characteristics, recognized a septate mycelium (one that could be classified as a *Rhizoctonia* type fungus, also responsible for orchid mycorrhizae) as well as another endophyte, similar to a phycomycete and responsible for VA infections. Following this clear distinction, VAM have since been identified as a type in themselves, distinct from other endomycorrhizae. Moreover, for the first time, Peyronel[7] claimed that the endophyte responsible for VAM formation belonged to the Endogonaceae family. (4) All the above-listed authors recognized mycorrhiza as a mutualistic, symbiotic association, since the host cell seemed to digest the fungus, thus obtaining nutrients.

Many of these definitions have survived the passage of time, others have been modified, and updated with the introduction both of new equipment that have made it possible to understand the physiological characteristics of a VAM infection, and of new techniques, that have allowed more sophisticated morphological analyses.

The aim of this chapter is to give (1) an anatomical and cytological description of VAM, as they are in angiosperm roots, for most of the available data concern plants from this phylum, and (2) a description and comparison of colonization processes occurring in bryophytes, pteridophytes and gymnosperms, which are traditionally said to form VAM infections.

II. FUNGAL INFECTION IN ANGIOSPERMS

In most cases the anatomical and cytological changes caused by VAM fungi in the

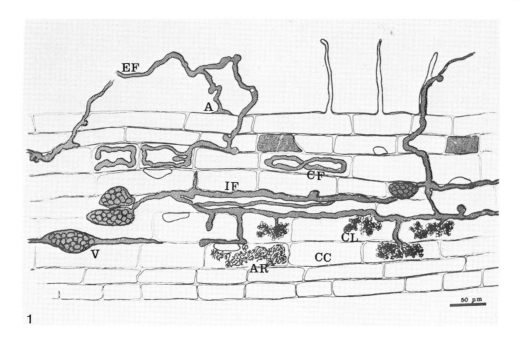

FIGURE 1. The drawing summarizes all the different phases of a typical VAM infection in a root. The endophyte forms an appressorium on the root surface, or penetrates the root by root hair. In the outer layer of the parenchymal cortex it forms intracellular coils; while in the inner layer it creates intercellular hyphae and arbuscules. Vesicles are scattered throughout the root. (EF) Extramatrical hyphae; (A) appressorium; (CF) coiled hyphae; (IF) intercellular hyphae; (CL) fungal clump; (V) vesicle; (CC) cortical cell; (AR) arbuscle.

host do not induce root alterations recognizable with the naked eye. Only in plants such as onions or other Liliaceae and maize, can mycorrhizal roots be recognized by their yellow color as compared to the uninfected white ones; according to Gerdemann,[10] the yellow color is usually soluble in water. Another characteristic which can sometimes be an indication whether VAM fungi are present is the size and morphology of the roots, and in particular the development of root hairs. For example, primitive angiosperm roots, typified by those of the order Magnoliales, are coarsely branched with rootlets (diameter generally more than 0.5 mm).[11] The latter usually lack root hairs and are heavily mycorrhizal, while graminoid roots have fine rootlets (less than 0.1 mm in diameter), which are covered with long root hairs and become mycorrhizal only when phosphorus availability is extremely low. Recent observations on tropical plants in Brazil and on some Liliaceae in Scotland have confirmed the correlation between the frequency and length of root hairs and the ability of the plant to grow without mycorrhizal fungi.[12,13]

The observation that plant species without fine root systems and intense root hair development are frequently more infected and more mycorrhizal-dependent, could reflect the fact that the symbiotic fungus replaces certain root hair functions such as mineral nutrient absorption.

Light microscopy is needed to grasp the complexity of the plant/mycorrhizal fungus interaction. Mycorrhizal infections occur solely in the epidermis or esoderm and the cortical parenchyma of roots which present a primary structure. The infection does not penetrate the endodermis and is therefore not present in the central vascular cylinder, nor is it present in the meristematic regions. As was observed by pioneers in this field, the infection develops in stages: (1) an extramatrical phase with extramatrical hyphae and external vesicles or spores scattered in the surrounding soil, and (2) an intraradical phase with intracellular unbranched hyphae, intercellular hyphae, branched intracellular hyphae (arbuscules), and vesicles.

A. The Extramatrical Phase

The development and spread of the extramatrical phase of VAM fungi differs greatly according to the type of soil, plants, and fungi. In some cases fungal growth in soil may be 80 to 134 times the length of the subtending mycorrhizal root.[14,15] In other cases it appears less developed. Morphologically, the extramatrical mycelium is continuous with the intraradical one, thus forming one infection unit.

The morphology of the extramatrical mycelium may vary considerably. Nicolson,[16] working on Gramineae, observed hyphae ranging from 2 to 27 μm in diameter. Hyphae can be classified as thick- or thin-walled; thick-walled hyphae, yellowish in color, have typical unilateral angular projections and are nonseptate.[7,9,16,17] However, many septate thin-walled hyphae develop as lateral branches from the thick-walled hyphae. Extramatrical fungal hyphae with angular projections and an irregular orientation have been observed in the scanning electron microscope (SEM) associated with *Vitis vinifera* roots infected by *Glomus fasciculatum;* the hypha bifurcates giving rise to two infecting branches (Figure 2). Powell[18] observed that laminate and yellow vacuolate spores germinate giving rise to straight thick-walled, aseptate hyphae. However close to the roots, they branch, forming fan-like structures of septate preinfection hyphae, which grow toward the root. Powell[18] has suggested that the presence of these structures is limited to the first phases of the host-fungus interaction and are seldom found in well-established mycorrhizal infections.

Vesicle-like, globose to obovate spores are regularly associated to the extramatrical mycelium. They are nearly always borne terminally or on short lateral branches. The external vesicles largely vary from 20 to 150 μm in diameter, and are thick-walled, with a dense cytoplasmic content rich in oil globules. With age they become vacuolated.[16,17]

Information on the ultrastructural organization of the extramatrical phase is still scarce, but according to recent data (limited to the external hyphae) they possess a cytoplasm with nuclei, mitochondria, and an endoplasmic reticulum rich in ribosomes.[19] Frequently there is a well-developed vacuolar system wherein the smaller vacuoles contain electron-dense granulese much like those resembling polyphosphate granules in intraradical hyphae (Figure 3). Most of the extramatrical hyphae have a highly vacuolated cytoplasm.

In *Glomus fasciculatum*, the hyphal wall has a rather complex two-layer structure, 0.3 to 0.5-μm thick. The outer layer is electron transparent and fibrillar at higher magnification, but its ultrastructure is not constant since it tends to separate from the inner underlying layer to swell and in the end it appears as a clear space limited by a sloughing line. The inner electron-dense layer is 0.15- to 0.40-μm thick and organized in alternate, light and dark, stacked lamellae. Ultrastructural and cytochemical reactions have shown that the wall of extraradical hyphae are formed by proteins and polysaccharides, some of which are alkali-insoluble and probably chitin[19] (Figure 4).

When in contact with the root surface, the extramatrical hypha swells apically and increases in size forming a more or less pronounced appressorium-like structure (Figure 5 and 6). Fan-like appressoria have been observed by Gianinazzi-Pearson et al.[20] on the root surface of wild raspberry colonized by the fine endophyte, identified as *Glomus tenue* according to Hall.[21] As Mosse[17] showed with her experiments on apple trees and strawberries, the extramatrical mycelium can give rise to a variable number of entry points in the root, according to the season and to the age of the root (from 2.6 to 21.1 entry points per millimeters of root length). In *Ammophyla arenaria*, only six penetrating hyphae per centimeter of root have been reported.[16] As has already been mentioned, hyphae do not penetrate meristems; in root organ cultures too, the infection takes place at 0.5 to 1 cm behind the root tip.[22]

Hyphae can penetrate the root in more than one way:

1. The infecting branch directly penetrates the wall of a root hair or of an epidermal or exodermal (in older roots) cell by means of a mechanism which is not yet fully understood, but which must involve mechanical and/or enzymatic actions. Light microscopy observations on *Gigaspora margarita* and *Glomus monosporum* illustrate how the penetrating hypha greatly reduces its diameter during wall penetration.[23,24] The same feature can be seen in a *Glomus* sp. infecting *Ornithogalum umbellatum* roots (Figure 6), where SEM analysis shows the hypha returning to its previous diameter immediately after penetration and forming a loop (Figure 7).
2. The infecting hyphal branch originating from the appressorium can pass through the empty spaces between the sloughing cells in the outer layer of the root and then enter the first intact layer of cortical cells, forming an intracellular loop (Figure 8).[25]
3. The infecting hypha penetrates the root between the epidermal cells and spreads intercellularly from the entry point.[26-27]

The different ways of root penetration seem to be linked to the anatomy of the plant root, depending most probably on the wall-thickening pattern of the outer cells.

B. The Intraradical Phase

When fungus is inside the root, the way in which it spreads varies, depending on the plant and fungus involved. Recently it has been claimed that it is possible to distinguish some of the infecting fungi on the basis of their specific infection pattern.[24,28,29] Although this method is of unquestionable value to diagnose infections, it has so far been tested on too limited a number of plants to be applied generally. Fungal structures will be described here in relation to their position within the root and when possible the characteristics of the endophyte will also be given.

1. Intracellular Hyphae in the Outer Cortical Layers of the Root

The outer cortical root layers are often colonized by intracellular hyphae, characterized by a linear or more often a looped arrangement, without any signs of branching. The infecting hypha of the VAM fungi can form intracellular coils in the first cell to be infected, with similar coils being subsequently formed in neighboring cells (Figure 9). Alternatively, the infecting branch penetrates the first cell without coiling and becomes organized in coils only in neighboring cells. Coil formation is common to the mycorrhizal roots of many plants including, for example, the Gramineae, grapevine, Star of Bethlehem, and wild raspberry.[7,20,30-32] Coils have been formed by *Glomus epigaeum* and *G. fasciculatum* in *Vitis vinifera* (Figure 10), by *Gigaspora gigantea* and *Acaulospora laevis* in subterranean clover, by *Gigaspora margarita* in *Trachymene anisocarpa* and *Leptospermum juniperinum*.[23,24]

On the contrary, *Glomus fasciculatum* and *G. monosporum* do not form coils in subterranean clover roots and neither do *G. mosseae* in onion roots nor *G. tenue* in wild raspberry roots.[20,24,26] According to Abbott,[29] the size of the intracellular unbranched hyphae (3 to 7 μm) depends on the type of fungus involved, while the amount and the behavior of the hyphae are probably influenced by the host.

SEM of cortical cells in *Liriodendron tulipifera* show that the coiled hyphae can fill a large part of the cell lumen.[33,34] Transmission electron microscopy (TEM) of host organelle distribution shows that host nuclei always appear to be in close association with hyphae. The latter contain small nuclei, cytoplasm with vacuoles enclosing electron-dense granules, lipid droplets, glycogen granules, and a thick osmiophilic wall (Figures 10 and 11). Throughout the intracellular colonization, the host plasmalemma

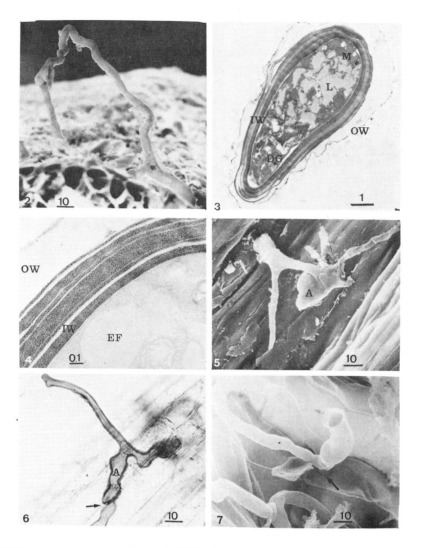

FIGURE 2. Scanning electron micrograph (SEM) of the extramatrical mycelium of *Glomus fasciculatum* on the root surface of *Vitis vinifera*. The fungus shows a wave-like pattern. It bifurcates at the point of contact with the root. Bar = 10 μm. FIGURE 3. Transmission electron micrograph (TEM) of an extramatrical hypha of *Glomus fasciculatum*. The cytoplasm has small mitochondria, large lipid globules, and electron-dense granules inside vacuoles. The wall consists of two clearly separated layers, i.e., one outer electron transparent layer and an electron-dense multilamellate inner layer. (M) Fungal mitochondrion; (L) fungal lipids; (IW) fungal inner wall; (OW) outer fungal wall; (DG) dense granule. Bar = 1 μm. FIGURE 4. Higher magnification of the wall of an extramatrical hypha of *Glomus fasciculatum* (TEM). After a silver proteinate test for localization of polysaccharides, the inner layer (IW) shows a clear lamellar structure. The cytochemical reaction together with chemical extractions suggests that the stained lamellae consists of alkali-insoluble polysaccharides. (OW) Outer fungal wall; (IW) fungal inner wall; (EF) extramatrical hyphae. Bar = 0.1 μm. FIGURE 5. A large appressorium of a *Glomus* sp. adhering to the outer root cells of *Vitis vinifera* giving origin to different penetrating hyphae (SEM). (A) Appressorium. Bar = 10 μm. FIGURE 6. Light microscope micrograph (LM) of an appressorium (A) of *Glomus* sp. on the root surface of *Ornithogalum umbellatum*. The extramatrical hypha bifurcates, giving rise to two larger branches. The infecting branch directly penetrates an epidermal cell, showing constriction of its diameter at the point of penetration (arrow). Bar = 10 μm. FIGURE 7. Intraradical Hypha of *Glomus* sp. immediately after the penetration of an epidermal cell of *Ornithogalum umbellatum*. A clear halo (arrow) can be seen around the hypha as well as the enlargement of the fungal diameter. The fungus immediately organizes a coil (SEM).

and tonoplast are invaginated around the fungus. Host plasmalemma and fungal walls are always separated by an osmiophilic fibrillar layer of matrix material (Figure 12), which is continuous with the host wall and morphologically similar.[30,32,34]

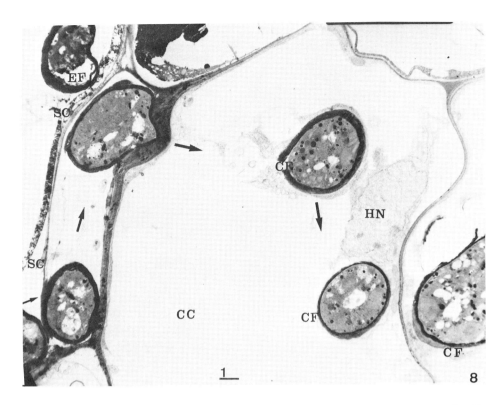

FIGURE 8. *Vitis vinifera* x *Glomus fasciculatum*. The extramatrical hypha passes between the sloughing cells, then enters the first intact layer of cortical cells, forming a loop. Arrows indicate the direction of penetration (TEM). (EF) Extramatrical hyphae; (SC) host sloughing cells; (CF) coiled hyphae; (HN) host nucleus; (CC) cortical cell. Bar = 1 μm.

Intracellular hyphae with coiled or linear orientation can display signs of deterioration, characterized by degeneration of the fungal cytoplasm and subsequently by hyphal wall collapse.[34] On passing from one cell layer into the next, the fungus penetrates another cell forming a new coil (Figure 10). Some authors have observed that this cell-to-cell passage of the fungus caused hyphal constriction at the point of penetration across the host cell walls, after which hyphae revert to their previous size (Figure 10).[32,34] During the cell-to-cell passage from the outer cortical layers to the inner ones, host cell plasmalemma appears to be continuous. This suggests a rather complex sequence and, in agreement with Grippiolo,[32] can be summarized as follows: (1) as the intracellular fungus reaches the periphery of the cell, the host plasmalemma which is invaginated around the fungus and limiting the interfacial zone becomes continuous with that adhering to the host cell wall; (2) the matrix material becomes continuous with the primary wall; and (3) the fungus passes through this and the middle lamella by both a mechanical and probably an enzymatic mechanism. The fungus then penetrates the underlying cell wall and ends up by causing the invagination of the plasmalemma in the next cell (Figure 13).

On reaching the middle area of the cortical parenchyma the fungus becomes intercellular due to a mechanism which partly resembles that suggested above. The infection subsequently spreads along the root by intercellular hyphae running parallel to the root axis.

2. Intercellular Hyphae

Intercellular hyphae produced by coils or directly by the penetrating hyphal branches are usually found in the intermediate layers of the cortical parenchyma. Their diameter

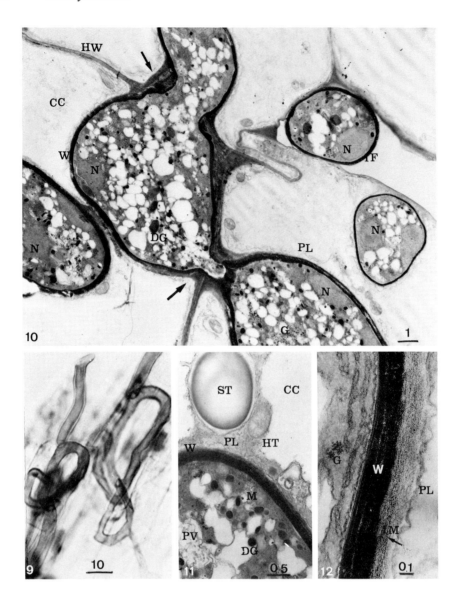

FIGURE 9. Coil of a vesicular-arbuscular fungus in neighboring cells of the outer cortical layer in an *Ornithogalum umbellatum* root (LM). Bar = 10 µm. FIGURE 10. Coiled hyphae of *Glomus fasciculatum* passing from a cortical cell of *Vitis vinifera* to two underlying ones. Hyphal constriction at the point of penetration is evident (arrow). Fungal wall is strongly electron dense. Small nuclei can be observed as well as electron-dense granules inside vacuoles. Host plasmalemma surrounds all the fungal hyphae (TEM). (HW) Host wall; (CC) cortical cell; (W) fungal wall; (N) fungal nucleus; (IF) intercellular hyphae; (DG) dense granule; (PL) host plasmalemma; (G) glycogen. Bar = 1 µm. FIGURE 11. Detail of the coil ultrastructure: electron-dense granules, glycogen deposits, mitochondria, and polyvesicular bodies. Host cytoplasm is limited by the host plasmalemma and host tonoplast. A large starch grain is also present (TEM). (ST) Starch; (CC) cortical cell; (W) fungal wall; (PL) host plasmalemma; (HT) host tonoplast; (M) fungal mitochondrion; (PV) fungal polyvesicular bodies; (DG) dense granule. Bar = 0.5 µm. FIGURE 12. Higher magnification of the contact area between the fungal wall and the host plasmalemma after a cytochemical reaction for polysaccharide localization. Matrix material shows a fine deposition of silver grains while deposition is very heavy on the fungal wall (TEM). (G) Glycogen; (W) fungal wall; (PL) host plasmalemma; (M) fungal mitochondrion. Bar = 0.1 µm.

ranges from 2 to 6 µm, with the exception of those of the fine endophytes, which are smaller (less than 2 µm). These hyphae dilate the intercellular spaces and are sometimes in bundles of three or four (Figure 16). They run in the parenchyma for considerable

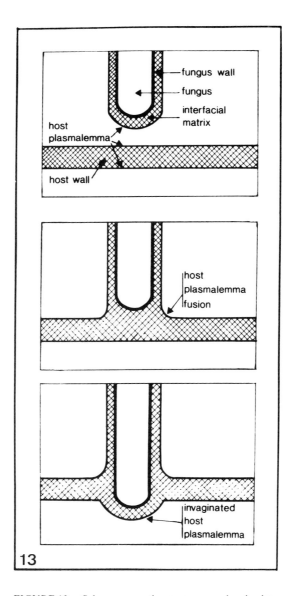

FIGURE 13. Scheme suggesting a sequence, showing how host plasmalemma appears continuous during the fungal passage from cell to cell (see text for a complete explanation).

distances (up to several millimeters) and sometimes have a wavy form as they follow the outline of the host plant cells.[31]

They often have intermittent projections and are at times swollen.[35] Abbott and Robson[35] have described H-connections among hyphae running parallel; a perpendicular branch arising from a longitudinal hypha probably divides in two directions forming yet another long hypha parallel to the original. Y-junctions are also to be found; they are formed by a longitudinal hypha branching to give two that are parallel. These features seem to be fungus-specific; the junctions have been observed in roots colonized by certain endophytes (e.g., *Glomus* spp.) but appear to be absent in other cases (e.g., *Gigaspora* and *Acaulospora*). The intercellular hyphae are septate when empty, but septa are rare in the active hyphae. The above characteristics can be easily observed in mycelium that has been extracted from a mycorrhizal root submitted to an enzymatic digestion, which digests the host cell walls.[36] In this way, well-defined Y- and H-junctions, septa, and sometimes anastomose-like bridges were observed in a *Glomus* sp. in *Ornithogalum umbellatum* roots (Figures 14 and 15).

TEM of intercellular hyphae (Figures 10, 16, 17, and 21) shows that they contain small nuclei (0.5 to 1.5 μm), mitochondria, globose electron-dense granules closely limited by membranes, electron-dense granules within larger vacuoles, α-glycogen granules, and lipid droplets.[25,26,30] The latter are common in both intercellular hyphae and in arbuscular branches, as Nemec[37] demonstrated using histochemical methods. Triglycerides appear to be the main components of the lipids occurring in the fungal structures of *Glomus etunicatum* in rough lemon roots. According to Harley[38] and Cox et al.,[39] lipids could be synthesized by the fungal endophyte as an alternative storage sink of transferred host photosynthate.

In more mature hyphae, the vacuolar system (formed by a large number of small vacuoles, each one separated from the next by tonoplast) becomes dominant and shows a characteristic reticulate pattern. In such hyphae, walls are constantly osmiophilic but vary in thickness. Many VAM endophytes have walls with only one layer and are rich in polysaccharides (Figure 17) and proteins.[19] Two layers have been observed in the fine endophyte *G. tenue*.[27]

Since the fungus spreads by dilating the host intercellular spaces, its wall can come into direct contact with the primary host cell wall and sometimes traces of the middle lamella may remain (Figure 16).[27,30,40] This morphological picture suggests that the fungus penetrates by an enzymatic mechanism, for example by producing hydrolases such as pectinase. However, data concerning such enzyme activities in VAM fungi are not yet available.

Intercellular hyphae as well as arbuscular trunks can sometimes harbor bacterium-like organelles (BLOs)[25,41] (Figure 18). The organisms can be free in the fungal protoplasts or contained within the fungal vacuoles (for a complete list of the papers on the subject, see Chapter 7).

In the various fungi so far examined *(Glomus caledonicum, G. fasciculatum, G. epigaeum, G. mosseae and Gigaspora margarita)*, BLOs vary in shape and size. One of these, BLO/1 is characterized by an irregular coccal shape, by abundant ribosome-like particles, an apparently homogenous cell wall and evident divisions, suggesting a possible cytobiotic state.[42] However, the bacterial nature and the role of these organelles in the mycorrhizal symbiosis remains debatable, until they can be cultured or physically extracted from the fungi.[42]

3. Arbuscules

In the inner layers of the cortical parenchyma, intercellular hyphae penetrate the cortical cells giving rise to a complex hyphal branching system, like "small bushes" which are called arbuscules.[43] The arbuscule (Figure 19) is the most significant structure in the VAM complex, in particular from a functional viewpoint; in fact, recent interpretations agree that the arbuscule is the preferential site for fungus/plant metabolite exchanges.[25,39]

The presence of arbuscules is a *sine qua non* to identify a VAM infection in a root. All endophyte fungi, recognized by Gerdemann and Trappe[44] and belonging to the genera *Glomus, Gigaspora, Acaulospora,* and *Sclerocystis,* form arbuscules. The initial classification of VAM, once known simply as endotrophic mycorrhizae, was also carried out on the basis of arbuscules being present and according to the type of degeneration they underwent.[6,45] According to these classic descriptions, branches were said to swell at their ends creating roundish bodies called "sporangioles" which represented the senescent phase of the infection. The whole arbuscule-sporangiole complex was said to end by disintegrating in the cells, digested by the host plant.

It was only with the use of electron microscopy that development and subsequent

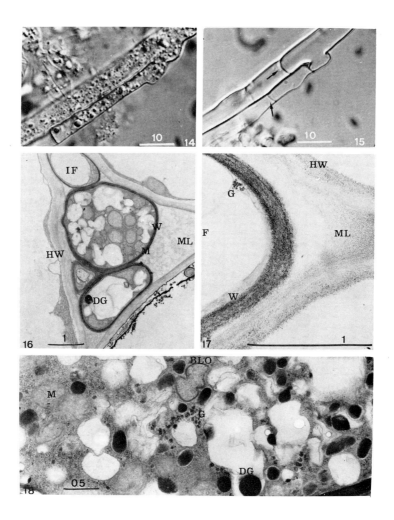

FIGURE 14. Intercellular hyphae of a *Glomus* sp. isolated from segments of mycorrhizal roots after enzyme maceration. Y-junctions show active intercellular hypha forming a parallel branch. Cytoplasmic structures are highlighted using Nomarski differential-interference contrast (LM). Bar = 10 μm. FIGURE 15. As in Figure 14, Y-junction of an empty intercellular hypha with frequent septa. Nomarski differential-interference contrast (LM). Bar = 10 μm. FIGURE 16. Bundles of four intercellular hyphae of a *Glomus* sp. in a naturally infected grapevine root. One can see electron-dense globules, mitochondria (M), and vacuoles in the fungal protoplasm limited by an osmiophilic wall. In the dilated intercellular spaces, remnants of the middle lamella (ML) are detectable (TEM). (IF) Intercellular hyphae; (W) fungal wall; (HW) host wall; (DG) dense granule. Bar = 1 μm. FIGURE 17. Higher magnification of the contact between an intercellular hypha and a host wall (HW). The fungal wall (FW) is osmiophilic and strongly marked by silver grains for polysaccharide localization. The deposition of silver grains is also evident on the primary host wall as well as on the middle lamella (ML). An electron-transparent space is evident between the fungal apex and the loose middle lamella. Glycogen (G) particles are strongly stained in the fungal cytoplasm (TEM). (F) Fungus. Bar = 1 μm. FIGURE 18. Intercellular hyphae of *Glomus fasciculatum* colonizing grapevine roots showing a bacterium-like organelle (BLO), free in the fungal protoplasm. The BLO has an electron-dense wall and a protoplasm rich in small granules. In addition, fungal protoplasm contains electron-dense globules inside vacuoles or closely limited by membrane, abundant glycogen (G) granules, and numerous mitochondria (M) (TEM). (DG) Dense granules. Bar = 0.5 μm.

degeneration of the arbuscule were fully understood (Figures 19-21). Many research workers have since provided very similar descriptions of this process in different plants colonized by different fungi.[26,27,30,40,46-49]

The hyphal branches arising from the intercellular hyphae enter host cells by stretching the host wall and by causing the host plasmalemma to invaginate. The arbuscule

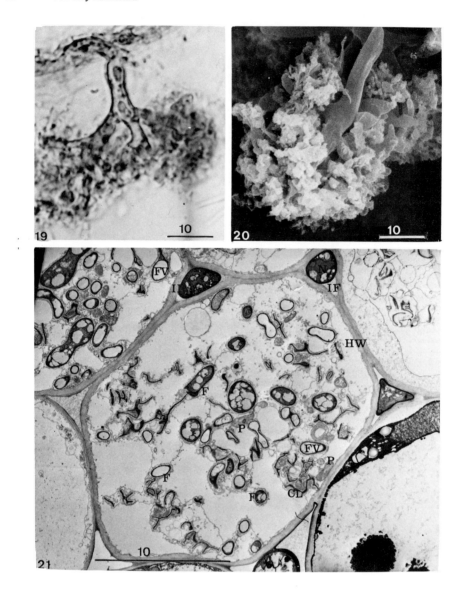

FIGURES 19—21. Arbuscules of a *Glomus* sp. in *Ornithogalum umbellatum* (19-20) and in *Vitis vinifera* (21) roots observed at LM, SEM, and TEM levels, respectively. (In Figure 21: (FV) fungal vacuole; (N) fungus nucleus; (IF) intercellular hyphae: (HW) host wall; (F) fungus; (P) plastid; (CL) fungal clump.) Bar = 10 μm.

trunk resembles the original intercellular hypha in size (3 to 4 μm in the coarse endophytes, under 2 μm in the fine endophytes) and ultrastructure; it is surrounded by a thick apposition layer continuous with the host wall material near the point of penetration (Figure 22). The arbuscule trunk bifurcates repeatedly inside the cell, thus giving rise to smaller branches (Figures 21 and 22). The latter range from 1 μm down to 0.3 to 0.5 μm in diameter and in the smallest branches fine and coarse endophytes are identical anatomically.

The pattern of branching can be clearly observed using SEM analysis; the arbuscular trunk is seen to branch dichotomously, giving rise to slender branches (Figure 20). The latter proliferate into smaller hyphae with short bifurcate terminals.[46,48] TEM shows that the arbuscular hyphae (always surrounded by the invaginated host plasmalemma) (Figures 21—23, and 42) contain numerous nuclei, mitochondria, glycogen particles, lipid globules, abundant polyvesicular bodies, and electron-dense granules inside small vacuoles.

These vacuoles are of great interest, for energy dispersive X-ray analysis has recently shown the presence of high levels of phosphorus and calcium within the electron-dense granules of *Glycine max* and *Allium cepa* mycorrhizae and they are consequently thought to be rich in polyphosphates.[50-51] Furthermore, they are the site of intense alkaline phosphatase (Figures 35 and 36) and possibly ATPase (Figure 34) activities.[52-53]

In the thinner branches the vacuoles become dominant and the electron-dense granules disappear (Figure 21). The pattern of distribution of electron-dense granules — present in the coils (Figure 11), intercellular hyphae (Figure 18), vesicles and trunk (Figure 22), and large arbuscular branches but absent in the highly vacuolated and collapsed hyphae is in agreement with the suggestion of Callow et al.[54] that the fungus can take up the phosphate from the soil, accumulate it as granules of polyphosphate, and transfer it in this form along the intercellular hyphae by cytoplasmic streaming to the active arbuscule. Here the granules are broken down and phosphorus released to the host. Polyphosphatases necessary for such a release of phosphorus have in fact been localized in the fungal mycelium within mycorrhizal roots.[36]

Morphologically, arbuscular walls are very distinctive. They are osmiophilic and starting from the arbuscule trunk they gradually become thinner and thinner and finally reach 20 nm in thickness. Cytochemical tests have revealed the presence of proteins and polysaccharides within the wall (Figure 23).[19,27,37] These polysaccharides, however, differ from those observed in the external hyphae and coils because they are mostly alkali-soluble.[19] Moreover, in the mycorrhiza *Ornithogalum umbellatum* x *G. fasciculatum,* ultrathin frozen sections have shown that the arbuscular wall texture is amorphous (Figure 24), compared to the fibrillar architecture shown by the host wall (Figure 25). Fibrils are not found within the fungal arbuscular wall, even after enzymatic digestions, proving that chitin fibrils are not present.[55] However, N-acetyl-glucosamine residues (chitin monomers) can be localized on the fungal wall, suggesting that during the arbuscular phase, chitin polymerization is not accomplished.[55] All the above described observations show that arbuscular branch walls are so thin they can be compared to those found in the apical regions of cultured mycelia.[19]

The arbuscule life span is limited to a few days (4 or 5, according to Cox and Tinker[56]), the arbuscular branches then deteriorating and collapsing. The latter processes have been described in great detail at the ultrastructural level by numerous research workers and can be summarized as follows: the smallest arbuscular branches show disorganized cytoplasmatic contents, loss of membrane integrity, no discernible organelles, and finally appear as an amorphous mass.[25-27,47-49] Sometimes the arbuscular branches become empty and differentiate a subapical septum which separates empty portions from functional ones (Figure 26).[40,57] Septum formation has not been observed in the degenerating hyphae of *Glomus tenue* in wild raspberry and *G. fasciculatum* in *Zea diploperennis*.[27,49] The walls of the empty zones collapse and then aggregate into clumps (Figure 27).

Roots collected from the field often show large arbuscular clumps, formed by the aggregation of smaller ones, filling the host cell.[40] In nature and in older plants this situation of senescent arbuscules is more frequent than that of the active arbuscules easily observed within young mycorrhizal roots obtained under controlled conditions.

4. Vesicles

Vesicles are globose bodies caused by an intercalary or terminal swelling of a hypha of the VAM fungus. Vesicles found within roots can be intercellular or intracellular, differ in size (30 to 50 μm — 80 to 100 μm), and may be found in both the inner and

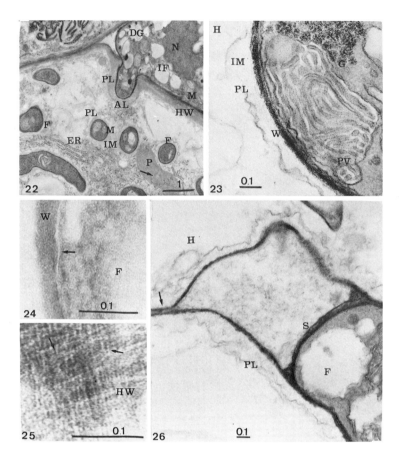

FIGURE 22. A hyphal branch arising from an intercellular hypha (IF), penetrating a host cell by stretching the host wall (HW) and by causing the host plasmalemma (PL) invagination. A thick apposition layer (AL) continuous with the host wall surrounds the fungal penetrating hypha. Smaller branches are observable inside the host cells, characterized by plastids (P) with dense globules and abundant rough reticulum (ER) (TEM). (DG) Dense granules; (N) fungal nucleus; (M) fungal mitochondrion; (F) fungus; (IM) interfacial material. Bar = 1 μm. FIGURE 23. Higher magnification of a thin arbuscular branch, showing a dense deposition of silver grains for polysaccharide localization on glycogen granules and on the thin wall (W). The fungus is surrounded by the host plasmalemma (PL); loose silver grains are evident in the interfacial area (IM) (TEM). (H) Host; (PV) host polyvesicular bodies. Bar = 0.1 μm. FIGURES 24, 25. Details of the host and fungus wall architecture after cryoultramicrotomy. (24) Arbuscular wall showing an amorphous texture; the fungal plasmalemma appears as an electron transparent line, see arrow (TEM), (W) fungal wall; (F) fungus. Bar = 0.1 μm. (25) The host wall (HW) displays fibrillar units embedded in an amorphous matrix (TEM). Bar = 0.1 μm. FIGURE 26. A septum (S) separates the empty portion of the arbuscular hypha from the active one underlying. The apical part of the empty portion is collapsed (arrow) giving rise to an isolated clump. Host plasmalemma (PL) is continuous around the fungal branches (TEM). (H) Host; (F) fungus. Bar = 0.1 μm.

the outer layers of the cortical parenchyma (Figure 28). Not all VAM fungi form vesicles within roots; for example *Gigaspora margarita* does not form them and *Acaulospora laevis* forms lobed or irregular intracellular vesicles. *A. trappei* forms smaller, unlobed vesicles and the genus *Glomus* gives rise to elliptical, inter- or intracellular ones.[24,29,44] The number of vesicles formed depends on the fungal species, for example it is high in *G. fasciculatum*, but low in *G. monosporum* and *G. caledonicum*.[28] The fine endophyte *(G. tenue)* usually produces many small vesicle-like structures about 0.5 μm in diameter.

From a cytological point of view, vesicles are the least studied of the VAM fungal

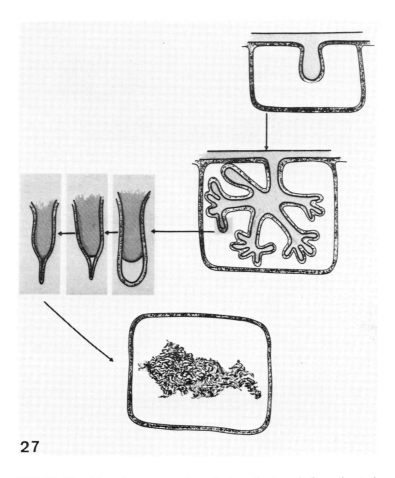

FIGURE 27. The scheme summarizes all steps of arbuscule formation and degeneration, i.e., trunk formation, dichotomous branching, septum formation, and clumps.

structures; their different phases of development have not been fully determined. However, from information gathered so far, it appears that during the initial phase the protoplasm contains many nuclei, glycogen granules, and small vacuoles with electron-dense granules, while mature vesicles are engorged with large lipid droplets (Figure 29) and organelles are no longer discernible. The outer vesicle surface appears smooth without ornamentation (Figure 30). The walls appear to be trilaminate, formed by layers of varying electron density. Intercellular vesicles and host walls are in direct contact, while intracellular vesicles are usually enclosed by a layer of condensed host cytoplasm.[25,30,34,48,58,59]

In the past, different hypotheses on the role of the vesicles have been proposed. For example, Peyronel[7] and Otto[60] suggested their function was that of reproductive organs, like sporangia. In McLennan's[8] opinion, vesicles are organs in which the usual function is to store large quantities of fat which are later transferred to cells in which digestion has proceeded. However, the cytological organization of the vesicles (mostly rich in lipids) and the fact that their numbers frequently increase in old or dead roots (Figure 30) suggest that they are mainly resting organs.

C. Host Response to Fungal Colonization

Organs such as stems and leaves undergo noticeable anatomical modifications following mycorrhizal infection.[61,62] More precisely, Daft and Okusanya[61] demonstrated that infection increased the amount of vascular tissue, the lignification of the xylem, and the number of vascular bundles. Krishna et al.[62] observed an increase in the leaf thickness, the size of the midrib vein, the mesophyll cells, and the number of plastids. On the contrary, such changes do not appear to occur in roots, but reports are very scanty on this subject. Further and more detailed investigations are needed on the anatomical organization of apical meristems and vascular cylinders, as had been done with other mycorrhizal associations, i.e., ectomycorrhizae and ericoid mycorrhizae, where fungal effects on the root anatomy have been investigated.[63,64]

Noticeable cytological variations have been observed, however, within VAM cells, mostly in those containing arbuscules. The host increases its volume of protoplasm in response to fungal colonization.[56] Host plasmalemma proliferates and surrounds all the fungal branches (Figures 21—23). Plasmalemma is typically three-layered, wavy-edged, and rich in polyvesicular bodies. The host nucleus is irregularly lobed, at times polyploid, and Golgi bodies are sometimes hyperactive. All these are signs of a metabolically active host cell.[25]

The primary host wall does not appear damaged during the fungal infection, while one of the most evident effects is the enlargement of the intercellular spaces, caused by the disappearance of the middle lamella (Figure 16). Moreover, wall-bound peroxidase activities decrease in the presence of the fungus.[25,65] Plasmodesmata usually occur between mycorrhizal cells.

Plastids are often modified and in fact, they can be blocked at the protoplastid phase (Figures 21 and 22), contain only small starch granules, or turn into chromoplasts.[27,34,40,66] The plastid morphology suggests modification in the carbohydrate metabolism, namely a reduction or disappearance in starch. The fact could be due, at least in part, to starch solubilization and transfer to the mycorrhizal fungus or more likely to a utilization of host photosynthate, preventing starch formation and storing.

During the process of mycorrhizal infection, the large central host vacuole (characteristic of differentiated cortical cells) disappears as it is split up by the developing hyphae (Figures 21 and 22). Cox and Tinker[56] calculated that the tonoplast increases in length by up to 2.2 times. At the end of the infection, the host cell reverts to its normal state and appearance (Figure 31). In fact, the amount of cytoplasm and number of organelles decreases and the large vacuole reforms.[49] In this phase large starch granules are found near arbuscular clumps (Figure 32).

In the cells containing coils, the cytological response is similar but weaker. There is a certain increase of cytoplasm, and a proliferation of the plasma membrane; the vacuole remains important and keeps its central position, while organelles (nucleus, mitochondria, endoplasmic reticulum) occur in the cell periphery or along the fungal coil (Figure 10). Plastids appear unchanged and often show large starch grains (Figure 11), suggesting that in coil containing cells, the modification of carbohydrate metabolism is not so important as the one observed in arbuscule-containing cells.

D. Host-Fungus Relationships

The understanding of fungus/plant interaction during the mycorrhizal association raises a series of problems of general biology, namely recognition between partners, specificity, symbiont growth, compatibility, and cellular interaction. These problems have not been analyzed sufficiently as far as mycorrhizae are concerned.[25,67] Problems such as host-fungus cellular interaction and the respective contact patterns have only received attention in the last few years. Here, a satisfactory morphological description of the contact patterns could represent an important contribution to understanding the exchanges which occur between the two partners.

In agreement with Bracker and Littlefield,[68] an interface (defined as a three dimensional zone associated with the interaction zone) is formed wherever partners of any

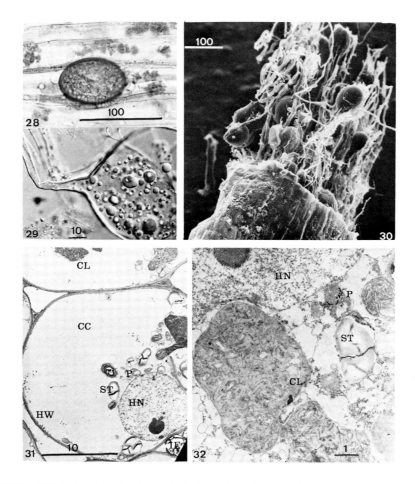

FIGURE 28. Elliptical vesicle of *Glomus fasciculatum* in the inner cortical layer of *Ornithogalum umbellatum* root (LM). Bar = 100 μm. FIGURE 29. Details of the subtending hypha and vesicle showing a cytoplasmatic continuity between them. Large droplets, probably of lipid nature are observed inside the vesicle, under Nomarski differential-interference contrast (LM). Bar = 10 μm. FIGURE 30. Old roots of *Allium cepa* filled up with elliptical *Glomus fasciculatum* vesicles in the exposed cortex (SEM). Bar = 100 μm. FIGURE 31. General view of a cortical cell (CC) of a grapevine root at the end of the arbuscular infection. Vacuole is large, filling up the cell lumen; large starch (ST) granules are present as well as a globose nucleus (HN). The fungal endophyte is reduced to clumps (CL) (TEM). (P) Plastid; (HW) host wall; (IF) intercellular hyphae. Bar = 10 μm. FIGURE 32. Detail of a cortical cell of a grapevine root at the end of the arbuscular infection showing a lobed nucleus (HN), starch (ST) granules, and fungal clumps (CL) (TEM). (P) Plastid. Bar = 1 μm.

symbiotic or pathogenic association come into direct contact. The interface consists of different components (walls of the partners, their respective plasma membranes, wall-derived materials) and the structure and organization of these may change during the development of the symbiotic interaction.

In all VAM studied so far, host and fungus have a wall-to-wall contact whenever the fungus is intercellular (Figure 17). However, when the fungus is intracellular, the interface is formed by the fungal wall and by the host plasmalemma, which invaginates, proliferates, and surrounds the intracellular fungus.[25] This appears to be the rule in many symbiotic or pathogenic associations characterized by biotrophic relationships between partners, but is not always true in ericoid endomycorrhizae.[69]

In VAM, the interface of intracellular coils which is limited by the fungal wall and host plasmalemma, contains a thick osmiophilic material (Figure 12). This interfacial material, laid down by the host, is rich in polysaccharides. It resembles primary wall material and seems to separate the two partners by acting as a permanent physical barrier. This picture changes completely when the intracellular fungus forms arbuscules. All arbuscular branches are also surrounded by host plasmalemma which increases its length.[56]

The interfacial zone, ranging from 150 to 750 nm in width, contains polysaccharide and protein materials of host origin,[70,71] the ultrastructural and cytochemical organization of which changes during the arbuscular development.[72] Electron-dense material is continuous with the primary host wall around the penetrating hypha (Figure 22), and its cytochemical features and architecture resemble the primary host wall. Less wall material is deposited by the host around the younger and finer fungal branches (Figure 23), even though the surrounding host plasmalemma does not show any significant morphological cytochemical modifications.[72] Nothing but scattered membranous vesicles (Figure 33) and scarce polysaccharide and protein material can be observed around the younger branches characterized by very thin walls (as already described), and well-developed polyvesicular bodies. After collapse and clumping of the arbuscular branches the quantity of disorganized material in the interfacial area increases.[72] At this stage the host origin of the interfacial material is not so clearly definable. In fact, sometimes fungal walls show a certain degree of degradation, and a fungal component in the interfacial material cannot always be completely excluded.

These morphological observations clearly show that the physical barrrier to nutrient exchange between host and fungus is minimal around the fine arbuscular branches, implying that in VAM the arbuscular interface is of primary importance in the active exchanges of nutrients between the living host and living fungus.[56,72] Numerous histological, cytochemical, and cytoenzymological results support this hypothesis.

As previously mentioned, the host plasmalemma surrounding the arbuscule branches is morphologically and cytochemically similar to normal plasmalemma. Furthermore, in mycorrhizal cells, ATPase activities in the host plasmalemma are specifically localized around the young arbuscular branches (Figure 34).[53] This considerably strengthens the idea that active phosphorus transfer occurs from the fungus to plant, since it is known that plasmalemma-bound ATPases are involved in active transmembrane ion transport in higher plants.[74]

Alkaline phosphatase activity, specific to VAM roots and correlated with a large number of active arbuscules and plant growth stimulation is localized at polyphosphate granules inside the fungal vacuoles of intercellular hyphae and mature arbuscules (Figures 35 and 36).[52,75] Consequently, the suggestion is that this enzyme activity may be involved in the active mechanism of P-transport, by VAM fungi to the host plant. On the contrary, acid phosphatase, also observed at light microscope level by Nemec[37] and MacDonald and Lewis[76] is associated with organelles in the young arbuscular branches, suggesting that it is probably involved in hyphal elongation and growth. Acid phosphatase activity has also been reported in the interfacial area, however, its role is not clearly defined.[25]

In conclusion, it appears likely from recent morphological studies that during VAM interactions the most important relationship is that of living host with living fungus and that the arbuscular phase is the most appropriate stage for nutrient exchanges between the symbionts, as compared to the other intraradical phases of the VAM fungus.

Fungal digestion, which many research workers once considered very important for nutrient transfer processes in VAM (for example, Scannerini[77] and Strullu and Gourret[78]) seems in fact to be limited to the terminal phases of the interaction, when the dead fungus, reduced to clumps, is eliminated inside the host cells which subsequently revert to their previous state. We still do not know how this happens!

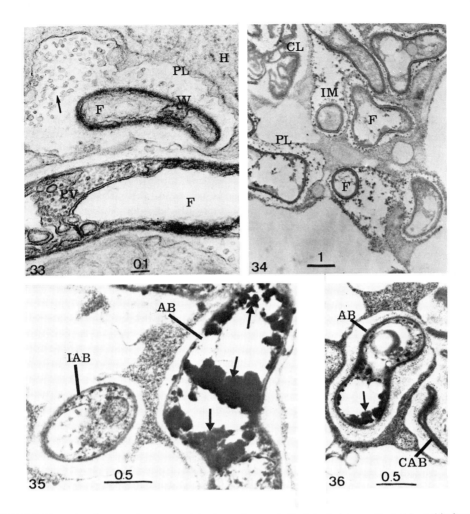

FIGURE 33. Abundant membranous vesicles (arrow) occur in the interfacial zone limited by a thin fungal wall (W) and the host plasmalemma (PL) around the young arbuscular branches (TEM). (H) Host; (F) fungus; (PV) fungal polyvesicular bodies. Bar = 0.1 μm. FIGURE 34. ATPase activities in a cortical cell of onion roots infected by *Glomus fasciculatum*. Part of an arbuscule showing living and collapsed hyphae. ATPase activity is localized along the host and fungal plasmalemmas (PL) and within the fungal vacuole. A diffuse lead phosphate precipitate is also present in the interfacial matrix (IM) (TEM). (F) Fungus; (CL) fungal clump. Bar = 1 μm. (From Marx, C., et al., *New Phytol.*, 90, 37, 1982. With permission.) FIGURES 35, 36. Alkaline phosphatase activity (arrow) (substrate: glycerophosphate) within the vacuoles of mature arbuscular branches (AB) of *Glomus mosseae* in onion roots. Activity is absent from the immature, small vacuolated branches (IAB) and from the collapsed senescent branches (CAB) (TEM). Bar = 0.5 μm. (From Gianinazzi, S. et al., *New Phytol.*, 82, 127, 1979. With permission.)

III. FUNGAL INFECTION IN BRYOPHYTES

It is well known that fungal endophytes can be associated with some bryophytes giving rise to associations sometimes called mycothalli.[43,79]

However, the only available descriptions of the anatomy of infection occurring in liverworts date back to 1949, when Stahl[80] noticed different degrees and patterns of infection in liverworts such as *Pellia, Monoclea,* and *Marchantia*. The penetrating hypha forms a typical infection unit with arbuscules, coils, and small intracellular vesicles inside the rhizoids and in the lowest layer of the thallus. Arbuscules usually collapse forming clumps, or so-called sporangiole. In other liverworts, coils, which in the end collapse, represent the most common structure.

Recently, Gourret and Strullu[81] investigated the fungus/liverwort association from an ultrastructural point of view. They observed endophytic hyphae in the cortical cells of thallus of *Pellia epiphylla* and surrounded by the invaginated host plasmalemma (Figure 37). In the interfacial area a thick osmiophilic layer was present. The hyphae were endowed with the usual cytoplasmic organelles and with electron-dense granules which in X-ray spectrometric analysis displayed phosphorus and calcium contents.[82]

The arbuscular hyphae emptied and collapsed giving rise to clumps, similar to those described in angiosperm mycorrhizal roots. However, the most striking feature was the presence of the endophyte inside cells containing true chloroplasts with thylacoids and grana. Large starch grains were observed only in uninfected cells.

Reports on the mycotrophic status of mosses are even scantier. However, according to recent studies, the association between identified VAM fungi and mosses may occur, though it is not a true symbiosis. Rabatin[83] observed fine hyphae and spores of *Glomus tenue* occurring along stems and leaves, but no fungal penetration. In another experiment, Parke and Linderman[84] inoculated *Funaria hygrometrica,* asparagus plants, and combinations of the two with *Glomus epigaeum* spores. Hyphae, vesicles, and spores were seen in moss plants, but only in those grown with asparagus "companion" plants. The fact that the fungal structures occurred only in the parenchymatous stem and leaf tissues of older moss gametophytes and that arbuscules were never observed probably means that the relationship between *G. epigaeum* and moss is not a mutualistic one.

IV. FUNGAL INFECTION IN PTERIDOPHYTES

During their life cycle, ferns may display a fungal infection in their thalli, rhizomes, or roots. There are not many reports about the occurrence of such infections in Pteridophytes.[43,79,85-88] Generally the infection process differs not only from species to species, but also from specimen to specimen in the same species. Moreover, Boullard[87] suggested that a close correlation exists between fungal colonization and fern evolution. In fact, fungal endophytes pass from an obligate state in Psilotaceae and Lycopodiaceae to a constant one in eurosporangiate ferns, a facultative one in some leptosporangiate ferns (Polypodiaceae, Pteridaceae, etc.), and in others (those belonging to aquatic families Azollaceae, Marsilaceae, etc.), they are completely absent.

Following this scheme, early Pteridophytes with fleshy perennial prothalli are obligate mycotrophs, while modern Pteridophytes, characterized by a reduced gametophytic phase, usually have uninfected gametophytes and a more reduced level of colonization in sporophytes.

Boullard[87] has recently provided an anatomical description of fungal colonization in ferns. Some characteristics constantly occur: the apical region, the stele, and the endodermis are uninfected in sporophytes; root hairs are sometimes absent in infected regions and gametangia in gametophytes are always free of fungi. Starch tends to disappear in infected cells and to reappear at the end of the infection. However, different patterns of colonization have been recognized in the various classes of Pteridophytes. Hyphae penetrate cortical cells directly or via root hairs, assuming a linear or, more often, a looped appearance; coils are sometimes the most common structure to be found. Arbuscules can easily be observed in sporophytes, such as *Pteridium aquilinum, Polysticum aculeatum,* or *Osmunda regalis.*[85,87] They degenerate and collapse forming clumps. Boullard claims that the frequent presence of clumps implies a short arbuscule life span. Spheroid or elliptical vesicles are found in intercalary or terminal positions.

Athyrium roots can harbor both a coarse endophyte (mostly coil forming), and a fine endophyte (closely resembling *Glomus tenue*), characterized by fan-like structures on the root surface, small vesicles, and fine arbuscules (Figures 38 and 39). Boullard[87] claims that the fungus/host association is mutualistic when and where the host "di-

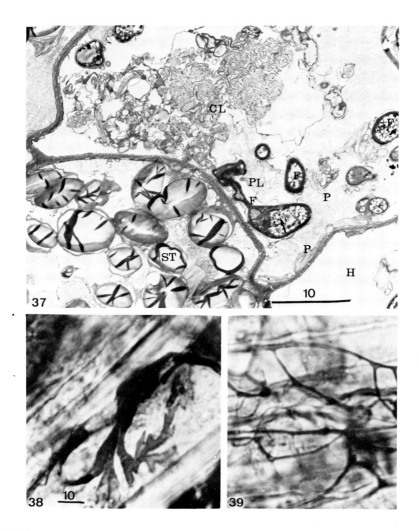

FIGURE 37. Mycothallus of *Pellia*. The phycomyceous endophyte forms arbuscular branches and clumps (CL) inside the green cells of the thallus. Plastids (P) show stacked grana. Starch (ST) is present only in the uninfected cells. Electron-dense globules are detectable inside the fungal vacuoles (TEM). (F) Fungus; (PL) host plasmalemma; (H) host. Bar = 10 μm. (Micrograph courtesy of Dr. D. S. Strullu.) FIGURE 38. Fanlike structures on the root surface of *Athyrium* (LM), formed by a fine endophyte. Bar = 10 μm. FIGURE 39. Small vesicles and fine intercellular hyphae colonizing *Athyrium* cortical cells (LM).

gests'' the fungus. This process follows different patterns, classified by Burgeff[45] as tolypophagy, thamniscophagy, and thamniscophysalidophagy, depending on whether coils, arbuscules and vesicles are "digested". These definitions, which Harley[43] described as "very uncomfortable words" are no longer in use.

Available cytological data are too limited to confirm the above interpretation. The presence of a fungal endophyte in *Psilotum nudum* gametophyte cells has been recently reported.[89] The unidentified fungus colonizes rhizoids and cortical cells with coarse and aseptate hyphae which have large nuclei, mitochondria, small vacuoles, thick walls, and are surrounded by an amorphous interfacial material and by the invaginated host plasmalemma. With time, the aged fungal hyphae vacuolate and collapse. Infected cortical cells appear vacuolated with a restricted cytoplasmic layer.

Cytological and physiological data are not yet sufficient to determine whether Pteridophytes are only colonized by typical VAM fungi, or whether the kind of interaction the ferns establish with the endophytes is a mutualistic one. However, as can be seen from the above descriptions, some features of fungal development and degeneration are similar in some respects to that occurring in VAM symbiosis in angiosperms.

V. FUNGAL INFECTION IN GYMNOSPERMS

Aseptate phycomycetous hyphae are commonly found in gymnosperm roots; Harley[43] has listed some in which VA-like mycorrhizae have been observed. Detailed descriptions are nevertheless still very rare and *Taxus baccata* mycorrhizae, carefully described by Prat[90] in 1926 remain one of the few examples to this day. In this conifer, fungal hyphae colonize only the feeder roots heavily. They form appressoria on the root surface and coils, arbuscules, and vesicles inside the cortical cells. Feeder roots grow for more than one season; in spring a new growing point develops isolated by dead tissue from the cortex of the parent root. At first this point remains uninfected and fungal hyphae, surviving in the humus surrounding the root, colonize the young branch in a second stage. Strullu[91] studied the ultrastructure of this natural association in great detail, although observations were limited to arbuscule-infected cells and to small intracellular vesicles. Intercellular hyphae are not evident, and spread of the infection occurs directly from cell to cell. On the contrary, other features resemble those described in mycorrhizal angiosperm roots. Arbuscular branches contain all the usual organelles and vary in size. The fungal wall is thin and osmiophilic electron-dense granules, commonly found inside the vacuoles, have been recognized as polyphosphates.[92] Direct microanalysis with a wavelength dispersive X-ray spectrometer has shown that these have a high phosphorus and calcium content.

All arbuscular branches are surrounded by the invaginated host plasmalemma, and the interfacial area, limited by fungal wall and host plasmalemma, contains electron-dense floccular matrix material, probably of host origin. The fungal branches vacuolate and collapse, giving rise to clumps. The host cell is characterized by a central nucleus, with a prominent nucleolus, endoplasmic reticulum, mitochondria, plastids with starch grains, and other modified plastids.

A similar pattern of colonization has been described at the light microscope level in *Sequoia gigantea, S. sempervirens,* and *Ginkgo biloba*.[93,94] In addition, in *G. biloba* (which is sometimes identified as a Praephanerogamic plant), Fontana[94] investigated mycorrhizal roots collected in the field as well as samples obtained under controlled conditions after inoculation with *Glomus epigaeum* spores (Figures 40 and 41). We can thus be certain that we are talking about a typical VAM fungus which gives a well-defined growth stimulation of the plant; the first ultrastructural investigations of the association *Ginkgo biloba/Glomus epigaeum* show the characteristic VAM pattern, namely coils, rare intercellular hyphae, and arbuscules similar to those already described by Strullu[92] in *Taxus* (Figure 42).

In conclusion, we still lack detailed studies on gymnosperms correlating the morphology of the infection, realized using identified VAM fungi, with its physiological effects. However, it is possible to say that as in angiosperms, VAM infections occurring in gymnosperms do not change the root morphology and infected regions often lack root hairs. On the contrary, colonization patterns do vary; coiled or linear hyphae are very common, arbuscules are formed by spreading directly from cell to cell, intercellular hyphae are scarce, and so are vesicles which tend to be minute, if present.

Host/fungus relationships seem to be cytologically homogeneous. Fungal hyphae are always surrounded by host plasmalemma with the interposing of the interfacial matrix material. Moreover, they always occur inside living and active host cells, a fact which suggests the presence of a biotrophic relationship between the two partners.

VI. CONCLUSIONS

Morphological and functional criteria emerging from recent cytological data as well

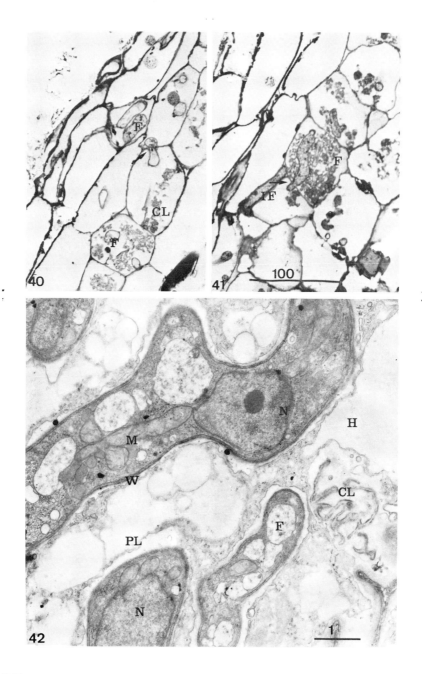

FIGURES 40, 41. *Ginkgo biloba* roots colonized by *Glomus epigaeum*. Longitudinal sections of the root show hyphae colonizing the inner cortical layer. Hypha are organized as coils, arbuscules, and intercellular hyphae. Cell to cell passages are evident (see arrows) (LM). (F) Fungus; (CL) fungal clump; (IF) intercellular hyphae. Bar = 100 μm. FIGURE 42. Active arbuscular hyphae of *Glomus epigaeum* in *Ginkgo biloba* roots characterized by the nucleus (N) with dense nucleoli, mitochondria (M), and small vacuoles. The host plasmalemma (PL) surrounds all the fungal branches. Collapsed hyphae (fungal clumps—CL) are evident (TEM). (H) Host; (W) fungal wall; (F) fungus. Bar = 1 μm.

as from the classical anatomical descriptions now enable us to identify VAM with more accuracy than in the past. The criteria that have been considered here are

1. Identity of the symbiotic fungus

2. Presence of the fungal structures, among which arbuscules are particularly important
3. Cytological changes in the root (or thallus)
4. Physiological relationships between the partners, justifying the use of the term "mutualistic association"
5. A cytological organization of the association which allows interactions between the two symbionts

On the basis of present-day knowledge, it is reasonable to say that mycorrhizal associations in angiosperms are the one case where all these morphological criteria can be applied. As far as the other groups are concerned:

Mosses: There are no characteristic mycorrhizae, with a mutualistic interaction between host and fungus.

Liverworts: The infection resembles that formed by VAM fungi in certain respects. However, the above-listed criteria are not satisfied as we do not yet have knowledge on their infective fungi, nor do we sufficiently understand the physiology or the cytological details of the relationship.

Ferns: The morphology of the infective process is definitely very much like that found in the VAM associations in angiosperms. However, the infection is not ubiquitous and seems to be dependent on the systematic position of the fern species. Therefore, at present the identification of VAM associations in ferns is purely based on a morphological resemblance, mostly at the level of the light microscope. No cytofunctional evidence is yet available nor have their endophytes been identified.

Gymnosperms: Although data are scarce and concern a limited number of species, the infection is morphofunctionally well characterized and is known to be formed by identified VAM fungi; it can therefore be considered as a characteristic VAM infection.

Many problems still remain, even when considering VAM associations in angiosperms, the most important of which are (1) The mechanisms of recognition between plants and VAM fungi. The plant is normally able to recognize the difference between "self" and "nonself", and as a result it can reject most of the microorganisms with which it interacts.[95] This means that VAM plants (80% of angiosperms) recognize all VAM fungi, which belong to no more than four or five genera of a single family, the Endogonaceae. The system therefore, has a low degree of specificity for the host and a high degree for the fungus. Generally speaking, recognition is probably established by the interaction of complementary macromolecules present on the surfaces of the wall of the two organisms which come into contact.[96] Since this same hypothesis has been applied with good evidence to other symbiotic associations, in particular in the *Rhizobium*/legume system, we can hypothesize that in VAM associations a recognition based on the interaction between polysaccharides and proteins present on the walls of the partners also exists.[19,97] (2) The complex morphogenesis of the VAM fungi within the root. This is one of the more typical features of VAM fungi compared to other mycorrhizal mycelia. Morphogenesis, which is also accompanied by conspicuous changes in cell wall structure and composition,[19] is probably regulated both by the fungal genome and by the host which intervenes in controlling and regulating development and growth of the symbiotic fungus by means of its plasmalemma that always surrounds the fungus. (3) The striking morphological similarities of different VAM systems. Present morphological analyses are rather limited considering the wide range of plants (from a systematic viewpoint) which can form VAM. The picture may therefore change following the input of new data, but even at this stage the range remains an exceptionally wide one, compared to other mycorrhizal types, most of which are found in one or a few families. Even roots with marked anatomical differences harbor fungal infections, the morphological characteristics of which are almost always the same.

Although recent observations suggest that anatomical features of colonization can be different and fungus-dependent,[29] all ultrastructural investigations agree on the same pattern of host/fungus cytological interactions, i.e., of a high cytological compatibility between the two partners. The host is metabolically very active as shown by an increase in protoplasm, but at the same time it allows fungal growth throughout its root cells, as if the fungus were "an integral part of the plant."[98] Both morphological homogeneity and cytological compatibility can be, at least partially explained on an evolutionary basis. According to Pirozynski and Malloch[99] and Nicolson,[100] one could hypothesize that VAM represent a very ancient symbiotic event in which previously divergent genomes form a new coevolutionary unit.[101] This suggestion may be in agreement with the symbiotic theory put forward by Margulis[102] for the symbiotic origin of organelles (plastids and mitochondria). Among other things, the symbiotic theory claims that living organisms do not descend from a unique ancestral prokaryotic population by direct filiation and accumulation of one-step mutations, but from eukaryotic ancestors evolved from prokaryotes by symbiotic events.[102]

Although in the case of VAM associations two symbiotic eukaryotes are involved, a similar event could be hypothesized, since VAM-like associations with phycomycetous fungi like *Paleomyces asteroxilii* may well date back to the Devonian period. In fact, Kidston and Lang,[103] studying the famous Rhynie fossil plants preserved in a petrified form, observed in *Rhynia* and *Asteroxylon* rhizomes fungi which were remarkably similar to modern VAM fungi. *Rhynia* and *Asteroxylon* are generally thought to be the ancestral forms of tracheophytes.

According to Pirozynski,[104] the fungal endophyte which developed a very long time ago, is morphologically stable and has shown a low degree of speciation. The host, and its epigeous parts in particular, seems to have responded to selective pressures, while the fungus has not. The morphological stability of these VAM associations may have been assured by interactions between physical and biological factors during the evolutionary processes. The root-soil interface may have been a relatively stable situation for the extramatrical hyphae, in comparison to the changes occurring in the soil and the endocellular life may have been a stable condition for the endophytic hyphae. This means that in the course of time, the host root cell represented and still represents for the VAM fungus a safe and fairly constant habitat, irrespective of the systematic position of the host plant.

ACKNOWLEDGMENTS

I wish to thank Drs. Anna Fontana, Aldo Fasolo, Silvio Gianinazzi, and Silvano Scannerini for valuable discussions, and particularly Dr. Vivienne Gianinazzi-Pearson for her critical reading and linguistic revision of the manuscript.

REFERENCES

1. Frank, A. B., Ueber neue Mykorrhiza-formen, *Ber. Bot. Gesell.*, 5, 395, 1887.
2. Schlicht, A., Beitrag zur Kenntniss de Verbreitung under der Bedentung der Mykorrhizen, *Landwirtsch. Jahrb.*, 18, 499, 1889.
3. Janse, J. M., Les endophytes radicaux de quelques plantes Javanaises, *Ann. Jard. Bot. Buitenzorg*, 14, 53, 1896-1897.

4. Dangeard, P. A., Le *"Rhizophagus populinus"*, Dangeard, *Botaniste*, 7, 285, 1900.
5. Petri, L., Ricerche sul significato morfologico dei prosporoidi (sporangioli di Janse) nelle micorrize endotrofiche, *Nuovo G. Bot. Ital.*, 10, 541, 1903.
6. Gallaud, I., Etude sur les mycorhizes endotrophes, *Rev. Gen. Bot.*, 17, 5, 1905.
7. Peyronel, B., Prime ricerche sulle micorrize endotrofiche e sulla micoflora radicicola normale delle fanerogame, *Riv. Biol.*, 6, 1, 1924.
8. McLennan, E. I., The endophytic fungus of *Lolium*. II. The mycorrhiza on the roots of *Lolium temulentum* L. with a discussion on the physiological relationships of the organism concerned, *Ann. Bot.*, 40, 43, 1926.
9. Butler, E. J., The occurrences and systematic position of the vesicular arbuscular type of mycorrhizal fungi, *Trans. Br. Mycol. Soc.*, 22, 274, 1939.
10. Gerdemann, J. W., A species of *Endogone* from corn causing vesicular-arbuscular mycorrhiza, *Mycologia*, 53, 254, 1961.
11. Baylis, G. T. S., The magnolioid mycorrhiza and mycotrophy in root systems derived from it, in *Endomycorrhizas*, Sanders, F. E., Mosse, B., and Tinker, P. B., Eds., Academic Press, London, 1975, 374.
12. St. John, T. V., Root size, root hairs and mycorrhizal infection: re-examination of Baylis's hypothesis with tropical trees, *New Phytol.*, 84, 483, 1980.
13. Chilvers, M. T. and Daft, M. F. J., Mycorrhizas of the Liliiflorae. II. Mycorrhiza formation and incidence of root hairs in field grown *Narcissus* L., *Tulipa* L. and *Crocus* L. cultivars, *New Phytol.*, 89, 247, 1981.
14. Sanders, F. E. and Tinker, P. B. H., Phosphate flow into mycorrhizal roots, *Pest. Sci.*, 4, 385, 1973.
15. Tisdall, J. M. and Oades, J. M., Stabilisation of soil aggregates by the root systems of ryegrass, *Aust. J. Soil Res.*, 17, 429, 1979.
16. Nicholson, T. H., Mycorrhiza in the *Gramineae*. I. Vesicular-arbuscular endophytes, with special reference to the external phase, *Trans. Br. Mycol. Soc.*, 42, 421, 1959.
17. Mosse, B., Observations on the extra-matrical mycelium of a vesicular-arbuscular endophyte, *Trans. Br. Mycol. Soc.*, 42, 439, 1959.
18. Powell, C. L., Development of mycorrhizal infections from *Endogone* spores and infected root segments, *Trans. Br. Mycol. Soc.*, 66, 439, 1976.
19. Bonfante-Fasolo, P. and Grippiolo, R., Ultrastructural and cytochemical changes in the wall of a vesicular-arbuscular mycorrhizal fungus during symbiosis, *Can. J. Bot.*, 60, 2303, 1982.
20. Gianinazzi-Pearson, V., Trouvelot, A., Morandi, D., and Marocke, R., Ecological variations in endomycorrhizas associated with wild raspberry populations in the Vosges region, *Acta Oecol.*, 1, 111, 1980.
21. Hall, I. R., Species and mycorrhizal infections of New Zealand *Endogonaceae*, *Trans. Br. Mycol. Soc.*, 68, 341, 1977.
22. Mosse, B. and Hepper, C., Vesicular-arbuscular mycorrhizal infections in root organ cultures, *Physiol. Plant Pathol.*, 5, 215, 1975.
23. Sward, R. J., Infection of Australian heathland plants by *Gigaspora margarita* (a vesicular-arbuscular mycorrhizal fungus), *Aust. J. Bot.*, 26, 253, 1978.
24. Abbott, L. K. and Robson, A. D., Growth of subterranean clover in relation to the formation of endomycorrhizas by introduced and indigenous fungi in a field soil, *New Phytol.*, 81, 575, 1978.
25. Scannerini, S. and Bonfante-Fasolo, P., Comparative ultrastructural analysis of mycorrhizal associations, *Can. J. Bot.*, 61, 917, 1983.
26. Cox, G. and Sanders, F., Ultrastructure of the host-fungus interface in a vesicular-arbuscular mycorrhiza, *New Phytol.*, 73, 901, 1974.
27. Gianinazzi-Pearson, V., Morandi, D., Dexheimer, J., and Gianinazzi, S., Ultrastructural and ultracytochemical features of a *Glomus tenuis* mycorrhiza, *New Phytol.*, 88, 633, 1981.
28. Abbott, L. K., Hall, I. R., and Robson, A. D., The use of infection morphology for identifying inoculant VA mycorrhizal fungi, *Abstr. 5th N. Am. Conf. Mycorrhizae*, Universitè Laval, Quebec, Canada, 1981, 52.
29. Abbott, L. M., Comparative anatomy of vesicular-arbuscular mycorrhizas formed on subterranean clover, *Aust. J. Bot.*, 30, 485, 1982.
30. Bonfante-Fasolo, P. and Scannerini, S., A cytological study of the vesicular-arbuscular mycorrhiza in *Ornithogalum umbellatum* L., *Allionia*, 22, 5, 1977.
31. Fontana, A., Bonfante-Fasolo, P., and Schubert, A., Caratterizzazione morfologica della micorrizia vescicolo-arbuscolare nella vite, *Quad. Vit. Enol.*, Universita di Torino, 137, 1978.
32. Grippiolo, R., Il processo di un fungo vescicolo-arbuscolare. I. Osservazioni ultrastrutturali sulla penetrazione del fungo da cellula a cellula nella radice, *Allionia*, 24, 49, 1981.
33. Kinden, D. A. and Brown, M. F., Electron microscopy of vesicular-arbuscular mycorrhizae of yellow poplar. I. Characterization of endophytic structures by scanning electron steroscopy, *Can. J. Microbiol.*, 21, 989, 1975.

34. Kinden, D. A. and Brown, M. F., Electron microscopy of vesicular-arbuscular mycorrhizae of yellow poplar. II. Intracellular hyphae and vesicles, *Can. J. Microbiol.,* 21, 1768, 1975.
35. Abbott, L. K. and Robson, A. D., A quantitative study of the spores and anatomy of mycorrhizas formed by a species of *Glomus,* with reference to its taxonomy, *Aust. J. Bot.,* 27, 363, 1979.
36. Capaccio, L. C. M. and Callow, J. A., The enzymes of polyphosphate metabolism in vesicular-arbuscular mycorrhizas, *New Phytol.,* 91, 81, 1982.
37. Nemec, S., Histochemical characteristics of *Glomus etunicatus* infection of *Citrus limon* fibrous roots, *Can. J. Bot.,* 59, 609, 1981.
38. Harley, J. L., Problems of mycotrophy, in *Endomycorrhizas,* Sanders, F. E., Mosse, B., and Tinker, P. B., Eds., Academic Press, London, 1975, 1.
39. Cox, G., Sanders, F. E., Tinker, P. B., and Wild, J. A., Ultrastructural evidence relating to host-endophyte transfer in a vesicular-arbuscular mycorrhiza, in *Endomycorrhizas,* Sanders, F. E., Mosse, B., and Tinker, P. B., Eds., Academic Press, London, 1975, 297.
40. Bonfante-Fasolo, P., Some ultrastructural features of the vesicular-arbuscular mycorrhiza in the grapevine, *Vitis,* 17, 386, 1978.
41. MacDonald, R. M. and Chandler, M. R., Bacterium-like organelles in the vesicular-arbuscular mycorrhizal fungus *Glomus caledonius, New Phytol.,* 89, 241, 1981.
42. MacDonald, R. M., Chandler, M. R., and Mosse, B., The occurrence of bacterium-like organelles in vesicular arbuscular mycorrhizal fungi, *New Phytol.,* 90, 659, 1982.
43. Harley, J. H., *The Biology of Mycorrhiza,* 2nd ed., Leonard Hill, London, 1969, 263.
44. Gerdemann, J. W. and Trappe, J. M., The *Endogonaceae* in the Pacific North West, *Mycol. Mem.,* 5, 1974.
45. Burgeff, H., Mycorrhiza, in *Manual of Pteriodology,* Vol. 1, Martinus Nijhoff, The Hague, 1938, 159.
46. Kinden, D. A. and Brown, M. F., Electron microscopy of vesicular-arbuscular mycorrhizae of yellow poplar. III. Host-endophyte interactions during arbuscular development, *Can. J. Microbiol.,* 21, 1930, 1975.
47. Kinden, D. A. and Brown, M. F., Electron microscopy of vesicular-arbuscular mycorrhiza of yellow poplar. IV. Host-endophyte interactions during arbuscular deterioration, *Can. J. Microbiol.,* 22, 64, 1976.
48. Holley, J. D. and Peterson, R. L., Development of a vesicular-arbuscular mycorrhiza in bean roots, *Can. J. Bot.,* 57, 1960, 1979.
49. Kariya, N. and Toth, R., Ultrastructure of the mycorrhizal association formed between *Zea diploperennis* and *Glomus fasciculatus, Mycologia,* 73, 1027, 1981.
50. White, J. A. and Brown, M. F., Ultrastructure and X-ray analysis of phosphorus granules in a vesicular-arbuscular mycorrhizal fungus, *Can. J. Bot.,* 57, 2812, 1979.
51. Cox, G., Moran, K. J., Sanders, F., Nockolds, C., and Tinker, P. B., Translocation and transfer of nutrients in vesicular-arbuscular mycorrhizas. III. Polyphosphate granules and phosphorus translocation, *New Phytol.,* 84, 649, 1980.
52. Gianinazzi, S., Gianinazzi-Pearson, V., and Dexheimer, J., Enzymatic studies on the metabolism of vesicular-arbuscular mycorrhiza. III. Ultrastructural localization of acid and alkaline phosphatase in onion roots infected by *Glomus mosseae* (Nicol. Gerd.), *New Phytol.,* 82, 127, 1979.
53. Marx, C., Dexheimer, J., Gianinazzi-Pearson, V., and Gianinazzi, S., Enzymatic studies on the metabolism of vesicular-arbuscular mycorrhizas. IV. Ultracytoenzymological evidence (ATPase) for active transfer processes in the host-arbuscule interface, *New Phytol.,* 90, 37, 1982.
54. Callow, J. A., Capaccio, L. C. M., Parish, G., and Tinker, P. B., Detection and estimation of polyphosphate in vesicular arbuscular mycorrhizas, *New Phytol.,* 80, 125, 1978.
55. Bonfante-Fasolo, P., Cell wall architectures in a mycorrhizal association as revealed by cryoultramycrotomy, *Protoplasma,* 111, 113, 1982.
56. Cox, G. and Tinker, P. B., Translocation and transfer of nutrients in vesicular-arbuscular mycorrhizas. I. The arbuscule and phosphorus transfer: a quantitative ultrastructural study, *New Phytol.,* 77, 371, 1976.
57. Scannerini, S., Bonfante, P. F., and Fontana, A., An ultrastructural model for the host-symbiont interaction in the endotrophic mycorrhizae of *Ornithogalum umbellatum* L., in *Endomycorrhizas,* Sanders, F. E., Mosse, B., and Tinker, P. B., Eds., Academic Press, London, 1975, 393.
58. Protsenko, M. A. and Shemakhanova, N. M., On the ultrastructure of external and intraroot vesicles of mycorrhizal fungi, *Mikol. Fitopatol.,* 8, 441, 1974.
59. Scannerini, S. and Bonfante-Fasolo, P., Dati preliminari sull'ultrastruttura di vescicole intracellulari nell'endomicorriza di *Ornithogalum umbellatum, Atti Accad. Sci. Torino Cl. Sci. Fis. Mat. Nat.,* 109, 619, 1975.
60. Otto, G., Beitrag zur frage der funktionellen Bedeutung der Vesikel der endotrophen Mykorrhiza an Samlingen von *Malus communis* L., *Arch. Mikrobiol.,* 32, 373, 1959.

61. Daft, M. J. and Okusanya, B. O., Effect of *Endogone* mycorrhiza on plant growth. VI. Influence of infection on the anatomy and reproductive development in four hosts, *New Phytol.*, 72, 1333, 1973.
62. Krishna, K. R., Suresh, H. M., Syamsunder, J., and Bagyaraj, D. J., Changes in the leaves of finger millet due to VA mycorrhizal infection, *New Phytol.*, 87, 717, 1981.
63. Clowes, F. A. L., Cell proliferation in ectotrophic mycorrhizas of *Fagus sylvatica* L., *New Phytol.*, 87, 547, 1981.
64. Berta, G. and Bonfante-Fasolo, P., Apical meristems in mycorrhizal and uninfected roots of *Calluna vulgaris*, *Plant Soil*, 71, 285, 1983.
65. Grippiolo, R., Peroxidase activity in the wall of Star of Bethlehem *(Ornithogalum umbellatum)* VA mycorrhizae, *Carologia*, 35, 384, 1982.
66. Scannerini, S. and Bonfante-Fasolo, P., Unusual plastids in an endomycorrhizal root, *Can. J. Bot.*, 55, 2471, 1977.
67. Molina, R. and Trappe, J. M., Lack of mycorrhizal specificity of the Ericaceous hosts *Arbutus menziesii* and *Arctostaphylos uva ursi*, *New Phytol.*, 90, 495, 1982.
68. Bracker, C. E. and Littlefield, L. J., Structural concepts of host-pathogen interfaces, in *Fungal Pathogenicity and the Plant's Response*, Byde, R. J. W. and Cutting, C. V., Eds., Academic Press, London, 1973, 159.
69. Bonfante-Fasolo, P. and Gianinazzi-Pearson, V., Ultrastructural aspects of endomycorrhiza in the Ericaceae. III. Morphology of the dissociated symbionts and modifications occurring during their reassociations in axenic culture, *New Phytol.*, 91, 691, 1982.
70. Dexheimer, J., Gianinazzi, S., and Gianinazzi-Pearson, V., Ultrastructural cytochemistry of the host-fungus interfaces in the endomycorrhizal association *Glomus mosseae - Allium cepa*, *Z. Pflanzenphysiol.*, 92, 191, 1979.
71. Scannerini, S. and Bonfante-Fasolo, P., Ultrastructural cytochemical demonstration of polysaccharides and proteins within the host-arbuscular interfacial matrix in an endomycorrhiza, *New Phytol.*, 83, 87, 1979.
72. Bonfante-Fasolo, P., Dexheimer, J., Gianinazzi, S., Gianinazzi-Pearson, V., and Scannerini, S., Cytochemical modifications in the host-fungus interface during intracellular interactions in vesicular-arbuscular mycorrhizae, *Plant Sci. Lett.*, 22, 13, 1981.
73. Tinker, P. B., Effects of vesicular-arbuscular mycorrhizas on plant nutrition and plant growth, *Physiol. Veg.*, 16, 743, 1978.
74. Marre, E., Fusicoccin: a tool in plant physiology, *Annu. Rev. Plant Physiol.*, 30, 273, 1979.
75. Gianinazzi-Pearson, V. and Gianinazzi, S., Enzymatic studies on the metabolism of vesicular-arbuscular mycorrhiza. II. Soluble alkaline phosphatase specific to mycorrhizal infection in onion roots, *Physiol. Plant Pathol.*, 12, 45, 1978.
76. MacDonald, R. M. and Lewis, M., The occurence of some acid phosphatases and dehydrogenases in the vesicular-arbuscular mycorrhizal fungus *Glomus mosseae*, *New Phytol.*, 80, 135, 1978.
77. Scannerini, S., Le ultrastrutture delle micorrize, *G. Bot. Ital.*, 109, 1975.
78. Strullu, D. G. and Gourret, J. P., Donnees ultrastructurales sur l'integration cellulaire de quelques parasites on symbiotes de plantes. II. Champignons mycorhiziens, *Bull. Soc. Bot. Fr. Actual. Bot.*, 127, 97, 1980.
79. Boullard, B., *Les Mycorhizes*, Masson et Cie, Paris, 1968, 47.
80. Stahl, M., Die Mykorrhiza der Leber moose mit besonderer Berucksischtigung der thallosen Formen, *Planta*, 37, 103, 1949.
81. Gourret, J. P. and Strullu, D. S., Etude cytophysiologique et ecologique des symbioses racinaires: nodules fixateurs d'azote et mycorhizes, *C. R. Sci. A. T. P. Rennes*, 28, 20, 1979.
82. Strullu, D. G., Gourret, J. P., and Garrec, J. P., Microanalyse des granules vacuolaires des ectomycorhizes, endomycorhizes et endomycothalles, *Physiol. Veg.*, 19, 367, 1981.
83. Rabatin, S. C., The occurrence of the vesicular-arbuscular mycorrhizal fungus *Glomus tenuis* with moss, *Mycologia*, 72, 191, 1980.
84. Parke, J. L. and Linderman, R. G., Association of vesicular-arbuscular mycorrhizal fungi with the moss *Funaria hygrometrica*, *Can. J. Bot.*, 58, 1898, 1980.
85. Fontana, A., Ricerche sulla simbiosi micorrizica nelle Pteridofite e sui microorganismi normalmente presenti nelle loro radici, *Allionia*, 5, 27, 1959.
86. Cooper, K. M., Endomycorrhizas affect growth of *Dryopteris filix-mas*, *Trans. Br. Mycol. Soc.*, 69, 161, 1977.
87. Boullard, B., *Considerations sur la Symbiose Fongique chez les Pteridophytes*, Syllogeus No. 19, National Museum of Natural Science, Ottawa, 1, 1979.
88. Laferriere, J. and Koske, R. E., Occurence of VA mycorrhizas in some Rhode Island Pteridophytes, *Trans. Br. Mycol. Soc.*, 76, 331, 1981.
89. Peterson, R. L., Howarth, M. J., and Whittier, D. P., Interactions between a fungal endophyte and gametophyte cells in *Psilotum nodum*, *Can. J. Bot.*, 59, 711, 1981.

90. Prat, H., Etude des mycorhizes du *Taxus baccata, Ann. Sci. Nat. Bot. Biol. Veg.,* 8, 141, 1926.
91. Strullu, D. G., Histologie et cytologie des endomycorhizes, *Physiol. Veg.,* 16, 657, 1978.
92. Strullu, D. G., Gourret, J. P., Garrec, J. P., and Fourcy, A., Ultrastructure and electron-probe microanalysis of the metachromatic vacuolar granules occurring in *Taxus* mycorrhizas, *New Phytol.,* 87, 537, 1981.
93. Mejstrik, V. and Kelley, A. P., Mycorrhizae in *Sequoia gigantea* Lindl. et Gard. and *Sequoia sempervirens* Endl., *Ceska Mykol.,* 33, 51, 1979.
94. Fontana, A., Mycorrhizal status in *Ginkgo biloba* roots, *Abst. 5th N. Am. Conf. Mycorrhizae,* Quebec, Canada, 1981, 43.
95. Sequeira, L., Lectins and their role in host-pathogen specificity, *Annu. Rev. Phytopathol.,* 16, 453, 1978.
96. Reisert, P., Plant cell surface structure and recognition phenomena with reference to symbioses, *Int. Rev. Cytol. Suppl.,* 12, 71, 1981.
97. Bonfante-Fasolo, P., Interaction entre plante-hote et champignon mycorhizogene. Role des polysaccharides et des proteines des parois cellulaires, in *Les Mycorhizes, partie integrante de la plante: Biologie et perspectives d'utilisation,* Gianinazzi-Pearson, V. and Trouvelot, A., Eds., Colloque de l'Institut National de la Recherche Agronomique, Paris, 1982, 41.
98. *Les Mycorhizes, partie integrante de la Plante: Biologie et perspectives d'utilisation,* Colloque de l'INRA, No. 13, 1982.
99. Pirozynski, K. A. and Malloch, D. W., The origin of land plants: a matter of mycotrophism, *Biosystems,* 6, 153, 1975.
100. Nicolson, T. H., Evolution of vesicular-arbuscular mycorrhizas, in *Endomycorrhizas,* Sanders, F. E., Mosse, B., and Tinker, P. B., Eds., Academic Press, London, 1975, 25.
101. Taylor, F. J. R., Symbionticism revisited: a discussion of the evolutionary impact of intracellular symbioses, in The Cell as a Habitat, Richmond, M. H. and Smith, D. C., Eds., *Proc. R. Soc. London,* 204, (Spec. Ed.), 267, 1979.
102. Margulis, L., *Symbiosis in Cell Evolution,* Freeman & Company, San Francisco, 1981, 37.
103. Kidston, R. and Lang, W. H., On old red sand stone plants showing structure from Rhynie chert bed, Aberdeenshire. V. The Thallophyta occurring in the peat bed, the succession of the plants throughout a vertical section of the bed, and the conditions of accumulation and preservation of the deposit, *Trans. R. Soc. Edinburgh,* 52, 855, 1921.
104. Pirozynski, K. A., Interactions between fungi and plants through the ages, *Can. J. Bot.,* 59, 1824, 1981.

Chapter 3

ECOLOGY OF VA MYCORRHIZAL FUNGI

Barbara A. Daniels Hetrick

TABLE OF CONTENTS

I.	Introduction	36
II.	Dispersal	36
	A. Active Dissemination	36
	B. Passive Dissemination	37
	1. Rodent Mycophagy	37
	2. Dispersal by Worms, Birds, Insects, Etc.	38
	3. Dispersal by Wind	38
III.	Spore Germination	39
	A. The Influence of Soil Microflora	39
	B. The Influence of Environmental Conditions	40
	1. Soil Water Potential	40
	2. Nutrient Content of Soil	40
	3. Host and Nonhost Plants	42
IV.	Colonization and Sporulation	42
	A. Fungal Species	43
	1. Inoculum Density	43
	2. Competition Between Fungal Species	43
	B. Environment	44
	1. Temperature and Light	44
	2. Soil Fertility	44
	3. Host Plant	45
V.	Survival	46
	A. Cropping Schedule	47
	B. Soil Condition	48
	C. Influence of Other Soil Organisms	48
	D. Season	50
VI.	Conclusions	50
References		51

I. INTRODUCTION

The occurrence and ecological importance of vesicular-arbuscular mycorrhizal (VAM) fungi have been extensively studied in certain plant communities such as the tropical rain forest[1-3] and sand dunes,[4,5] but much less extensive information is available on the ecology of the fungal symbionts themselves. The factors which influence their survival, germination, root colonization, and sporulation in nature are only partially understood. As we move closer to using mycorrhizal fungi to increase agricultural plant yield, a more thorough understanding of these factors will be critical.

In this chapter our knowledge of the factors which influence the colonization and life cycle of VAM fungi will be summarized. Extensive research has been published on certain aspects of this subject, making a complete review of all literature beyond the scope of a book chapter. In these cases only representative studies are discussed. More often, however, the literature cited is a fairly complete representation of the state of this science. Far more research is needed on the ecology of VAM fungi, and these areas are discussed in the following chapter as well. Where less complete information was available I have taken the liberty of speculating, in hopes of stimulating some thought on the part of the reader. Hopefully, more research in VAM ecology will follow, even if it is just to prove me incorrect.

II. DISPERSAL

VAM fungi are indigenous to soils throughout the world. In fact, many VAM species are represented on most continents. As an explanation for their remarkably widespread distribution, Trappe[6] proposed that VAM fungi were disseminated intercontinentally prior to continental drift. The super continent Gondwanaland is thought to have begun to break apart and drift north about 125 million years ago.[7] Fossil records of plants containing VAM-like structures[8] have been dated to about 370 million years ago,[9] so that distribution of VAM fungi could have occurred as Trappe suggested. Alternatively, the widespread distribution of VAM fungi may simply reflect the millions of years over which dispersal of these fungi has occurred.

Later in this section the means by which VAM fungi are known to be dispersed will be discussed. These include active dissemination (growth of mycelium through soil) and passive dissemination where VAM fungi are moved by wind, water, or by soil microorganisms.

A. Active Dissemination

VAM fungi may be disseminated in a variety of ways. Active dispersal occurs as mycelia grow through the soil. Powell[10] studied the rate of spread of VAM fungi through fumigated sandy loam soil, which had been inoculated with mycorrhizal fungi or planted with mycorrhiza-infected plants, and projected that an efficient mycorrhizal fungus might move only 65 m in 150 years or 0.43 m/year. Powell[10] further demonstrated that VAM fungal species differed in their rate of spread and in their ability to retain possession of colonized plants under encroachment by other VAM fungal species. The rate of spread was reduced in inoculated soil preplanted with seedlings already mycorrhizal but was increased in soil previously cropped with nonmycorrhizal plants. Apparently, introduced VAM fungi spread more readily through soils which already contain low populations of VAM fungi, particularly if the population has been depressed by cultivation to nonhost crops.

More recently it has been shown that plant species and root density may significantly influence the rate of VAM fungus spread.[11] In clover, the greatest rate of spread of *Glomus fasciculatum* was 1 cm/week while in fescue, *G. fasciculatum* spread at only

0.7 cm/week. These experiments showed that root density is most critical when plants are young and root density is low. In fact, supraoptimal root densities were achieved in fescue (a grass which develops an extensive root system) and rate of fungal spread was reduced as plant size increased. A similar supraoptimal root density was not achieved in clover because of the less extensive root system.

The previous experiments were conducted in the greenhouse in fumigated soil, and it is difficult to project the rate of mycelial growth and VAM fungus spread through field soils. Colonization by morphologically similar indigenous VAM fungi in nonsterile field soils makes this type of study difficult. However, using sporulation by nonindigenous VAM fungi as proof of spread, Mosse et al.[12] demonstrated that *Glomus caledonicum* was able to spread 7 to 13 cm from an inoculation point after 13 weeks. No correlation was observed between rate of spread and plant size, but spread rate was greater in nonsterilized plots than in those receiving formalin treatments. In these experiments host species also significantly affected spread rate.

Higher rates of fungal spread (1.5 to 3.4 m/year) have been reported for certain soilborne plant pathogenic fungi in nonsterile soils,[13] although factors such as soil fertility, seasonal fluctuation in moisture, temperature, and microbial activity will influence the rate of spread of VAM fungi. The mathematical model suggested by Smith and Walker[14] may be used for determining colonization frequencies and rate of fungal growth through the plant.

Whether VAM fungi grow in a directed way, i.e., toward a root stimulus or randomly in soil has been debated. Directed growth would most likely make optimum use of energy supplies in the spore and would increase the number of infective hyphal strands which reach a host.[13] Powell,[15] using the buried slide technique in partially sterilized soil, demonstrated little or no attraction of VAM hyphae to roots until random contact occurred, except with hyphae from honey-colored spores *(Acaulospora laevis)* which frequently grow toward the roots. Koske[16] has demonstrated chemotactic attraction of hyphae of *Gigaspora margarita* to host roots in vitro. Since hyphae would pass through the air to reach the host roots suspended above germinating spores, the attractant is probably a volatile substance. Whether such chemotactic substances are produced under field conditions and can direct mycelial growth in the field has not been studied.

B. Passive Dissemination
1. Rodent Mycophagy
Many soil-borne fungi have developed highly specialized methods of dispersal. This is particularly true of the hypogeous fungi which fruit in sporocarps below ground. These hypogeous fungi are protected from climatic stresses which might prevent sporulation of epigeous fungi. As spores mature these hypogeous fungi frequently emit an odor which, by becoming increasingly strong, attracts rodents. The rodents eat the sporocarp, digest the peridium or glebal mycelial constituents, and defecate the spores which remain intact. These spores are thus packaged in a fertile environment and are somewhat protected. In addition, adaptation to dispersal by rodents may increase the probability that spores will be deposited on or near roots of susceptible host plants.[17]

Species of Endogonaceae sporulate singly in soil and also in hypogeous or epigeous sporocarps, but so far, only sporocarpic species of Endogonaceae have been found in rodent stomachs. Since Fogel and Trappe[18] observed a relationship between mammal size and the size of sporocarps ingested, it is not surprising that endogonaceous sporocarps, generally smaller (1 to 10 mm in diameter) than those formed by Ascomycetes and Basidiomycetes, are ingested by smaller mammals such as Soridae (shrews), Zapodidae (jumping mice), Cricetidae (mice, rats, lemmings, voles), and Ochotonidae (pikas). Despite the large spore size (up to 400 µm), these spores pass through the rodent digestive system and remain germinable.[19] Furthermore, the defecated spores are capable of initiating typical VAM fungal infections.[20] Mycophagy remains the only known dispersal method for spores of Endogonaceae formed in hypogeous sporocarps.

It is curious that Endogonaceae sporocarps, even when mature, do not emit a strong odor as produced by Basidiomycete and Ascomycete hypogeous sporocarps. Although the bright color of epigeous sporocarps might visually attract rodents, the means by which hypogeous sporocarps are detected is unclear. Unless ingestion of these hypogeous sporocarps is entirely random, they are probably still detected by smell.

2. Dispersal by Worms, Birds, Insects, Etc.

A number of other vectors of VAM fungal spores have been described. As early as 1922, endogonaceous spores were observed in the digestive tracts of millipedes[21] and more recently they have been found in grasshoppers and crickets,[22] and in earthworm and ant casts.[23] These vectors probably ingest single endogonaceous spores but not whole sporocarps.[18] It is clear from the research of McIlveen and Cole[23] that endogonaceous spores remain viable following earthworm ingestion since earthworm casts give rise to typical VAM colonization when inoculated onto soybean plants.

The major contribution of some of these vectors may be that soil containing spores or spores themselves are brought to the soil surface, thus favoring further dispersal by wind if that occurs.[23] Soil containing spores can also be brought to the soil surface by the activity of mud dauber wasps, robins, or sparrows in whose nests spores have been found.[23] Spores from swallow nests were also able to initiate typical VAM colonizations.

The importance of digging or burrowing animals which do not themselves ingest fungi but bring soil to the surface has recently been dramatically illustrated by MacMahon.[24] He observed that very few plants have re-established themselves on Mount St. Helens following the 1981 volcanic eruption, except on gopher mounds. The soil from these gopher mounds contained 24 to 80 VAM fungal spores per gram of soil as compared with adjacent ash above the entrained organic soil layer which contained 0.1 to 1.7 spores per gram of soil. In stripmined areas, rodent dissemination of VAM fungal propagules has been shown to increase the rate of revegetation.[20] At Mount St. Helens, burrowing and digging animals which bring inoculum to the surface may be equally vital to revegetation.

3. Dispersal by Wind

Spores brought to the soil surface by a variety of vectors are theoretically available for wind dissemination. To date, there were no reports that spores are wind disseminated or that these spores which might be dispersed during wind or dust storms remain viable. However, there is increasing circumstantial evidence which strongly implies that wind dissemination is effective and occurs regularly.

As already mentioned, Hansen and Ueckert[22] reported that VAM fungal spores had been ingested by grasshoppers and crickets. Similarly, Ponder[25] reported VAM fungal spores in grasshopper and rabbit droppings. While occasional bits of soil might be ingested by these animals, they are primarily leaf feeders. The presence of VAM fungal spores in their digestive tracts or feces, therefore implies that spores were present on leaves prior to feeding, probably as a result of wind dispersal.[26] The presence of viable VAM fungal spores in rabbit droppings certainly implies that the spores can survive wind dissemination.[25] Taber[27] has observed VAM fungal spores in *Portulaca* seed capsules which are oriented on the plant toward the wind. VAM fungal spores appear trapped on the mucilaginous surface of seeds within the capsule, and were probably wind disseminated.

The turbulence and wind velocity necessary to disseminate spores of VAM fungi and the distance they could be transported have not been studied. The unusually large size of VAM fungal spores (up to 400 μm in diameter) makes comparison with dissemination of other fungal spores difficult. Tommerup and Carter,[28] in devising a method to separate spores from soil, have demonstrated that velocities of 0.10 to 0.55 m/sec would transport spores. The maximum velocity used was that necessary to move 100-μm quartz particles. Spores exposed to these wind velocities were also demonstrated to be viable. Thus, while direct evidence for wind dissemination is still lacking, this form of dispersal is certainly strongly implied from the studies mentioned above.

The influence of man in transporting equipment or plants with adherent soil has not been discussed. Certainly VAM-colonized plants would show no visible symptoms which would cause them to be discarded as might occur with pathogen-infected plants. This form of dissemination may be of greatest importance in the dispersal of VAM fungal spores but there are no data on which this can be evaluated.

III. SPORE GERMINATION

The germination of soil fungi, particularly soil-borne plant pathogenic fungi, has been the subject of considerable research, because induced failure to germinate or germination in the absence of a host plant can result in satisfactory pathogen control. Many soil-borne pathogenic fungi have developed higher degrees of host specificity and germinate only or most frequently in the presence of a susceptible host plant. Thus, host root exudates are often the "triggers" which initiate spore germination. Failure to infect rapidly after germination will frequently result in secondary spore formation or lysis of germ tubes by the soil microflora.[29,30] Although VAM fungi are soil-borne and infect living plant roots, these fungi exhibit little host specificity and have developed different "triggers" for germination. In general, the germination of spores of VAM fungi may be influenced more by soil microorganisms and the physical and chemical environments than by the presence or absence of host or nonhost plant roots.

A. The Influence of Soil Microflora

Germination of VAM fungal spores in vitro can be erratic. On agar media a dormancy factor which retards germination has been observed. Godfrey[31] reported that high levels of *Glomus microcarpum* spore germination occurred only after spores were incubated to remove a "dormancy factor". Daniels and Graham[32] observed reduced germination of freshly harvested spores of *Glomus mosseae*. Hepper and Smith[33] also noted the improved spore germination of *G. mosseae* after storage for several months. In contrast, Koske[34] demonstrated that spores of *Gigaspora gigantea* freshly collected from the field, germinated well without prior storage or cold treatment on sand or agar medium. This apparent contradiction may be explained if spores collected from the field are naturally aged and thus require no additional storage to remove a dormancy factor. It is also possible that spores of *Glomus* spp. require different stimuli for germination than spores of *Gigaspora* spp.

The nature of this dormancy factor which Godfrey[31] reports is unclear. Perhaps VAM fungal spores are not physiologically mature even when they appear morphologically mature. Alternatively, VAM fungal spores may contain self-inhibitors removed in the presence of soil microorganisms or by leaching.[35] Nevertheless, improved germination under nonsterile conditions or with nonsterile soil amendments has been demonstrated by several researchers.[32,35,36]

In 1959, Mosse[36] observed that spores of a *Glomus* sp. germinated poorly on water agar but exhibited much improved germination on water agar containing nonsterile soil overlaid with autoclaved cellophane disks. She proposed that a water soluble, heat labile, dialyzable, microbially produced soil constituent stimulated germination. Daniels and Graham[32] also reported stimulation of germination by a dialyzable soil constituent but attributed this to nutrient effects. Subsequently, Daniels and Trappe[35] found that spores of *Glomus epigaeum* failed to germinate in autoclaved, steamed, or gamma-irradiated soil, while high levels of germination occurred in nonsterile soils. This implies that germination is suppressed in sterilized soil because of some inhibitory compound released during sterilization. Alternatively, inhibitory levels of nutrient released following sterilization may also explain the stimulation of germination observed in nonsterile soil.

If a germination stimulus exists in nonsterile soil the nature of this stimulus has not been elucidated. It has been suggested, however, that self-inhibitors of spore germination might be removed by the activity of soil microorganisms.[35] In support of this hypothesis, germination did occur in autoclaved kaolin clay and activated charcoal, substances with high cation exchange values, perhaps capable of removing or adsorbing self-inhibitory compounds from the spore. The hypothesis of self-inhibition of germination is strengthened by the observation that hyphal growth following germination is also improved in the presence of activated charcoal, again suggesting the action of self-inhibitors.[37] That the germination of *Gigaspora* spp. spores may be regulated differently is again suggested by Koske,[34] who found no difference between germination in sterile or nonsterile sand. The low nutrient status of sand, however, might also explain these apparently contradictory results.

B. The Influence of Environmental Conditions
1. Soil Water Potential

The influence of soil water potential on VAM fungal spores has been studied by Daniels and Trappe[35] using *Glomus epigaeum* added to silt loam of varied moisture contents, and by Koske[34] using *Gigaspora gigantea* placed on sand to which concentrations of polyethylene glycol were added. *Glomus epigaeum* spores germinated best at moisture contents between field capacity and soil saturation. Below field capacity, germination declined with no germination occurring below -31 bars.[34] In contrast, *Gigspora gigantea* germination was strongly inhibited at -10 bars but higher levels of germination could eventually be obtained at low water potentials if spores were incubated longer.[34] Koske further observed that germ tube length was reduced at low water potentials. Thus, low water potential can, at least with *Gigaspora* spp., delay germination and reduce the hyphal growth from germinated spores. Since spores of *Gigaspora* spp. are capable of re-germination,[38] this inhibition or reduced germination may be energy saving.

Further research will be necessary to determine whether *Glomus* spp. also exhibit a reduced germination rate in low water potential conditions as do *Gigaspora* spp. Because spores of *Gigaspora* spp. germinate by production of a new germ tube which penetrates the spore wall, their germination is more easily observed. This germ tube, because of its cylindrical shape, is easily distinguished from the bulbous, gametangial hyphal attachment which characterizes this genus. In contrast, spores of *Glomus* spp. germinate by regrowth of their hyphal attachment or new germ tubes grow through the old hyphal attachments on which the spores were formed. It is far more difficult, therefore, to ascertain whether spores of *Glomus* spp. have indeed germinated, and for these reasons it is extremely difficult to observe a reduced germination rate which might occur at low water potentials.

2. Nutrient Content of Soil

The germination of VAM fungal spores does not appear to be greatly influenced by soil fertility. Koske[34] observed no difference in germination of *Gigaspora gigantea* spores regardless of phosphorus concentrations. Similarly, Daniels and Trappe[35] observed that additions of nitrogen or potassium did not appreciably stimulate or inhibit germination, and although certain levels of phosphorus amendment gave a statistically significant increase in germination, this stimulation was probably not biologically significant. However, Siqueira et al.[39] have also observed that a phosphorus amendment increased spore germination on water agar but nitrogen and potassium amendments had no effect.

Though inorganic nutrients such as nitrogen and phosphorus have little effect on VAM fungal spore germination, addition of glucose to soil depressed the germination of *Gigaspora gigantea*.[34] Both germ tube length and number per spore were reduced. Similarly, Siqueira et al.[39] reported decreased germination of *G. margarita* spores and reduced germ tube growth following amendment of water agar with a range of organic substrates. It would be interesting to know whether organic substrate amendment is fungicidal or only fungistatic in nature. In soil, amendment of glucose probably results in a flush of microbial activity resulting in the glucose being rapidly metabolized. If, under high nutrient conditions, germination of some spores is prevented, germination should resume soon when the glucose level is depleted, however, Koske[34] reports that germination was still lower in glucose-amended soil after 7 days than that observed in unamended soil. Alternatively, the increased microbial activity in glucose-amended soil might result in increased spore lysis or hyperparasitism of VAM fungal spores, thereby reducing the level of germination. The mechanism by which organic substrates reduce spore germination on agar medium is unclear. The fact that organic substrates reduce germination both on agar medium and in nonsterile soil may indicate that these substrates directly influence spore germination rather than stimulate microbial activity which then indirectly inhibits spore germination.

The pH optimum for spore germination will probably differ with each VAM species and the environment to which each is indigenous.[40] For example, *Glomus mosseae*, common in alkaline flatland soils,[41] germinated well on water or soil extract agar at pH 6 to 9. In comparison, *Gigaspora coralloidea* and *G. heterogama* from more acidic Florida soils germinated best at pH of from 4 to 6.[40] Spore germination of *Glomus epigaeum* occurred over a wide range of soil pH, with optimum germination occurring between 6 and 8. Thus, it appears that pH can influence the germination of VAM fungal spores, but germination seems to occur within a range which is still acceptable for plant growth.

From the literature it is difficult to interpret how pH influences spore germination. In soil, nutrient concentration, excepting phosphorus,[35] has little effect on spore germination. Thus, it seems unlikely that pH-induced differences in nutrient availability are responsible for stimulation or inhibition of VAM fungal spore germination. However, Siqueira et al.[39] report a significant interaction between pH and nutrient which influences spore germination on agar medium. Apparently, the optimum pH for spore germination may depend not only on the fungal species but also on the nutrient content of the germination medium.

Just as the optimum pH for germination appears dependent on the environmental adaptation of a VAM fungal species, the temperature range over which germination occurs may also depend on the species of VAM fungi and environments to which they are ecologically adapted.[42] Florida isolates of *Gigaspora* spp. germinated best on soil extract agar incubated at 25 to 35°C while *Glomus mosseae* from a cooler Washington State environment germinated best when incubated at 18 to 20°C. Koske[34] found optimum germination for *Gigaspora gigantea* from Rhode Island to be 30°C, while Daniels and Trappe[35] observed that *Glomus epigaeum* from Oregon germinated best at 22°C. While these differences in temperature optima may reflect ecological adaptations, *Gigaspora* spp. from Rhode Island and Florida had similar temperature requirements even though the climate of their sources were quite dissimilar. The similar pH and temperature requirements of *Glomus* spp. as compared with *Gigaspora* spp. may imply that differences between the two genera must also be considered.

3. Host and Nonhost Plants

As already mentioned, it is unusual for root-infecting fungi to germinate in the absence of host plant roots. VAM fungi seem to be an exception. Using the buried slide technique in partially sterile soil, Powell[15] demonstrated that VAM fungal spores germinated similarly whether or not onion roots were present. That spores will germinate well in nonsterile soil in the absence of host roots is well documented.[34-36] In fact, Daniels and Trappe[35] observed no additional stimulation of germination in the presence of host roots, and germination occurred equally well in the presence of nonhost or ectomycorrhizal plant roots which could lead to reduced populations of VAM fungi in soils. In contrast, Graham[43] observed that germination of *Glomus epigaeum* spores was increased, and that germ tube lengths were greater, when spores were exposed to root exudates. The germ tubes of these treated spores also branched more frequently than nontreated spore germ tubes.

VAM fungi are remarkably nonhost-specific. Certain VAM species may be more efficient in stimulating the growth of certain plant species, but each VAM fungus is generally able to colonize every VAM host species.[44] Further research may reveal exceptional VAM fungal species adapted to a narrow host range within a particular ecological environment, but this will certainly be the exception rather than the rule. Fungi which infect few or single plant species might require, for their survival, a plant stimulus for germination. The apparent lack of influence of plant roots on VAM spore germination, therefore, probably reflects the adaptation of these fungi to extremely wide host ranges.

Spore germination of VAM fungi appears to be controlled by levels of soil temperature, moisture, and pH which also induce plant seed germination. Thus, spores of VAM fungi germinate when newly formed growing roots are likely to be present. The extremely large spore size of these fungi (approximately 35 to 400 μm) and energy contained therein may permit the VAM fungi to grow through the soil for long distances in "search" of host roots. The ability to germinate more than once[38] probably also increases the chance that VAM fungi will successfully penetrate a host. However, in cases such as monoculture of nonhost plants, germination in the absence of host roots could be detrimental and result in reduced populations of VAM fungi in the soil.[10]

IV. COLONIZATION AND SPORULATION

Root colonization and subsequent spore production by VAM fungi are influenced by a wide range of environmental, host, and fungal effects. Colonization is used here as an alternative to the somewhat confusing term "root infection". In many cases, the factors which stimulate or inhibit colonization probably also stimulate or inhibit sporulation since these two phenomenon are often closely related.[45,46] For example, Hayman,[45] in sampling wheat plants through a growing season, observed that root colonization and spore production increased through the season, peaking just prior to harvest. Application of nitrogen fertilizer not only reduced root colonization but suppressed spore formation as well.

Although the close relationship between colonization and sporulation led Daft and Nicolson[46] to suggest that spore numbers might be an effective measure of root colonization, these two phenomena are not necessarily correlated. In temperate climates where root growth by perennial plants is more or less continuous, Baylis[47] found that few spores were produced despite relatively high levels of root colonization. He suggested that no evolutionary stimulus for spore production existed if root growth was not intermittent. In this section the influence of soil environment, host plant, and fungal species on colonization and sporulation of VAM fungi will be considered.

A. Fungal Species
1. Inoculum Density
The influence of inoculum density on root colonization and subsequent sporulation has been carefully studied by researchers seeking to increase plant growth stimulation or inoculum production. From these studies it appears that increased inoculum dosage results in increased percentage root colonization,[48-50] but eventually an upper limit of root colonization is reached and greater inoculum dosages will result in no further increase in root colonization.[49] Both Daft and Nicolson[51] and Ferguson[50] have examined the influence of inoculum density on rate of colonization. By examining plants with varied inoculum dosage at different growth stages they determined that increased inoculum dosage results in increased colonization rate, i.e., plants become more highly infected earlier, but the final colonization rates were similar regardless of inoculum dosage. However, in annual crops with short growing seasons, high inoculum dosages may be required if maximum plant growth stimulation is to be achieved. For inoculum production purposes as well, optimum spore production is achieved sooner if the necessary inoculum dosage is applied.[50]

2. Competition Between Fungal Species
Numerous attempts have been made over the years to determine the influence of mixed VAM fungal species on root colonization and sporulation. Whether VAM fungal species have synergistic or competitive effects on each other has been difficult to determine, however, because of the morphological similarity of mycorrhizal structures formed by the various VAM endophytes. Biochemical or genetic labeling of a particular fungus to follow its colonization in a plant, and thus distinguish it from other VAM fungal species has not yet been accomplished. Only with *Glomus tenue* and a few other such species which produce morphologically distinct structures have such studies been possible.[53] For example, Ross and Ruttencutter[54] were able to compare (on the same plant) colonization caused by *Gigaspora gigantea* with that resulting from *Glomus macrocarpum* because the former species fails to form vesicles while the latter does. So distinctive are the colonization structures of certain VAM fungal species that Abbott and Robson[55] proposed a dichotomous key for their identification based on root colonization structures. The species included in this key were *Gigaspora gigantea, Glomus tenue, G. monosporum, G. fasciculatum,* and *Acaulospora laevis.* The applicability of VAM fungal species identification based on root colonization structures to other species has not been tested (see Chapter 4).

Because of the experimental difficulties, the effect of multiple VAM fungal species colonizations has only been assessed in terms of the resulting plant growth response. There are numerous reports[57-61] of increased plant growth from inoculation with nonindigenous VAM species even into nonsterile soils containing indigenous species. It is assumed that these nonindigenous species are more efficient though this increased plant growth could result from uneven inoculum levels. It is also possible that increased plant growth reflects better competitive ability of the introduced fungus.

In support of this latter hypothesis, Ocampo et al.[62] found that less VAM colonization occurred in barley, lettuce, potato, or onion from the mixed endophyte population in nonsterile soil than occurred when *Glomus fasciculatum* alone was inoculated onto these crops in sterilized soil. It is unclear, however, whether inoculum potential for each species was equal in these experiments. Crush[63] has proposed that colonization by other VAM fungi may be "shed" in favor or *G. tenue* colonization because the latter is a superior competitor for phosphorus. The superior competitive ability of *G. tenue* was also demonstrated by Powell and Daniel[61] who observed increased plant growth when *G. tenue* was added to host pot cultures already colonized by other VAM fungi. Further evidence for competition between VAM fungal species came from Ross and Ruttencutter,[54] who found that less colonization occurred in *G. macrocarpum*-inoculated peanuts and soybeans than when these hosts were inoculated with *Gigaspora gigantea* and *Glomus macrocarpum* together. The percentage of *G. macrocarpum* colonization, as measured by vesicle formation, was reduced when combined with *Gigaspora gigantea*, suggesting the superior competitive ability of *G. gigantea*. Although many differences in efficiency of VAM fungi may reflect differences in competitive ability among species for host root, it is also possible that these differences may be attributed to the adaptation of VAM fungi to particular soil conditions.[64]

B. Environment
1. Temperature and Light

Both temperature and light have been shown to have a significant influence on colonization and sporulation by VAM fungi under greenhouse conditions. Higher temperatures generally result in greater root colonization,[65,66] and increased sporulation.[65] Studying the effects of temperature on VAM establishment, Schenck and Schroder[67] observed that maximum arbuscule development occurred near 30°C but that mycelial colonization of the root surface was greatest between 28 and 34°C. Sporulation and vesicle development were greatest at 35°C. Periods of cold stress followed by maintenance of high soil temperature have also been shown to increase colonization and sporulation.[50]

Although increased light intensity generally increases percentage colonization,[50,66,68-73] longer daylengths also increase root colonization.[74-78] In fact, a photoperiod of 12 hr or more may be more important than light intensity in providing high levels of root colonization,[66,77] but if suitable daylength is provided, increased light intensity may still increase colonization.[79] Though low light intensity can significantly reduce root colonization, its effect on sporulation may be less pronounced.[50] All of these studies have been conducted under greenhouse conditions, and more research is necessary to see whether the results apply to the field as well.

2. Soil Fertility

It is widely accepted that maximum root colonization and sporulation occur in soil of low fertility. Both phosphorus[45,51,80-82] and nitrogen[83-85] may significantly reduce root colonization if present at high levels, and a delicate balance between these two elements appears to exist. Bevege[86] found that root colonization increased as nitrogen content increased if phosphorus levels were moderate. At higher levels of phosphorus, however, nitrogen applications were inhibitory. A similar balance has been observed for zinc and phosphorus.[87]

The effect of high soil fertility on root colonization depends on the host plant grown. Strezemska[88] observed that root colonization of rye, wheat, barley, and oats was reduced after years of cropping in highly fertilized soils, but colonization of bean roots was not similarly reduced under these conditions. The differing sensitivity of crop plants to soil fertility may explain Gerdemann's[89] observation that many crops in the midwest U.S. remained highly mycorrhizal despite high soil fertility.

Research by Menge and co-workers[90] has shown that much of the influence of soil fertility on root colonization is plant mediated. Using a split root technique, phosphorus fertilizer was applied to half of a sudangrass root system.[90] Sporulation even in the unfertilized half was reduced. High levels of root colonization and sporulation were also observed in roots exposed to high levels of soil phosphorus if the bulk of the root system was exposed to lesser amounts of phosphorus. Thus the internal phosphorus concentration of plants controls the level of root colonization and sporulation by VAM fungi which will occur. Ratnayake et al.[91] and Graham et al.[92] have further demonstrated that the net leakage of root exudates is significantly greater under low phosphorus levels. High levels of exudation were correlated with decreased phospholipid levels and increased permeability of root membranes. They suggested that root colonization by VAM fungi is inhibited at high phosphorus levels because of the decreased root exudation. Thus, the content of phosphorus in the root can mediate root colonization by VAM fungi.

That it is the internal phosphorus content of roots which controls root colonization by VAM fungi is also suggested by the research of Sieverding[93] and Nelsen and Safir,[94] who observed that greater root colonization occurs in drought-stressed plants than in plants receiving adequate water. Nelsen and Safir[94] suggested that high levels of root colonization can occur in drought-stressed plant even in highly fertile soils because low moisture levels can reduce the diffusion rate of nutrients such as phosphorus and decrease the availability of these nutrients to the plants.

3. Host Plant

The presence or absence of a host plant obviously plays a large role in whether or not colonization and subsequent sporulation will occur. Nonhost plants such as Chenopodiaceae and Cruciferae species can become minimally colonized by VAM fungi,[95] particularly when grown in the presence of host plants.[62,96] The influence of nonhost plants on the colonization of host plants has been studied with contradictory results. For example, the presence of nonmycorrhizal plants has resulted in reduced colonization of mycorrhizal host plants,[97-99] possibly because of toxic nonhost-root exudates[97,98] or seed coat components.[99] In contrast, Ocampo et al.[62] detected no reduced colonization of mycorrhizal plants cropped together with nonmycorrhizal hosts. In fact, onions became more colonized when grown with nonhost swedes than when grown alone and similar results were observed in barley cropped together with rape. These experiments were conducted in pots in the greenhouse. If planting host and nonhost species together results in more rapid utilization of soil nutrients because there is more root volume per pot, an increase in root colonization might not be unexpected.

There is also evidence that competition can arise between host species. Colonization of weakly mycorrhizal plants can be increased when grown in the presence of strongly mycorrhizal "nurse plants".[100] In contrast, Ocampo et al.[62] demonstrated that less colonization occurred in onions grown in combination with barley or potato than alone. Similarly, maize and onions grown alone were more colonized than when grown together. Once again, however, it is possible that the reduced colonization occurring when two host plants are cropped together reflects an increased amount of plant root biomass available for colonization. Clearly, more research is needed to understand the impact of host plant competition on root colonization.

The affinity of host plants to VAM endophytes will also determine the degree of colonization or sporulation which occurs. Schenck and Kinloch[101] observed that the incidence of VAM fungal species (determined as spore numbers in soil) depended upon the plant species which was colonized. A woodland site was newly planted with 6 agronomic crops and grown in monoculture for 7 years. Spore number of *Gigaspora* spp. were most numerous around soybean plants, but *Glomus* and *Acaulospora* spp. predominated around monocotyledonous crops. This influence of host plant on incidence of VAM fungi has also been observed by Kruckelmann[95] on a site where 6 crops were grown in monoculture for 16 years. It appears that the host plant can affect sporulation and possibly survival of VAM fungi.

Although VAM fungi have extremely wide host ranges,[1] the existence of host preference has been suggested by many researchers.[76,86,102-104] This preferential association between certain plant and fungal species can be evaluated with respect to combinations which provide the greatest plant growth stimulation and the greatest root colonization or maximum sporulation, but these three factors are not necessarily correlated.[104] There are no reports at present of exclusive colonizations between a VAM fungus and host plant. The factors which determine host-symbiont affinities have not been studied and no doubt are of considerable importance.

In general, the factors which result in maximum plant growth may result in maximum sporulation.[46,85,105] Reduction in plant size by pruning had no influence on root colonization but inhibited sporulation by 3-month-old plants.[50] Defoliation of plants[79] can also decrease root colonization and sporulation. Similarly, logging of sugar maple trees resulted in reduced sporocarp formation as compared with sporulation in unlogged stands.[10]

V. SURVIVAL

Both spores of VAM fungi and colonized root pieces can be effective propagules, initiating typical VAM colonization in host plants,[11,50,57,107-109] although the ability of VAM fungi to persist in soil may depend partly on the type of propagule formed. It is clear that hyphal fragments and root pieces colonize host roots more rapidly than spore inoculum, thus producing a growth response sooner,[57,110,111] but the ability of this inoculum form to survive in the absence of an actively growing host is questionable.

In New Zealand bush soils, Baylis[47] observed a scarcity of spores as compared with soils carrying native grasses or those in cultivation. He suggested that the latter soils, subjected to intermittent root growth and drying, might stimulate sporulation or select for isolates of VAM fungi which sporulate. Alternatively, in the bush soils, where actively growing roots are always present because of year-round adequate soil moisture and temperature, VAM fungal isolates which do not sporulate may have been favored because sporulation is unnecessary in these soils.

The greater ability of spore inoculum to withstand harsh environmental conditions has been suggested by Hall.[111] The amount of infective fungal material in pelleted inoculum (composed of spores, colonized root pieces, and hyphae) decreased rapidly when dried for only 2 weeks. Presumably, VAM hyphal fragments were less able to survive drying than were spores. Similarly, Gould and Liberta[112] compared the inoculum level contained in stored topsoil with the level found in undisturbed soil. While spore numbers in each sample were similar, the stored topsoil had a lower inoculum potential (initiated fewer colonizations) than did undisturbed soil. Although the heterogeneity of soil samples may account for some of these differencs it is also possible that this variation reflects the greater survivability of spore inoculum.

In contrast, Tommerup and Abbott[109] demonstrated that root pieces colonized by several VAM fungi, when stored in soil dried to -50 MPa, remained infective even after 6 months. It is unclear, however, whether the inoculum potential of these pieces was reduced during storage, i.e., whether the energy for growth of the fungus available for host colonization[29] had declined. It is also possible that storage under moisture conditions more conducive to root decomposition or to root dessication might result in loss of viability for hyphal fragments contained in root pieces.

To summarize, colonized root pieces can remain as viable propagules for extended periods in partially dried soil,[109] but rapidly lose viability in moist soils under storage.[111,112] For this reason spores of VAM fungi are probably the primary propagule for survival during periods of intermittent root growth or under field conditions in the absence of a host plant except under very dry soil conditions. Spore populations in soil probably also decline, but rates of decline and factors affecting decline have received little attention. The inoculum potential assays of Porter[113] and Powell[114] will make this type of study possible though these most probable number techniques have yet to be applied to research on spore survival.

If spores are the predominant survival structure of VAM fungi, the inoculum potential of a soil could be directly related to the number of spores produced. The nutritional conditions under which spores of pathogenic fungi are formed influence the inoculum potential of those spores.[115-117] If this is true for VAM fungi as well, then the environmental conditions prior to and during sporulation would also influence the VAM fungal inoculum potential of a soil. Therefore, the factors which influenced sporulation by controlling the amount of initial inoculum may indirectly influence VAM fungus survival in a soil.

Survival of propagules of VAM fungi is probably directly influenced by hyperparasites, unsuccessful attempts at colonization, proximity of host plants, etc. Throughout the remainder of this section only those factors which directly influence propagule survival will be discussed. The factors which affect colonization and sporulation have been discussed in another section of this chapter, but their indirect influence on survival should be considered.

A. Cropping Schedule

The influence of cropping sequence on survival of VAM fungi is not well understood. The number of growing seasons in the absence of a host plant apparently influences VAM survival whether the soil is fallowed or cropped to a nonhost. Since new spores are not produced in these situations, the population of VAM fungi can only decline. Soil storage or fallowing appears to reduce the inoculum potential level in soil. Gould and Liberta[112] demonstrated reduced inoculum potential in stored topsoil compared with undisturbed soil even though spore populations did not differ greatly. Similarly, Black and Tinker[118] found that spore levels generally decreased by the end of a barley growing season, but a greater rate of decrease was evident in fallowed or nonhost-cropped soil. A year of fallow or nonhost cropping reduced by half the colonization level obtained in the subsequent barley crop. Ocampo[119] however, observed no difference in colonization levels whether plants were grown in soil kept fallow for 10 weeks prior to planting or in soil amended with inoculum stored under refrigeration. These results suggest that the length of time a host is absent may influence the amount of inoculum which survives.

As discussed in Section IV (Colonization and Sporulation), of this chapter, nonhost plants belonging to the Chenopodiaceae and Cruciferae are reported to become minimally colonized by VAM fungi.[95] However, these nonhost plants have also been observed to reduce host plant colonization although Ocampo[119] observed no such deleterious effects and, in fact, reported increased colonization of host plants in close proximity to nonhost plants. Ocampo et al.,[62] therefore, suggested that VAM fungi may be able to derive some benefit from nonhost plants and that nonhost cropping may favor the development of VAM fungi more than fallowing. This hypothesis was tested subsequently in a series of crop rotation experiments.[119] While greatest initial colonization occurred in soil precropped with a host plant, more colonization occurred in soil precropped with a nonhost plant than in soil previously fallow. After 8 weeks, however, the colonization level was similar in all plants. These results are similar to those of Daft and Nicolson[120] who demonstrated lower initial colonization levels in tomatoes inoculated with low inoculum numbers than in plants which received large inoculum dosages. Regardless of inoculum level, similar colonization levels were achieved by the end of the experiments.

Thus it appears that inoculum levels, which decrease during fallow or nonhost cropping, can after one season's cultivation of a host crop overcome the negative effect of fallowing or nonhost cropping. After many years of deleterious cropping schedules, however, VAM populations may be markedly reduced.[95] Whether at such extremely low levels of inoculum more than one host cropping season might be necessary to restore VAM inoculum potential to levels which benefit plant growth has not been studied. Alternatively, if VAM inoculum can in effect be eliminated from soils by overfertilization or deleterious cropping schedule, reinoculation with VAM fungi may be necessary.

B. Soil Condition

Both soil fertility and pH significantly influence spore production of VAM fungi[95] and thus (indirectly) influence VAM survival. Although high levels of phosphorus and nitrogen in soil and artificial media are known to inhibit or reduce root colonization[82,84,107,121] and subsequent spore formation,[83,95] it is the nutritional status of the plant, not the soil fertility, which determines the degree of colonization and spore formation which will occur.[90] As previously discussed, spore germination occurs independently of soil fertility. However, in fertile soils root exudation may be decreased, and root colonization by VAM fungi may be inhibited.[91,92] If, as Koske[38] suggests, spores are able to germinate more than once, then each unsuccessful attempt at colonization probably results in reduced inoculum potential, but not spore death. The number of such unsuccessful attempts and the energy cost of each attempt has not been studied.

Soil temperature influences root colonization and sporulation[65-67] and may indirectly affect VAM fungal propagule survival. The direct effects of temperature on spore survival have not been studied. Because soil temperature directly influences VAM fungal spore germination and plant root growth it may directly influence the number of successful root penetrations that can occur. In addition, the activity of the soil microflora which is influenced by soil temperature[30] may affect the rate of VAM hyphal lysis or hyperparasitism of spores.

Though soil conditions can influence the survival of VAM fungi in general, adaptation to edaphic factors can influence the performance and persistence of particular VAM fungi. Lambert and co-workers[64] demonstrated that indigenous fungi are often more efficient at increasing plant growth in soils to which they have become adapted. In contrast, introduced fungi were able to increase plant growth more than indigenous fungi initially but not after the first cutting of a forage crop. This strongly implies that VAM isolates adapted to particular edaphic conditions possess some survival advantage over introduced species in soils of similar conditions. Probably these well-adapted isolates are better able to colonize, spread through the plant, or sporulate, indirectly influencing survival. Alternatively, these fungi may have greater inoculum potential and competitive ability under certain soil conditions, but this has not been studied.

C. Influence of Other Soil Organisms

Fungal spores provide a large nutrient base within the soil. Soil animals such as mites or Collembola can feed on fungal spores or they may be ingested by worms. Nematodes are also capable of piercing and sucking the contents of fungal spores. Though these interactions are known to occur, the quantitative impact of this feeding on fungal populations is unknown.[30]

It has been suggested that bacteria can penetrate spores[122] but more commonly they remain on the spore surface while producing fungal cell wall degrading enzymes as well as enzymes to disintegrate the fungal protoplasm. Pigmented fungal spores containing melanin compounds are most resistant to such enzymatic lysis. In addition, bacteria deplete the nutrients around spores, thus increasing the leaching of nutrients from spores,[30] and can induce autolysis of fungal spores. Whether pigmented spores can be thus induced to autolyse is as yet unclear.

Old and Wong[122] have observed that pigmented spores are frequently perforated although the means of perforation was unclear. Using "nucleopore" membranes of varied pore sizes they initially determined that soil animals could not have passed through the pores to cause perforation. Neither were they able to isolate fungi or bacteria from perforated spores which reproduced the perforation symptom on healthy spores. In later studies, Anderson and Patrick[123] discovered that the perforations and depressions in spores they observed were caused by vampyrid amoebae which could constrict their bodies sufficiently to pass through the nucleopore filters. The ubiquitous nature of these amoebae and their ability to perforate hyphae and spores of many fungal species was subsequently demonstrated. Recently, Coley et al.[124] have observed amoebae associated with VAM spore walls, though these appeared to benefit the VAM fungi and not lyse them.

The ability of fungi to parasitize other fungi has been extensively studied. Many fungi are capable of direct penetration although enzymatic degradation is probably also used to gain entrance to the fungal protoplasm. Generally, mycoparasitism occurs more at high temperature under high nutrient conditions, but it must be remembered that most research on mycoparasitism has been accomplished in vitro. The importance of this phenomenon in vivo has not been determined but may not be of great ecological significance for most soil fungi.[30]

VAM fungi are, no doubt, subjected to the same microbial impediments to spore survival as other soil fungi. In fact, the large spore size of VAM fungi (35 to 400 μm in diameter) and their ubiquity in soils makes them a likely source of food, probably subject to intense microbial activity. With the exception of fungal hyperparasites, the microorganisms which lyse or inactivate VAM fungal spores have received little attention.

Warnock et al.[125] observed that leek plants colonized by *Glomus fasciculatum* failed to show a growth response if Collembola or springtails were present. Plants with a VAM fungus and Collembola were similar in size to noninoculated plants. They suggest that in the presence of Collembola, the VAM fungus may not be active or that Collembola graze on external VAM hyphae. Though not entirely conclusive, the presence of hyphae in the gut of Collembolas makes the latter explanation more plausible. Thus, insect grazing can also reduce the VAM fungal population of soils. The frequency of this type of grazing by other insect species has not been studied.

The fungi most frequently found to hyperparasitize VAM fungal spores belong to the Mastigomycotina, fungi with a zoosporic stage in their life cycle. These include *Rhizidiomycopsis* sp.,[126] *Phlyctochytrium* sp.,[54,127] and a pythium-like fungus.[54] A hydromyxomycete, *Labyrinthula* sp. was also recently described.[128] Daniels and Menge[127] have also described two hyphomycetous parasites, *Anguillospora pseudolongissima* and *Humicola fuscoatra*. The latter was observed to hyperparasitize *Phytophthora* and *Pythium* oospores as well.[129] Some of these hyperparasites probably also parasitize VAM fungal hyphae.[54] Undoubtedly many more hyperparasitic fungi will be described in the future.

As VAM fungal spores mature, the primary cell wall thickens and spores usually become darker and more melanized. This melanization corresponds to increasing resistance to hyperparasites, and spores of light-colored species were more susceptible to hyperparasites than were darker heavily melanized species.[127] This may mean that spores, once mature, will have greater survival ability or require penetration by mechanical means.

Comparative susceptibility of VAM fungal species to hyperparasites probably influences their survival in soil and may also influence the competitive ability of these fungi.[54] As susceptibility and resistance appear to depend in part on melanization, darker-spored species might be expected to dominate in natural ecosystems, but the influence of hyperparasites on species succession has not been explored. If hyperparasites can reduce or eliminate the population of VAM fungi in soils cropped to VAM-dependent plants, a reduction in plant growth may occur. In this situation hyperparasites might be considered as secondary plant pathogens.[127]

Though unexplained, it is interesting to note that VAM fungi frequently sporulate in protected areas in the soil. For example, *Glomus fasciculatum* spores form in old seed coats[27,130-132] or even within nematode cysts.[132] These cysts may be sites of VAM fungal sporulation whether or not they contain live nematodes. *G. deserticola* has also been observed to sporulate flatly around the inner cap surface of small plastic ultracentrifuge tubes.[133] The advantage of such sporulation or sporulation within roots is unclear unless these sites are more protected from parasitic microorganisms or predatory larvae. Alternatively, these sites may simply provide the otherwise unavailable space in which sporocarps can form, though this seems less likely.

D. Season

It is thought that sporulation occurs in response to intermittent root growth but spore production probably increases after periods of extensive root growth or as the host matures and senesces.[45,134,135] Many surveys have documented seasonal variations within VAM fungal populations[45,118,135,136] usually based on the spore numbers isolated. Whether peak spore production occurs in spring/summer or summer/fall seems to be related to climate and the crop but a significant decline in spore numbers occurs during the winter. Apparently, enough inoculum survives the winter to initiate colonization in spring. Spore levels increase during the growing season, then decline by the following spring.

The cause of this decline is unclear but it could be attributed to spontaneous germination or spore death, ingestion by soil fauna, destruction by soil fungi and other parasites, or stimulation of germination in the absence of a living host.[136] Whether the level of microbial activity in winter months is sufficient for spore destruction is undetermined, however.

VI. CONCLUSIONS

Although our understanding of the ecology of VAM fungi is increasing, there is still much to be learned. The composite life cycle of these fungi is becoming clearer, but the ecological adaptation of the various fungal species has received little attention. For example, we have no clear understanding of how the VAM fungal species differ from one another. Which fungal species are best adapted to certain soil conditions or host plants? How does the environment or host plant influence the symbiosis? Under what conditions are particular VAM fungi likely to be pathogenic rather than symbiotic? Which factors govern the distribution of certain VAM fungi, i.e., why are particular fungi somewhat cosmopolitan while others are generally more limited to their distribution?

In the past the aim of most research was to understand the general phenomenon of mycorrhizal symbiosis, i.e., how it worked and what factors influenced it. Representative VAM fungi were used to conduct experiments with little understanding of the influence the particular fungal species exerted on the experiment. It was enough then to know that the VAM fungal species differed in their distribution, their benefit to host plants, and their ability to sporulate or colonize roots. However, in the future we will need to study specifically how and why each VAM fungal species differs from the others and to which edaphic conditions and host each VAM fungus is best adapted.

Without a clear understanding the ecology of each VAM fungal species, our ability to manipulate the mycorrhizal symbiosis to the benefit of agriculture will be severely limited.

REFERENCES

1. Mosse, B., The role of mycorrhiza in phosphorus solubilization, in 4th Int. Conf. Global Impacts Appl. Microbiol., Furtado, J. S., Ed., Sao Paulo, Brasil, 1973, 543.
2. Janos, D. P., Mycorrhizae influence tropical succession, *Biotropica,* 12(Suppl.), 56, 1980.
3. Diem, H. G., Gueye, I., Gianinazzi-Pearson, V., Fortin, J. A. and Dommerques, Y. R., Ecology of VA mycorrhiza in the tropics: the semi-arid zone of Senegal, *Acta Oecologia Oecol. Plant.,* 2, 53, 1981.
4. Koske, R. E., Sutton, J. C., and Sheppard, B. R., Ecology of Endogone in Lake Huron sand dunes, *Can. J. Bot.,* 53, 87, 1975.
5. Sutton, J. C. and Sheppard, B. R., Aggregation of sand-dune soil by endomycorrhizal fungi, *Can. J. Bot.,* 54, 326, 1976.
6. Trappe, J. M., Biogeography of hypogeous fungi: trees, mammals, and continental drift, *Abstr. 2nd Int. Mycol. Congr. Inc.,* Bigelow, H. E. and Simmons, E. G., Eds., University of South Florida, Tampa, 1977, 675.
7. Raven, P. H., Evert, R. F., and Curtis, H., *Biology of Plants,* 2nd ed., Worth Publishers, New York, 1976, 685.
8. Kidston, R. and Lang, W. H., On Old Red sandstone plants showing structure from the Rhynie chert bed, Aberdeenshire. V. The Thallophyta occurring in the peat bed, the succession of the plants throughout a vertical section of the bed, and the conditions of accumulation and preservation of the deposit, *Trans. R. Soc. Edinburgh,* 52, 855, 1921.
9. Chaloner, W. G., The rise of the first land plants, *Biol. Rev. Biol. Proc. Cambridge Philos. Soc.,* 45, 353, 1970.
10. Powell, C. Spread of mycorrhizal fungi through soil, *N.Z. J. Agric. Res.,* 22, 335, 1979.
11. Warner, A. and Mosse, B., Factors affecting the spread of vesicular mycorrhizal fungi in soil. I. Root density, *New Phytol.,* 90, 529, 1982.
12. Mosse, B., Warner, A., and Clarke, C. A., Plant growth response to vesicular-arbuscular mycorrhiza. XIII. Spread of an introduced VA endophyte in the field and residual growth effects of inoculation in the second year, *New Phytol.,* 90, 521, 1982.
13. Wallace, H. R., Dispersal in time and space: soil pathogens, in *Plant Disease: An Advanced Treatise,* Vol. 2, Horsfall, J. G. and Cowling, E. B., Eds., Academic Press, New York, 1978. 181.
14. Smith, S. E., and Walker, N. A., A quantitative study of mycorrhizal infection in Trifolium: separate determination of the rates of infection and of mycelial growth, *New Phytol.,* 89, 225, 1981.
15. Powell, C., Development of mycorrhizal infections from *Endogone* spores and infected root segments, *Trans. Br. Mycol. Soc.,* 66, 439, 1976.
16. Koske, R. E., Attraction of germ tubes of *Gigaspora gigantea* to plant roots by a volatile mechanism, *Abstr. 5th North Am. Conf. Mycorrizae,* Universite Laval, Quebec, Canada 1981, 18.
17. Trappe, J. M. and Maser, C., Ectomycorrhizal fungi: interactions of mushrooms and truffles with beasts and trees, in Mushrooms and Man, An Interdisciplinary Approach to Mycology, Walters, T., Ed., USDA Forest Service, Washington, D.C., 1977, 165.
18. Fogel, R. and Trappe, J.M., Fungus consumption (mycophagy) by small animals, *Northwest Sci.,* 52, 1, 1978.
19. Trappe, J. M. and Maser, C., Germination of spores of *Glomus macrocarpus* (Endogonaceae) after passage through a rodent digestive track, *Mycologia,* 68, 433, 1976.
20. Rothwell, F. M. and Holt, C., Vesicular-arbuscular mycorrhizae established with *Glomus fasciculatus* spores isolated from the feces of cricetine mice, Forest Service Res. Note NE-259, USDA Forest Service, Washington, D.C., 1978, 1,
21. Thaxter, R., A revision of the Endogoneae, *Proc. Am. Acad. Arts Sci.,* 57, 291, 1922.
22. Hansen, R. M. and Ueckert, P. N., Dietary similarity of some primary consumers, *Ecology,* 51, 640, 1970.
23. McIlveen, W. D. and Cole, H., Jr., Spore dispersal by Endogonaceae by worms, ants, wasps, and birds, *Can. J. Bot.,* 54, 1486, 1976.
24. MacMahon, J. A., Mount St. Helens revisited, *Nat. Hist.,* 91, 14, 1982.

25. Ponder, F., Jr., Rabbits and grasshoppers: vectors of endomycorrhizal fungi on new coal mine spoil, Forest Service Res. Note NC-250 USDA Forest Service, Washington, D.C., 1980, 1.
26. MacMahon, J. A., Personal communication, 1982.
27. Taber, R. A., Personal communication, 1982.
28. Tommerup, I. C. and Carter, D. J., Dry separation of mcroorganisms from soil, *Soil Biol. Biochem.*, 14, 69, 1982.
29. Garrett, S. D., *Pathogenic Root Infecting Fungi*, Cambridge University Press, Cambridge, 1970, 1.
30. Griffin, D. M., *Ecology of Soil Fungi*, Syracuse University Press, Syracuse, N.Y., 1972, 1
31. Godfrey, R. M., Studies on British species of *Endogone*. III. Germination of spores, *Trans. Br. Mycol. Soc.*, 40, 203, 1957.
32. Daniels, B. A. and Graham, S. O., Effects of nutrition and soil extracts on germination of *Glomus mosseae* spores, *Mycologia*, 68, 108, 1976.
33. Hepper, C. M. and Smith, G. A., Observations on the germination of *Endogone* spores, *Trans. Br. Mycol. Soc.*, 66, 189, 1976.
34. Koske, R. E., *Gigaspora gigantea:* observations on spore germination of a VA-mycorrhizal fungus, *Mycologia*, 73, 288, 1981.
35. Daniels, B. A. and Trappe, J. M., Factors affecting spore germination of the vesicular-arbuscular mycorrhizal fungus, *Glomus epigaeus, Mycologia*, 72, 457, 1980.
36. Mosse, B., The regular germination of resting spores and some observations on the growth requirements of an *Endogone* sp. causing vesicular-arbuscular mycorrhiza, *Trans. Br. Mycol. Soc.*, 42, 273, 1959.
37. Watrud, L. S., Heithaus, J. J., and Jaworski, E. G., Evidence of production of inhibitor by the vesicular-arbuscular-mycorrhizal fungus *Gigaspora margarita, Mycologia*, 70, 821, 1978.
38. Koske, R. E., Multiple germination by spores of *Gigaspora gigantea, Trans. Br. Mycol. Soc.*, 76, 328, 1981.
39. Siqueira, J. O., Hubbell, D. H., and Schenck, N. C., Spore germination and germ tube growth of a vesicular-arbuscular mycorrhizal fungus, *in vitro, Mycologia*, 74, 952, 1982.
40. Green, N. E., Graham, S. O., and Schenck, N. C., The influence of pH on the germination of vesicular-arbuscular mycorrhizal spores, *Mycologia*, 68, 929, 1976.
41. Gerdemann, J. W. and Trappe, J. M., The Endogonaceae in the Pacific Northwest, *Mycol. Mem.*, 5, 1, 1974.
42. Schenck, N. C., Graham, S. O., and Green, N. E., Temperature and light effects on contamination and spore germination of vesicular-arbuscular mycorrhizal fungi, *Mycologia*, 57, 1189, 1975.
43. Graham, J. H., Effect of citrus root exudates on germination of chlamydospores of the vesicular-arbuscular mycorrhizal fungus, *Glomus epigaeum, Mycologia*, 74, 831, 1982.
44. Mosse, B., Advances in the study of vesicular-arbuscular mycorrhiza, *Annu. Rev. Phytopathol.*, 11, 171, 1973.
45. Hayman, D. S., *Endogone* spore numbers in soil and vesicular-arbuscular mycorrhiza in wheat as influenced by season and soil treatment, *Trans. Br. Mycol. Soc.*, 54, 53, 1970.
46. Daft, M. J. and Nicolson, T. H., Effect of *Endogone* mycorrhiza on plant growth. IV. Quantitative relationships between the growth of the host and the development of the endophyte in tomato and maize, *New Phytol.*, 71, 287, 1972.
47. Baylis, G. T. S., Host treatment and spore production by *Endogone, N.Z. J. Bot.*, 7, 173, 1969.
48. Johnson, P. N., Mycorrhizal endogonaceae in a New Zealand forest, *New Phytol.*, 78, 161, 1977.
49. Carling, D. E., Brown, M. F., and Brown, R. A., Colonization rates and growth responses of soybean *Glycine max* plants infected by vesicular-arbuscular mycorrhizal fungi, *Can. J. Bot.*, 57, 1769, 1979.
50. Ferguson, J. J., Inoculum Production and Field Application of Vesicular-Arbuscular Mycorrhizal Fungi, Ph.D. thesis, University of California, Riverside, 1981.
51. Daft, M. J. and Nicolson, T. H., Effect of *Endogone* mycorrhiza on plant growth. III. Influence of inoculum concentration on growth and infection in tomato, *New Phytol.*, 68, 953, 1969.
52. Daniels, B. A. and Menge, J. A., Evaluation of the commercial potential of the vesicular-arbuscular mycorrhizal fungus, *Glomus epigaeus, New Phytol.*, 87, 345, 1981.
53. Rabtin, S. C., Seasonal and edaphic variation in vesicular-arbuscular mycorrhizal infection of grasses by *Glomus tenuis, New Phytol.*, 83, 95, 1979.
54. Ross, J. P. and Ruttencutter, R., Population dynamics of two vesicular-arbuscular endomycorrhizal fungi and the role of hyperparasitic fungi, *Phytopathology*, 67, 490, 1977.
55. Abbott, L. K. and Robson, A. D., Growth of subterranean clover in relation to the formation of endomycorrhizas by introduced and indigenous fungi in a field soil, *New Phytol.*, 81, 575, 1978.
56. Mosse, B., Hayman, D. S. and Ide, G. J., Growth responses of plants in unsterilized soil to inoculation with vesicular-arbuscular mycorrhiza, *Nature (London)*, 224, 1031, 1969.
57. Powell, C., Mycorrhizal fungi stimulate clover growth in New Zealand hill country soils, *Nature (London)*, 264, 436, 1976.

58. Powell, C., Mycorrhizas in hill country soils. II. Effect of several mycorrhizal fungi on clover growth in sterilized soils, *N.Z. J. Agric. Res.*, 20, 59, 1977.
59. Powell, C., Mycorrhizas in hill country soils. III. Effect of inoculation on clover growth in unsterile soils, *N. Z. J. Agric. Res.*, 20, 343, 1977.
60. Powell, C., Mycorrhizas in hill country soils. V. Growth responses in ryegrass, *N.Z. J. Agric. Res.*, 21, 495, 1977.
61. Powell, C. and Daniel, J., Mycorrhizal fungi stimulate uptake of soluble and insoluble phosphate fertilizer from a phosphate-deficient soil, *New Phytol.*, 80, 351, 1978.
62. Ocampo, J.A., Martin, J., and Hayman, D. S., Influence of plant interactions on vesicular-arbuscular mycorrhizal infections. I. Host and non-host plants grown together, *New Phytol.*, 84, 27, 1980.
63. Crush, J. R., The effect of *Rhizophagus tenuis* mycorrhizas on ryegrass, cocksfoot and sweet vernal, *New Phytol.*, 72, 965, 1973.
64. Lambert, D. H., Cole, H., Jr., and Baker, D. E., Adaptation of vesicular-arbuscular mycorrhizae to edaphic factors, *New Phytol.*, 85, 513, 1980.
65. Furlan, V. and Fortin, J. A., Formation of endomycorrhizae by *Endogone calospora* on *Allium cepa* under three different temperature regimes, *Nat. Can. (Quebec)*, 100, 467, 1973.
66. Hayman, D. S., Plant growth responses to vesicular-arbuscular mycorrhiza. VI. Effect of light and temperature, *New Phytol.*, 73, 71, 1974.
67. Schenck, N. C. and Schroder, V. N., Temperature response of *Endogone* mycorrhiza on soybean roots, *Mycologia*, 66, 600, 1974.
68. Peyronel, B., Prime osservazione sui rapporti tra luce e simbiosi micorrizica, *Lab. Chanousia Giardi. Bot. Alpino Piccolo San Bernardo*, 4, 1, 1940.
69. Peuss, H., Untersuchungen zur Okologie und Bedeutung der Tabakmycorrhiza, *Arch. Mikrobiol.*, 29, 112, 1958.
70. Schrader, R., Untersuchungen zur Biologie der mycorrhiza der Erbsenmycorrhiza, *Arch. Mikrobiol.*, 32, 81, 1958.
71. Winter, A. G. and Meloh, K. A., Untersuchungen uber den Einfluss der endotrophen Mycorrhiza auf die Entwicklung von Zea mays l., *Naturwissenschaften*, 45, 319, 1958.
72. Meloh, K. A., Untersuchungen zur Biologie der endotrophen mycorrhiza bei *Zea mays* L. and *Avena sativa* L., *Arch. Mikrobiol.*, 46, 369, 1963.
73. Furlan, V. and Fortin, J. A., Effects of light intensity on the formation of vesicular-arbuscular endomycorrhizas on *Allium cepa* by *Gigaspora calospora*, *New Phytol.*, 79, 335, 1977.
74. Boullard, B., Premieres observations concernant l'influence du photoperiodisme sur la formation de mycorrhizes, *Mem. Soc. Sci. Nat. Math. Cherbourg*, 48, 1, 1957.
75. Boullard, B., Relations entra la photoperiode et l'abondance des mycorrhizes chez l'*Aster tripolum* L. (Composees), *Bull. Soc. Bot. Fr.*, 106, 131, 1959.
76. Tolle, R., Untersuchungen uber die Pseudomycorrhiza von Gramineen, *Arch. Mikrobiol.*, 30, 285, 1958.
77. McCool, P. M., Effect of Ozone Stress on Development of Vesicular-Arbuscular Mycorrhiza and Growth Response of Tomato and Citrus, Ph.D. thesis, University of California, Riverside, 1981.
78. Johnson, C. R., Menge, J. A., Schwab, S., and Ting, I. P., Interaction of photoperiod and vesicular-arbuscular mycorrhizae on growth and metabolism of sweet orange, *New Phytol.*, 90, 665, 1982.
79. Daft, M. J. and El-Giahmi, A. A., Effect of arbuscular mycorrhiza on plant growth. VIII. Effects of defoliation and light on selected hosts, *New Phytol.*, 80, 365, 1978.
80. Kruschcheva, E. P., Conditions favouring formation of maize mycorrhiza, *Agrobiologiya*, 4, 588, 1960.
81. Ross, J. P., Effect of phosphate fertilization on yield of mycorrhizal and nonmycorrhizal soybeans, *Phytopathology*, 61, 1400, 1971.
82. Khan, A. G., The effect of vesicular-arbuscular mycorrhizal associations on growth of cereals. I. Effects on maize growth, *New Phytol.*, 71, 613, 1972.
83. Porter, D. M. and Beute, M. K., *Endogone* species in roots of Virginia type peanuts, *Abstr. Phytopathol.*, 62, 783, 1972.
84. Hayman, D. S., The occurrence of mycorrhizas in crops as affected by soil fertility, in *Endomycorrhizas*, Sanders, F. E., Mosse, B., and Tinker, P. B., Eds., Academic Press, London, 1975, 495.
85. Redhead, J. F., Endotrophic mycorrhizas in Nigeria: some aspects of the ecology of the endotrophic mycorrhizal association of *Khaya grandifoliola*, in *Endomycorrhizas*, Sanders, F. E., Mosse, B., and Tinker, P. B., Eds., Academic Press, London, 1975, 461.
86. Bevege, D. I., Vesicular-Arbuscular Mycorrhizas of Araucaria: Aspects of Their Ecology and Physiology and Role in Nitrogen Fixation, Ph.D. thesis, University of New England, Armidale, New South Wales, 1972.
87. McIlveen, W. D. and Cole, H., Jr., Influence of zinc on development of the endomycorrhizal fungus *Glomus mosseae* and its mediation of phosphorus uptake by *Glycine max* "Amsoy 71", *Agric. Environ.*, 4, 245, 1979.

88. Strzemska, J., Mycorrhiza in farm crops grown in monoculture, in *Endomycorrhizas,* Sanders, F. E., Mosse, B., and Tinker, P. B., Eds., Academic Press, London, 1975, 545.
89. Gerdemann, J. W., The significance of V-A mycorrhizae in plant nutrition, in *Root Diseases and Soilborne Pathogens,* Tousson, T. A., Bega, R. V., and Nelson, P. E., Eds., University of California Press, Berkeley, 1970, 125.
90. Menge, J., Steirle, D., Bagyaraj, D. J., Johnson, E. L. V., and Leonard, R. T., Phosphorus concentrations in plants responsible for inhibition of mycorrhizal infection, *New Phytol.,* 80, 575, 1978.
91. Ratnayake, M., Leonard, R. T., and Menge, J. A., Root exudation in relation to supply of phosphorus and its possible relevance to mycorrhizal formation, *New Phytol.,* 81, 543, 1978.
92. Graham, J. H., Leonard, R. T., and Menge, J. A., Membrane-mediated decrease in root exudation responsible for phosphorus inhibition of vesicular-arbuscular mycorrhiza formation, *Plant Physiol.,* 68, 548, 1981.
93. Sieverding, E., Einflusse der Bodenfeuchte auf die Effektivitat der VA-mykorriza, *Angew. Bot.,* 53, 91, 1979.
94. Nelsen, C. E. and Safir, G. R., Increased drought resistance in onion plants by mycorrhizal infection, *Planta,* 154, 407, 1982.
95. Kruckelmann, H. W., Effect of fertilizers, soils, soil tillage and plant species on the frequency of *Endogone* chlamydospores and mycorrhizal infections in arable soils, in *Endomycorrhizas,* Sanders, F. E., Mosse, B., and Tinker, P. B., Eds., Academic Press, London, 1975, 511.
96. Hirrel, M. C., Mehravaran, H., and Gerdemann, J. W., Vesicular-arbuscular mycorrhizae in the Chenopodiaceae and Cruciferae: do they occur? *Can. J. Bot.,* 56, 2813, 1978.
97. Hayman, D. S., Johnson, A. M., and Ruddlesdin, I., The influence of phosphate and crop species on Endogone spores and vesicular-arbuscular mycorrhiza under field conditions, *Plant Soil,* 43, 489, 1975.
98. Iqbal, S. H. and Qureshi, K. S., The influence of mixed sowing (cereals and crucifers) and crop rotation on the development of mycorrhiza and subsequent growth of crops under field conditions, *Biologia (Lahore),* 22, 287, 1976.
99. Morley, C. D. and Mosse, B., Abnormal vesicular arbuscular mycorrhizal infections in white clover induced by Lupin, *Trans. Br. Mycol. Soc.,* 67, 510, 1976.
100. Trinick, M. J. and Mosse, B., in *Rothamsted Exp. Stn. Annu. Rep.,* 1, 253, 1975.
101. Schenck, N. C. and Kinloch, R. A., Incidence of mycorrhizal fungi on six field crops in monoculture on a newly cleared woodland site, *Mycologia,* 72, 445, 1980.
102. Fox, J. A. and Spasoff, L., Host range studies of Endogone gigantea, in *Proc. Va. J. Sci.,* 23, 121, 1971—72.
103. Crush, J. R., Significance of endomycorrhizas in tussock grasslands in Otago, New Zealand, *N.Z. J. Bot.,* 11, 645, 1973.
104. Mosse, B., Specificity in VA mycorrhizas, in *Endomycorrhizas,* Sanders, F. E., Mosse, B., and Tinker, P. B., Eds., Academic Press, London, 1975, 469.
105. Saif, S. R., The influence of stage of host development on vesicular-arbuscular mycorrhizae and Endogonaceous spore populations in field grown vegetable crops. I. Summer-grown crops, *New Phytol.,* 79, 341, 1977.
106. Kessler, K. J. and Blank, R. W., *Endogone* sporocarps associated with sugar maple, *Mycologia,* 64, 634, 1972.
107. Sparling, G. P. and Tinker, P. B., Mycorrhizas in Pennine grassland, in *Endomycorrhizas,* Sanders, F. E., Mosse, B., and Tinker, P. B., Eds., Academic Press, London, 1975, 545.
108. Read, D. J., Koucheki, H. K., and Hodgson, J., Vesicular-arbuscular mycorrhiza in natural vegetation systems. I. The occurrence of infections, *New Phytol.,* 77, 641, 1976.
109. Tommerup, I. C. and Abbott, L. K., Prolonged survival and viability of VA mycorrhizal hyphae after root death, *Soil Biol. Biochem.,* 13, 431, 1981.
110. Hall, I. R., Response of *Coprosma robusta* to different forms of endomycorrhizal inoculum, *Trans. Br. Mycol. Soc.,* 67, 409, 1976.
111. Hall, I. R., Soil pellets to introduce vesicular-arbuscular mycorrhizal fungi into soil, *Soil Biol. Biochem.,* 11, 85, 1979.
112. Gould, A. B. and Liberta, A. E., Effects of topsoil storage during surface mining on the viability of vesicular-arbuscular mycorrhiza, *Mycologia,* 73, 914, 1981.
113. Porter, W. M., The most probable number method for enumerating infective propagules of vesicular-arbuscular mycorrhizal fungi in soil, *Aust. J. Soil Res.,* 17, 515, 1979.
114. Powell, C., Mycorrhizal infectivity of eroded soils, *Soil Biol Biochem.,* 12, 247, 1980.
115. Toussoun, T. A., Nash, S. M., and Snyder, W. C., The effect of nitrogen sources and glucose on the pathogenesis of *Fusarium solani* f. phaseoli, *Phytopathology,* 50, 137, 1960.
116. Williams, F. J., Antecedent nitrogen sources affecting virulence of *Colletotrichum phemoides, Phytopathology,* 55, 333, 1965.

117. Trainor, M. J. and Martinson, C. A., Nutrition during spore production and the inoculum potential of *Helminthosporium maydis* Race T, *Phytopathology*, 68, 1049, 1978.
118. Black, R. and Tinker, P. B., The development of endomycorrhizal root systems. II. Effect of agronomic factors and soil conditions on the development of vesicular-arbusclar mycorrhizal infection in barley and on the endophyte spore density, *New Phytol.*, 83, 401, 1979.
119. Ocampo, J. A., Effect of crop rotations involving host and non-host plants on vesicular-arbuscular mycorrhizal infection of host plants, *Plant Soil*, 56, 283, 1980.
120. Daft, M. J. and Nicolson, T. H., Effect of *Endogone* mycorrhiza on plant growth. III. Influence of inoculum concentration on growth and infection in tomato, *New Phytol.*, 69, 953, 1968.
121. Mosse, B. and Phillips, J. M., The influence of phosphate and other nutrients on the development of vesicular-arbuscular mycorrhiza in culture, *J. Gen. Microbiol.*, 69, 157, 1971.
122. Old, K. M. and Wong, J. N. F., Perforation and lysis of fungal spores in natural soil, *Soil Biol. Biochem.*, 8, 285, 1976.
123. Anderson, T. R. and Patrick, Z. A., Mycophagous amoeboid organisms from soil that perforate spores of *Thielaviopsis basicola* and *Cochliobolus sativus*, *Phytopathology*, 68, 1618, 1978.
124. Coley, S. C., Baker, K. L., Hooper, G. R., and Safir, G. R., Spore wall development of the endomycorrhizal fungi *Glomus fasciculatus* and *Glomus mosseae* and a possible association with an amoeboid species, *Abstr. Proc. Am. Phytopathol. Soc.*, 12, 392, 1978.
125. Warnock, A. J., Fitter, A. H., and Usher, M. B., The influence of a springtail *Folsomia candidi* (insecta, collembola) on the mycorrhizal association of leek *Allium porrum* and the vesicular-arbuscular mycorrhizal endophyte *Glomus fasciculatus*, *New Phytol.*, 90, 285, 1982.
126. Schenck, N. C. and Nicolson, T. H., A zoosporic fungus occurring on species of *Gigaspora margarita* and other vesicular-arbuscular mycorrhizal fungi, *Mycologia*, 69, 1049, 1977.
127. Daniels, B. A. and Menge, J. A., Hyperparasitization of vesicular-arbuscular mycorrhizal fungi, *Phytopathology*, 70, 584, 1980.
128. Koske, R. E., *Labrinthula* inside the spore of a vesicular-arbuscular mycorrhizal fungus, *Mycologia*, 73, 1175, 1981.
129. Sneh, B., Humble, S. J., and Lockwood, J. L., Parasitism of oospores of *Phytophthora megasperma* var *sojae*, *P. cactorum*, *Pythium* sp. and *Aphanomyces euteiches* in soil by oomycetes, chytridimycetes, hyphomycetes, actinomycetes and bacteria, *Phytopathology*, 67, 622, 1977.
130. Daniels, B. A., Personal observation, 1980.
131. Menge, J. A., Personal communication, 1980.
132. Nigh, E., Personal communication, 1981.
133. Daniels, B. A., Unpublished data, 1980.
134. Mosse, B. and Bowen, G. D., The distribution of *Endogone* spores in some Australian and New Zealand soils and in an experimental field soil at Rothamsted, *Trans. Br. Mycol. Soc.*, 51, 485, 1968.
135. Mason, D. T., A survey of numbers of Endogone spores in soil cropped with barley, raspberry and strawberry, *Hortic. Res.*, 4, 98, 1964.
136. Sutton, J. C. and Barron, G. L., Population dynamics of *Endogone* spores in soil, *Can. J. Bot.*, 50, 1909, 1972.

Chapter 4

TAXONOMY OF VA MYCORRHIZAL FUNGI

I. R. Hall

TABLE OF CONTENTS

I.	Introduction	58
II.	Incidence, Collection, and Storage of Endogonaceae	58
	A. Incidence of Endogonaceae	58
	B. Collection and Extraction of Sporocarps and Spores	59
	C. Storage of Sporocarps and Spores	62
III.	Sectioning Endogonaceous Spores	62
IV.	Features of the Genera	62
	A. *Endogone*	63
	B. *Gigaspora*	63
	C. *Acaulospora*	67
	D. *Entrophospora*	68
	E. *Glomus*	68
	F. *Sclerocystis*	68
	G. *Glaziella*	68
	H. *Modicella*	68
	I. *Complexipes*	69
V.	Identifying and Describing Endogonaceae	69
VI.	The Keys	72
	A. Key to the Sporocarpic Species	72
	B. Key to the Nonsporocarpic Species — Subtending Hyphae Observable	79
	C. Key to the Nonsporocarpic Species — Subtending Hyphae not Observable	86
	D. Key to the Genera	88
VII.	Glossary	89
References		92

I. INTRODUCTION

With the possible exception of the work by Hawker and colleagues,[1-4] the fungal species responsible for the formation of vesicular-arbuscular mycorrhizae (VAM) have been placed in four of the genera (*Acaulospora, Gigaspora, Glomus,* and *Sclerocystis*) in the Endogonaceae,[5] an ancient group which has Devonian fossil representatives.[6,7] *Endogone sensu stricto* has a few members which are known to form ectomycorrhizae,[5] but the mycorrhiza-forming status of the rest of the species in *Endogone* and in the other four genera (*Complexipes, Entrophospora, Glaziella,* and *Modicella*) is unknown. I therefore feel that it would be imprudent to consider only those species known to form VAM and consequently all the genera and species in the Endogonaceae are covered in this chapter.

Thaxter[8] commented: "The true relationships of the group to other families of fungi have long been a matter of conjecture, as is evident from the terms — asci, sporangia, cysts, vesicles etc. — which have been applied by various authors to the chlamydospores alone." The latest contribution to this confusion came from Pirozynski and Malloch[9] who suggested that at least some of the genera in the Endogonaceae are Oomycetes. However, comparative analysis of hyphae of Endogonaceae, other Zygomycetes, Chytridiomycetes and Oomycetes[10] using gas-liquid chromatography (GLC) and Curie-point pyrolysis mass spectrometry has firmly placed the Endogonaceae in what has become their traditional position within the Zygomycetes.[8,11-13] However, it is still a matter of some discussion if *Modicella* and *Complexipes* are endogonaceous (see below) and whether the family is best placed in its own order, the Endogonales,[13] or along with, for example, the Mucoraceae and Mortierellaceae in the Mucorales.[14]

Since Gerdemann and Trappe[5] produced their much-needed revision of the Endogonaceae, the number of recognized species has more than doubled.[15-38] Despite this and the erection of the genera *Complexipes*[35] and *Entrophospora*,[16] the Gerdemann and Trappe system of classification is still adequate and a further revision is not required. However, because of the increased number of species and the need for clarification of the taxonomic status of some informally described species included in the Hall and Fish[39] key to the Endogonaceae, four new keys have been compiled. From time to time users of Hall and Fish's and others keys[5,17,26,31,37,39] have written to me asking for clarification of terms that have been used. Therefore, I have also included in this chapter what I hope is an adequate illustrated guide to the Endogonaceae. Because of their propensity to form mycorrhizae, Endogonaceae are at present receiving increased attention. However, those unfamiliar with the group often have difficulty locating papers describing useful techniques and obtaining instruction in preparing endogonaceous material for identification, etc. I have therefore attempted here to provide a summary of these useful techniques and a comprehensive bibliography.

II. INCIDENCE, COLLECTION, AND STORAGE OF ENDOGONACEAE

A. Incidence of Endogonaceae

There is a tendency for some species to fruit only at certain times of the year. For example, in a temperate rain forest in the south of the South Island of New Zealand, *Endogone pisiformis* Link ex Fries has only been found in February, while *Glomus convolutum* Gerdemann & Trappe tends to produce its sporocarps during May, and *Sclerocystis dussii* (Pat.) von Hohn spores during August.[40] Superimposed on this are seasonal fluctuations in total spore numbers,[24,41-45] the effects of fertilizers,[41,46] cultivation, etc.[47] There also seem to be good and bad spore-forming years,[40,43] and there is a tendency for some species to have extremely limited distributions.[24] Consequently, it is important to realize that a few collections of spores or epigeous sporocarps from an area will probably not represent all of the endogonaceous fungi present. Only after repeated forays over a number of years will an appreciation be gained of the number and perhaps importance of the spore-forming species. Even then it should be borne in mind that in some localities nonsporing or rarely sporing races may occur[24,48] where it might be disadvantageous for a fungus to produce spores. For example, in temperate rain forests there is a dense mat of roots and VAM fungal hyphae do not have to travel far from an infected root to a newly formed uninfected root. There is no need for the fungi to produce spores and indeed a fungus which produced them would seemingly be committing its resources to structures which would give it no ecological advantage. It is therefore not so surprising that in New Zealand temperate rain forests it is common to find few or no spores in soil samples, and the predominance of spores of a particular species does not indicate that ecologically it is the most important.[40]

If it is suspected that rarely sporing races are present in an area it may be possible to encourage spore formation using the following techniques:

1. Roots and the soil fractions between 100 and 1000 μm (see Section II.B.) can be used to inoculate a suitable uninfected host plant growing in a sterile soil in the greenhouse.[5]
2. Well-infected mycorrhizal plants from the field can be transplanted into sterilized soil in the greenhouse.[5,24]
3. Suitable uninfected host plants raised in sterilized soil can be transplanted into the field and brought back after they had become mycorrhizal.[24,49]
4. Uninfected host plants can be grown in field soils in the glasshouse and spore formation hopefully encouraged by altering the moisture, temperature, or fertility status of the soil.[40,50]

After a few weeks or months, soil cores can be removed from the pots, wet sieved (see Section II.B.), and the spores extracted and monoaxenic cultures established.[51] Fine endophytes *Glomus tenue* (Greenall) Hall and *Glomus pallidum* Hall were isolated from several New Zealand forest soils using the above techniques,[24,40] but as their spores are smaller than most other VAM fungi (7 to 15 μm and 28 to 68 μm in diameter, respectively) they cannot be picked out of soil sievings using forceps. However, cultures can be established by sieving out the appropriate soil fraction and inoculating seedlings with it. For this special sieves can be made from nylon monofilament fabrics (e.g., Nybolt, Swiss Silk Bolting Cloth Manufacturing Company, Ltd., Zurich). Pathogens introduced at the same time may be eliminated by growing the endophytes on a series of hosts on which the pathogens cannot survive, a technique which has been used with success by Menge et al.[52] in California.

B. Collection and Extraction of Sporocarps and Spores

Some epigeous sporocarps such as those formed by *Glaziella aurantiaca* (Berk. & Curtis) Cooke are conspicuous as they are both large and brightly colored. Others (*e.g., Glomus macrocarpum* var *macrocarpum* Tul. & Tul.) can be small and can often blend with the texture and color of the soil on which they are formed. Hypogeous sporocarps are even more difficult to find even if one has a "hypogeous instinct".[8] The only way I know of finding large specimens is to rake painstakingly through large quantities of litter and topsoil in the hope of finding some. While elutriation can be used to extract small sporocarps (less than approximately 1 mm) or ectocarpic spores,[53,54] the most commonly used technique is that of wet sieving devised by nematologists and adopted by Gerdemann and Nicolson.[55] This involves washing a slurry of soil through a graded series of soil sieves, e.g., 2000, 750, 300, 100, and approximately 60 μm. The material retained on the sieves can then either be washed from the sieves, decanted into

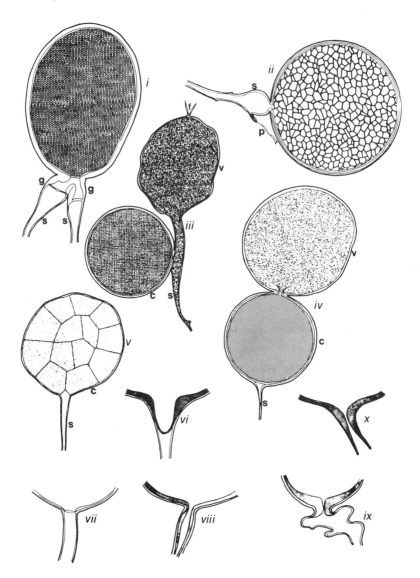

FIGURE 1. Important features of some of the genera. The drawings are not to scale. (i) *Endogone incrassata* zygospore subtended by two gametangia (g) attached to their suspensors (s). (ii) *Gigaspora decipiens* azygospore with bulbous subtending hypha (s) bearing a septate lateral projection (P). (iii) *Acaulospora laevis* mother hypha (s) terminating in a thin-walled mother vesicle (v) bearing two filiform processes (f) with the azygospore (c) formed laterally on the mother hypha. (iv) *Entrophospora infrequens* mother vesicle (v) formed terminally on a thin-walled subtending hypha (s) with the resting spore (c) formed within the subtending hypha. (v) *Modicella malleola* sporangium (c) containing polyhedral sporangiospores; formed on a subtending hypha (sporangiophore — s) without a columella. (vi) *Glomus mosseae* funnel-shaped subtending hypha with a cup-shaped septum. (vii) *Glomus pulvinatum* simple subtending hypha with a flat septum. (viii) *Glomus invermaium* simple, but constricted subtending hypha without a septum. (ix), (x) *Glomus macrocarpum* var *macrocarpum* complex, two examples of "simple" much occluded subtending hyphae of this difficult, heterologous, taxon.

petri dishes, and observed using a × 20 dissecting microscope, or processed further.[54,56-59] Of these further processing techniques I prefer to use the centrifugation technique developed by Tommerup and Kidby[59] at the University of Western Australia. In this technique, suspensions of soil sievings are shaken with finely ground kaolin (this is best done in approximately 75 mm × approximately 30 mm centrifuge tubes)

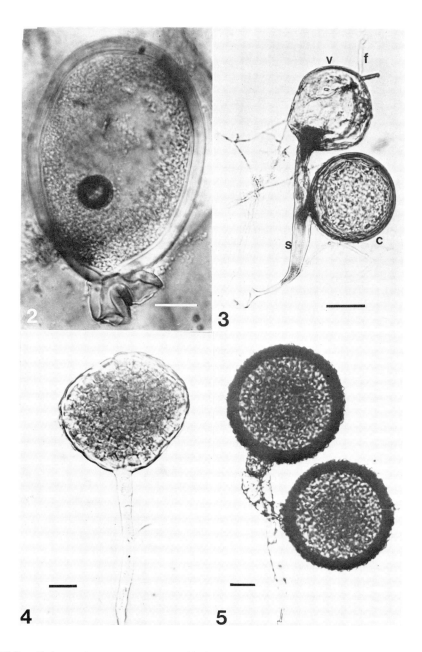

FIGURE 2. *Endogone incrassata* zygospore with the remains of its gametangia. A thick outer wall layer is present and a thin inner layer continues across the pore between the spore contents and the gametangia (Nomarski Interference). Scale = 20 μm. FIGURE 3. *Acaulospora laevis* mother hypha (s), mother vesicle (v) with filiform processes (f), and sessile resting spore (c) formed on the side of the mother hypha (vesicle and mother hypha stained with trypan blue). Scale = 50 μm. FIGURE 4. *Modicella malleola* sporangium containing about 25 sporangiospores. Scale = 20 μm. FIGURE 5. *Complexipes moniliformis* chlamydospores on septate subtending hyphae. Scale = 20 μm.

which are then spun at approximately 72 m·s^{-2} for 5 min. The supernatant, containing the light organic material is then poured off and the remaining soil particles and spores (with the kaolin) are dispersed in a 50% sucrose solution. This is again spun at 72 m·s^{-2} for about 45 sec, and the supernatant containing the spores is filtered through a Whatman No. 540 filter paper. A Buchner filter and pump speeds up this process. The spores, some of the kaolin, and a small proportion of the soil contaminant can then be washed off the filter paper onto a glass fiber filter pad, e.g., Whatman GF/A, and the remaining kaolin washed through with water. The pad can then be placed in a petri dish with a little water and the spores observed. It should be noted however, that air drying the soil prior to sieving, rubbing the soil through the sieves, the use of soil dispersants (e.g., sodium hexametaphosphate), or the use of Tommerup and Kidby's technique might lead to a change in spore color or spore damage. For example, the thin outer wall of *Glomus caledoncium* (Nicol. & Gerd.) Trappe & Gerdemann can be lost, which can make subsequent identification difficult. Consequently, when spores are to be extracted for taxonomic purposes such extraction processes should be avoided.

C. Storage of Sporocarps and Spores

Sporocarps larger than 1 mm can be stored dry in herbarium packets or small bottles, or in small bottles of lactophenol. Small portions of sporocarps, small sporocarps such as those of *Sclerocystis*, and single spores are best mounted on microscope slides. If lactophenol is used as a mounting medium, the edges of the cover slips will have to be sealed with nail varnish. Alternatively, "Miracle" Mounting Fluid, polyvinyl lactic acid, or polyvinyl lactophenol may be used[60,61] (see Glossary). However, it is important that a note of spore color and size be made before mounting as spore color can change in the mountant and the weight and surface tension produced by a coverslip can change a spore's diameter. As it may take several days or even weeks for some spores to clear adequately revealing details of, for example, spore wall structure, I prefer to use lactophenol as a mounting medium. If unmounted spores and sporocarps are to be sent through the mails, they are less likely to sustain damage if they are placed in sealed screw-topped vials filled with lactophenol. Also as lactophenol is a good preservative, the possibility of live pathogens contaminating the endogonaceous material is not a serious consideration for Port Agricultural Authorities.

III. SECTIONING ENDOGONACEOUS SPORES

Sometimes it is possible to gain useful information about spores from thin sections.[24] For transmission electron microscope (TEM) purposes, see papers by Mosse,[63-65] and Sward.[66-68] However, useful information can be obtained if thin sections are observed with a light microscope. For this technique, spores can be embedded in glycol methacrylate resin using 3% glutaraldehyde or 5% acrolein as fixatives (24 hr at 0°C).[62-65] Puncturing the spores with a needle prior to fixation and embedding usually causes some disruption of cytoplasmic details,[63-65] but this is sometimes necessary to get good infiltration. Sections can be cut between <0.1 and 5 μm on an ultratome and transferred from the knife edge to a large drop of water on a clean slide with a pair of No. 5 forceps or a fine brush. As the sections hydrate they tend to twist and fold and it may be necessary to flatten them out with a pair of needles. After air drying some can be stained with 0.5% aqueous Toluidine Blue buffered to pH 4.4 or 0.05% aqueous Trypan Blue before mounting in paraffin oil or a permanent mounting medium, e.g. Euparol® or "Miracle" Mounting Fluid. I obtained my best results when the glycol methacrylate resin was hardened for 48 hr at 60°C and the blocks stored at room temperature for several months before sectioning.[24]

IV. FEATURES OF THE GENERA

At present nine genera are recognized in the Endogonaceae: *Acaulospora, Endogone, Gigaspora, Glaziella, Glomus, Modicella, Sclerocystis*,[5] *Complexipes*,[35] and *Entrophospora*.[16] These produce zygospores, azygospores, chlamydospores, or sporangia

(Figure 1), either ectocarpically within the soil or in roots, or in hypogeous or epigeous sporocarps. Normally only one type of spore is formed in a sporocarp but both Thaxter[8] and Godfrey[69] have described one sporocarp each that contained both chlamydospores and zygospores. They attached great importance to these specimens as they apparently justified the inclusion of the zygosporic species (*Endogone*), the chlamydosporic species (*Glomus, Glaziella,* and *Sclerocystis*), and therefore the later-described genera (*Acaulospora, Gigaspora, Complexipes,* and *Entrophospora*) in the one family. However, Gerdemann[70] concluded that Thaxter's material (*Endogone fasciculata*) was probably a mixture of chlamydosporic and zygosporic species and selected the former as the lectotype. Gerdemann and Trappe[5] have subsequently suggested that both Thaxter's and Godfrey's collections were either chance mixtures of two species or due to hyperparasitism.

More evidence for the inclusion of the diverse forms in the one family was provided by Kanouse[71] who reported that in cultures of *Endogone sphagnophila* (= *E. pisiformis*)[5] chlamydospores, zygospores, and sporangia were produced which could justify the inclusion of chlamydosporic, zygosporic, and sporangiosporic (*Modicella*) genera in the one family. However, this evidence too was unsubstantial, as Gerdemann and Trappe[5] could not reproduce Kanouse's findings and it is possible that Kanouse had a *Mortierella* (an ubiquitous genus) contaminating the cultures. Consequently, while there is an overall similarity in the morphology of spores and sporocarps produced by the species in the different genera, there is no real evidence for the inclusion of chlamydosporic, zygosporic, and sporangiosporic species in the Endogonaceae. This family must therefore be considered as a collection of genera which are probably not closely related.

A. *Endogone*

In this genus (Figures 1, 2, and 11) are placed those species which are known to form zygospores in sporocarps. The sporocarps may be solid or hollow, lobed, stratified, with the spores formed acrogenously from a sterile pad, or aggregated into mats.[5] They may have distinctive odors of fish, onions, or burnt sugar.[5] The mode of formation and development of the zygospores parallels that in other members of the Mucorales. The gametangia may be equal or unequal in size, they may be close together and attached to the same part of the spore, or widely spaced and attached to different parts of the spore. In a few species, one or both of the gametangia may persist and are obvious on at least some of the zygospores in a mature sporocarp; however, it is more common for the gametangia to be ephemeral, or become obscured by a mantle or glebal hyphae. Most species have zygospores with characteristic even-sized oil globules and a two-layered wall with the inner layer completely closing the pore between the spore and the gametangia. Both of these features can be useful when attempting to identify zygosporic species which do not have obvious gametangia.

The mycorrhizal status of only one species, *Endogone flammicorona* Trappe & Gerdemann (previously designated *E. lactiflua* Berk. & Broome) has been published.[72-74] It was shown to form or to be closely associated with a characteristic ectomycorrhiza with *Pseudotsuga menziezii* (Mirb.) Franco and *Pinus strobus* L. However, Warcup[75] has also synthesized mycorrhizas with other species of *Endogone*. The germination of *Endogone* zygospores has not been observed.

B. *Gigaspora*

In this genus (Figures 1, 17, 26-28, and 30), spores are invariably formed ectocarpically on a single bulbous-shaped subtending hypha which usually has one or more short lateral projections of unknown function. Because of the resemblance of the single subtending hypha to the gametangia of *Endogone*, Nicolson and Gerdemann[76] and Ger-

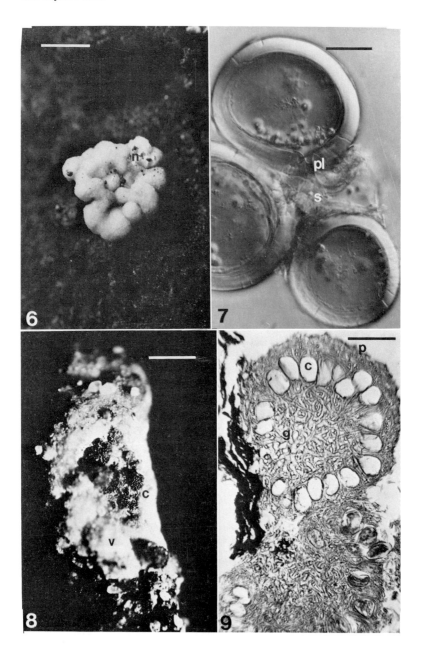

FIGURE 6. *Glomus pallidum* sporocarp damaged somewhat by the action of nematodes (n). Scale = 5 mm. FIGURE 7. *Glomus fuegianum* sporocarp containing three sessile chlamydospores on a common subtending hypha (s). There is a plug (pl) in the pore between the subtending hypha and the spore contents (Nomarski Interference). Scale = 50 μm. FIGURE 8. *Sclerocystis dussii* sporocarp with blackberry-like clusters of chlamydospores (c) surrounded by the shriveled remains of the thin-walled vesicles (v). Scale = 2 mm. FIGURE 9. *Sclerocystis coremioides* sporocarp sectioned to show the chlamydospores (c) surrounding a central plexus of glebal hyphae (g) and contained in an investing peridium (p). Scale = 100 μm.

demann and Trappe[5] termed the spores azygospores, though there is no real evidence that they are either formed parthenogenically or have evolved from zygospores via the reduction of one of the pair of gametangia.

The spore wall invariably has at least 2 layers but there may be as many as 20. The outer layer may be ornamented. Germination is by one of two methods. In those spe-

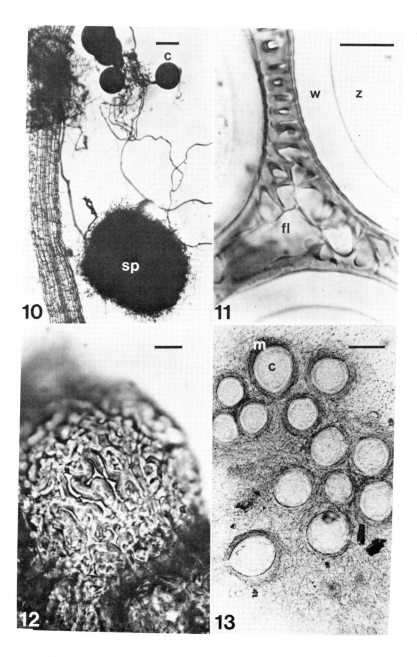

FIGURE 10. *Glomus mosseae* sporocarp (sp) containing a single chlamydospore, and six young chlamydospores (c) without peridial hyphae. Scale = 100 μm. FIGURE 11. *Endogone flammicorona* sporocarp sectioned to show the thick-walled hyphae of the mantle — Flammenkrone (fl) — between three thick-walled (w) zygospores (z). Scale = 10 μm. FIGURE 12. *Sclerocystis sinuosa*. Surface view of the sinuous peridial hyphae. Scale = 10 μm. FIGURE 13. *Glomus convolutum* sporocarp squashed to show the dark mantles (m) surrounding the chlamydospores (c). Scale = 100 μm.

cies which resemble *Gigaspora aurigloba* Hall,[24] a compartment or compartments develop between two inner layers of spore wall and germ tubes arising from the compartment egress by penetrating the outer layers of spore wall. Alternatively, as in *Gigaspora margarita* Becker & Hall,[17,66-68] compartments are not formed and the germ tubes arise from between warty material laid down on the inner surface of the spore wall.

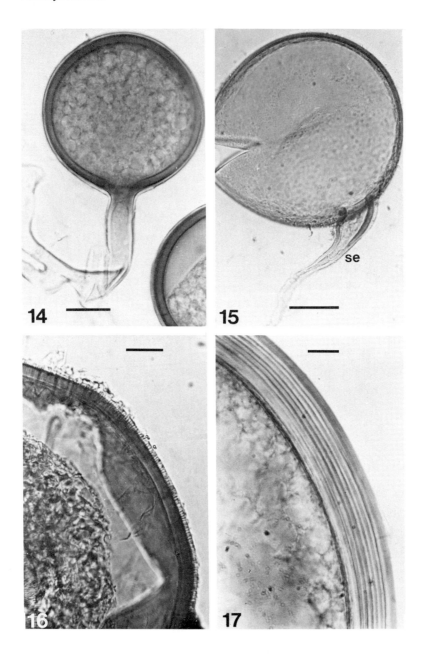

FIGURE 14. *Glomus pallidum* chlamydospores with a sightly constricted subtending hypha and a two-layered wall. Scale = 20 μm. FIGURE 15. *Glomus caledonicum* chlamydospore with a funnel-shaped subtending hypha. The outer layer of the spore wall extends down the subtending hypha as a loosely fitting sleeve. A septum (se) extends across the subtending hypha some distance below the point of attachment to the spore. Scale = 50 μm. FIGURE 16. *Glomus monosporum* chlamydospore wall with the thick inner layer and thin outer layer separated by minute echinulations. Scale = 20 μm. FIGURE 17. *Gigaspora decipiens* spore wall with about 14 distinct layers. (Nomarski Interference). Scale = 20 μm.

The majority of species have been shown to form endomycorrhizae with arbuscules, but without vesicles. However, vesicle-like structures are formed singly or in clusters on coiled hyphae in the soil. The shapes and ornamentation of these soil-borne vesicles can sometimes be used to identify the species producing them.

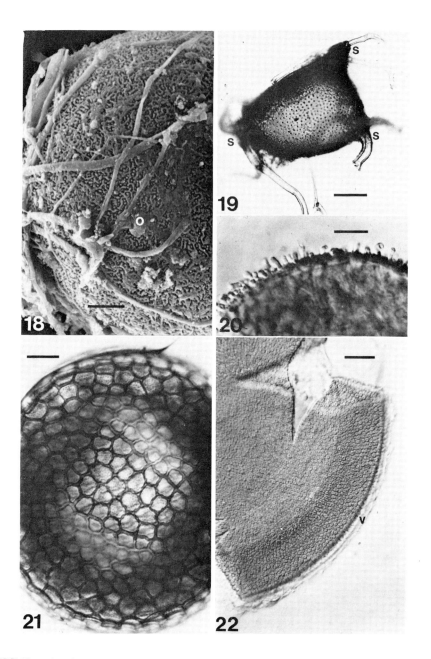

FIGURE 18. *Acaulospora spinosa* resting spore with the remains of the outer wall layer (o) covering the spines and ridges which ornament the second wall layer. (SEM) Scale = 20 μm. FIGURE 19. *Glomus multicaule* chlamydospore with three attached hyphae (s) and showing the rounded projections which cover the spore surface. Scale = 50 μm. FIGURE 20. *Complexipes moniliformis*. Club-shaped and reflexed projections which cover the surfaces of the spores and subtending hyphae (Nomarski Interference). Scale = 20 μm. FIGURE 21. *Acaulospora bireticulata* resting spore showing the network of the outer reticulum and the small pores within the inner reticulum. Scale = 20 μm. FIGURE 22. *Entrophosphora infrequens* ornamented resting spore surrounded by the remains of the hyaline subtending hypha (v) (Nomarski Interference). Scale = 20 μm.

C. *Acaulospora*

As in *Gigaspora*, the "azygospores" are formed singly and ectocarpically in the soil (Figures 1, 3, 18, and 21). Spore formation begins with the formation of a large thin-walled vesicle on the end of a wide hypha.[5,63,77] The contents of this vesicle then migrate back down the hypha and enter the sessile resting spore which forms on the side of the parent hypha. These resting spores invariably have two or more layered walls and in some spores, very complex ornamentation occurs on the outer surface. Germination is via peripheral compartments in *Acaulospora laevis* Gerdemann & Trappe similar to those in *Gigaspora aurigloba*.[63,65] VAM are formed which in *Acaulospora laevis* have characteristic lobed vesicles.

D. *Entrophospora*

As with *Acaulospora*, production of the resting spore is preceded by the formation of a terminal vesicle, but unlike *Acaulospora* the resting spore is formed inside the parent hypha just below the vesicle (Figures 1 and 22). The resting spore walls have a thick ornamented outer layer and a thin membranous inner layer. On many spores the wall of the parent hypha remains intact appearing as a thick, colorless outer wall layer. Only one species (*Entrophospora infrequens* (Hall) Ames & Schneider) has been described, which has polygonal projections on the outer surface of the resting spore wall. A second as yet undescribed species has been collected by Jehne in Australia. It has pores on the outer surface of the resting spore wall but in other respects resembles *E. infrequens*. Germination has not been observed in either species and their mycorrhizal status is unknown. Ames and Schneider[16] have suggested that *E. infrequens* may be hyperparasitic on mycorrhizal fungi, though not apparently on *Gigaspora*.

E. *Glomus*

Chlamydospores are formed in hypogeous or epigeous sporocarps, in sporocarps on twigs or leaves in the litter layer, singly in the soil, or in the cortex of roots (Figures 1, 6, 7, 10, 13-16, 19, 23-25, and 29). Except in one species (*Glomus multicaule* Gerdemann & Bakshi) there is usually only one subtending hypha. With the exception of *Glomus fuegianum* (Spegazzini) Trappe & Gerdemann, the spores are formed on the end of a hypha which can be constricted at the point of attachment to the spore, have parallel side walls, or become markedly wider at the point of attachment to the spore.

The spore walls can have one to many layers but ornamentation as complex as in some of the ornamented *Acaulospora* and *Gigaspora* does not occur. Germination is either via the old subtending hypha or more rarely through the spore wall. Endomycorrhizae with both vesicles and arbuscules have been formed by those *Glomus* species which have been successfully tested.

F. *Sclerocystis*

Chlamydospores are formed similar to those produced by *Glomus*, but are invariably formed in tight sporocarps with the spores arranged in a single layer around a central plexus of glebal hyphae (Figures 8, 9, and 12). In some species a peridium is present and in *Sclerocystis dussii* the sporocarps are interspersed with yellow, thin-walled vesicles. Endomycorrhizae with vesicles and arbuscules have been formed with those species tested.

G. *Glaziella*

This monotypic genus produces chlamydospores scattered randomly in the walls of hollow orange or red sporocarps up to 50 mm in diameter. *G. aurantiaca* (Berk & Curtis) Cooke has been recorded from the tropical areas of Brazil,[8] Cuba,[8] Jamaica,[8] Indonesia,[78] and the Marshall Islands.[79] Its mycorrhizal relationships and the mode of germination of its spores are unknown.

H. *Modicella*

Both species in this genus produce sporangiospores in thin-walled terminal sporangia formed in sporocarps (Figures 1 and 4). Germination of the sporangiospores has not been observed and the mycorrhizal status of the species are unknown. The genus may be better placed in the Mortierellaceae.[5,71,80]

I. *Complexipes*

Spores of this monotypic genus resemble *Glomus* chlamydospores and are produced singly or in loose clusters in the soil on the ends of septate hyphae (Figures 5 and 20).[24,35,81] The spore wall is two-layered and the outer surfaces of the spores and subtending hyphae are covered with club-shaped or reflexed projections. This fungus formed ectendomycorrhizas on *Pinus resinosa* Ait.[81] Neither Wilcox et al.[81] nor Walker[35] give information as to how the spores germinate. There is only scant evidence for the inclusion of this genus in the Endogonaceae.[35]

V. IDENTIFYING AND DESCRIBING ENDOGONACEAE

Gerdemann and Trappe[5] left some taxa as heterologous assemblages of relatively dissimilar collections, e.g., *Glomus macrocarpum* var *macrocarpum* and *Glomus fasciculatum* (Thaxter sensu Gerd.) Gerdemann & Trappe. Also, in some taxa, spore morphology may be inherently variable, e.g., the presence or absence of a thin outer wall layer in *Glomus mosseae* (Nicol. & Gerd.) Gerdemann & Trappe, or it may change markedly as the spores develop, e.g., *Glomus merredum* Porter & Hall. This variability can create problems when it comes to identification. For example, immature spores of *Gigaspora decipiens* Hall & Abbott and *Gigaspora margarita* can easily be mistaken. It is therefore important to be certain that collections of species contain spores showing the full range of morphology as well as anything else that conceivably could be of use, e.g., soil-borne vesicles and samples of mycorrhizae, if formed. In the past, these precautions have not always been taken. For example, Hall[24] described *Glomus infrequens* from a few spores collected from one locality, and Ames and Schneider[16] subsequently showed that the spores described by Hall were formed in a manner more complex than those of *Glomus* and erected the genus *Entrophospora*. The taxonomist presented with inadequate material therefore faces the dilemma of: (1) describing a species formally in the hope that someone else will recognize it and add to the information known about it, or (2) doing nothing in the hope that he will eventually unearth more material. Some have attempted to circumvent the problem by informally describing their "spore types" and assigning numbers, letters, or colloquial names to them. In essence this is a good idea, unfortunately these descriptions are often inadequate[39] which has led to problems. For example, I have seen spores of at least four distinct species which were all called "white reticulate"[77] by their respective users.

Recently, there has been an upsurge in the use of the electron microscope in the study of the fungi. Fortunately to date, no one has seen fit to use features which can only be seen with SEM or TEM to describe Endogonaceae, though SEM and TEM have been useful in clarifying features such as wall structure.[63-68] I would prefer to see that this state of affairs continued. My reasons for this should be obvious: not everyone who needs to identify a species has the time, facilities, or the expertise required to mount material for SEM studies or to embed it for ultrathin sectioning. This is particularly so if the identification of the fungus is incidental to the main line of the research as it is with much of the applied work on VAM.

With few exceptions (e.g., *Glomus tenue*), VAM fungi have been described and are identified from the morphology of their spores. Generally no reference has been made in the species descriptions to the remaining portions of their life cycles. This is partly historical as many species were described before Mosse[82,83] showed that a species later to be named *Glomus mosseae* was responsible for the formation of a VAM. Since then, a large number of new species have been described but apart from recording that they

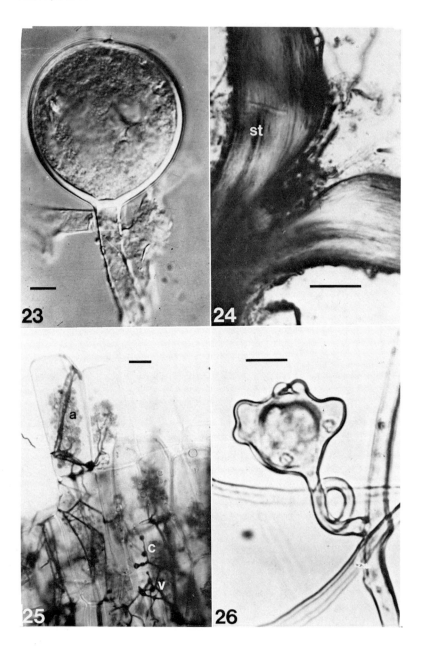

FIGURE 23. *Glomus pulvinatum* chlamydospore with a distinct flat septum at the point of attachment of the subtending hypha to the spore. (Nomarski Interference). Scale = 10 μm. FIGURE 24. *Glomus macrocarpum* flaking spore.[24] Section through the base of a chlamydospore showing the striated spore wall (st) and a plug occluding the pore. Scale = 10 μm. FIGURE 25. *Glomus tenue* (the fine endophyte) arbuscles (a), vesicles (v), spores (c), and very fine hyphae (as this root has been stained with trypan blue the hyphae are wider than is normal). Scale = 20 μm. FIGURE 26. *Gigaspora coralloidea* soil-borne lobed vesicle formed on a helically coiled hypha. Scale = 10 μm.

did or did not form mycorrhizas, few details of the morphology of their mycorrhizae have been recorded. There are a few exceptions. *Glomus tenue*, for example, can only be identified from the very fine hyphae and very small vesicles it produces inside roots as its spores are too small (7 to 15 μm) to collect using standard soil sieves. Additionally, the complete absence of internal vesicles but the presence of soil borne vesicles is

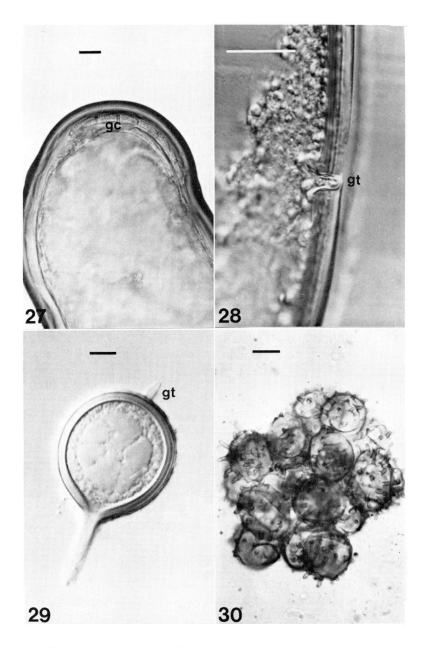

FIGURE 27. *Gigaspora aurigloba* spore with germination compartments (gc) formed between two inner layers of spore wall (Nomarski Interference). Scale = 20 μm. FIGURE 28. *Gigaspora margarita* spore wall with a germination tube (gt) formed between warts on the inner spore wall and without the prior formation of germination compartments. (Nomarski Interference). Scale = 10 μm. FIGURE 29. *Glomus pallidum* chlamydospore containing a vacuolate cytoplasm with its germination tube (gt) opposite the subtending hypha (Nomarski Interference). Scale = 10 μm. FIGURE 30. *Gigaspora decipiens*. Cluster of soil-borne echinulate vesicles. Scale = 20 μm.

a sure indication of a *Gigaspora* sp. However, Abbott and Robson[84-86] can now identify some species from a variety of characteristics of the anatomy of infections. They[85] recommend that future descriptions of endogonaceous species should include a description of the anatomy of its mycorrhiza produced under a defined set of conditions. While this is the ideal it only makes the taxonomist's dilemma worse if his fungus fails to form mycorrhizae.

VI. THE KEYS

In 1979, Hall and Fish[39] provided a single key to the Endogonaceae which, unlike most other keys, did not require a user to determine which genus he was dealing with before identifying the species. Examination of the key of Hall and Fish[39] reveals that all the sporocarpic species key out in the first section, and the nonsporocarpic in the second. Therefore, to reduce the size of a key a user has to handle, a separate key was compiled for the sporocarpic species (Key A). Most people who have looked for endogonaceous spores in soil sievings will have encountered the ectocarpic spore without a subtending hypha and have been frustrated when it came to identification. Two keys were therefore compiled for the nonsporocarpic species. Only those which produce and retain obvious subtending hyphae on their spores are included in the first of these (Key B). The second key deals with ectocarpic spores without subtending hyphae which may be difficult to identify conclusively and gives instruction on what a user should look for if he is to identify a species with this type of spore (Key C). However, for the benefit of workers who are not keen in identifying their fungi up to species level, but interested in knowing the genus, a simple key to the genera is provided (Key D).

All those species that had been published by the middle of 1982[15-18,21-39] have been included in the keys except for *Sclerocystis indicus* Bhattacharjee & Mukerji[20] and *Glomus reticulatum* Bhattacharjee & Mukerji[19] which were brought to my attention too late to be included. As with Hall and Fish's key,[39] species which were excluded from the Endogonaceae by Thaxter[8] and Gerdemann and Trappe[5] have again been omitted. Similarly, informally but incompletely described "species" have not been included; i.e., only those informally described species which have been adequately described are included in the keys. However, those formally described species which I consider to be synonymous with previously described taxa have not been omitted and consequently key out as well as can be expected. Species referred to as WUM 1—8 in the key refer to species described by Hall and Abbott[25] and Porter and Hall.[17] Computer techniques[87] identical to those used by Hall and Fish[39] were again employed to produce the keys. The list of diagnostic characters and their reliability weightings were slightly changed. This list of characters and the enlarged character state listings, i.e., the computer input, which form a summary of the features of endogonaceous fungi, are much too large to include in this chapter (but can be obtained by writing the author).[88]

Because of the variability in spore morphology exhibited by some species (see above) some flexibility has been provided in the keys which allows for minor user "errors". For example, it is not critical for a user to see the thin outer spore wall layer or sporocarps of *Glomus mosseae* in order to identify this species. Gross errors such as misunderstanding the terminology could not and have not been allowed for.

In the keys the number in the brackets following the dichotomy number is merely the position(s) in the key which leads to that particular dichotomy. It was a feature of the computer program and provides a useful check with long keys. The superscript numerals next to the species binomials in the keys are references to articles which contain recent useful descriptions of the species. These may not necessarily be the original description nor a description of the type.

A. Key to the Sporocarpic Species

1. (0) Spore wall of one layer or if two layered, outer difficult to discern; if one layered wall may be striated 2
   Spore wall of two or more distinct layers 31
2. (1) Sporocarps with a covering of large thin-walled vesicles or spores interspersed with large thin-walled vesicles 3

	Sporocarps not as above 4
3.	(2) Spores in sporocarp arranged in a single layer around a central plexus of sterile hyphae; pore at point of attachment of spore and attached hypha usually with a distinct septum; maximum dimension of sporocarp <1 mm.. *Sclerocystis dussii* (Pat.) von Höhn[5]
	Spores in sporocarp not as above; pore at point of attachment of spore and attached hypha never or rarely with a distinct septum; sporocarp >1 mm diameter .... *Glomus vesiculiferum* (Thaxter) Gerdemann & Trappe[5]
4.	(2) Spores with thin walls and always containing few to many thin-walled sporangiospores.. 5
	Spores not as above but internal spores may be present 6
5.	(4) Sporangiospores 4—12 in number, rarely more, 12—38 μm in diameter... *Modicella reniformis* (Bres.) Gerdemann & Trappe[8]
	Sporangiospores greater than 20 in number, 7—17 μm in diameter-.. *Modicella malleola* (Harkn). Gerdemann & Trappe[5]
6.	(4) Spores and sporocarp often exuding a sticky latex when cut; spores containing a white latex, spore wall grading in color from dark at outer surface to subhyaline near inner surface .... *Glomus melanosporum* Gerdemann & Trappe[5]
	Spores and sporocarp not exuding a sticky latex when cut; spores not containing white latex but hyphae in sporocarp may exude latex when cut.. 7
7.	(6) Spores clavate, much longer than wide 140—185 × 20—50 μm, with a very thick walled tip .... *Sclerocystis clavispora* Trappe[33]
	Spores not as above 8
8.	(7) Outer surface of sporocarp a thick gelatinous wall, the sporocarp with a central cavity; sporocarps up to 50 mm diameter ... *Glaziella aurantiaca* (Berk. & Curtis) Cooke[8,78,79]
	Sporocarp not as above 9
9.	(8) Maximum dimension of sporocarp <1 mm 10
	Maximum dimension of sporocarp >1 mm 16
10.	(9) Sporocarps containing only one spore or if more than one spore, then within the sporocarp spores arranged randomly; peridium of thin-walled interwoven hyphae...11
	Within the sporocarp spores arranged in a single layer around a central plexus of sterile hyphae, spores may be interspersed with thin-walled hyaline vesicles; peridium of thick-walled interwoven hyphae..........13
11.	(10) Sporocarp covered with pubescent hyphae clustered into tapered fasicles; number of spores in a sporocarp generally greater than 12; subtending hypha always hyaline .... *Glomus pubescens* (Sacc. & Ellis) Trappe & Gerdemann[5]
	Sporocarp without a covering as above; number of spores in a sporocarp <12; subtending hyphae if visible colored, though may be hyaline in immature spores..12
12.	(11) Spores often filled with hyphae; sporocarps containing 1—3 spores; diameter of subtending hypha at widest part 8—26 μm; outer surface of inner wall of mature spores ornamented with minute echinulate pro-

	jections; projections and thin outer wall not always obvious.. *Glomus monosporum* Gerdemann & Trappe[5,24,25,85]
	Spores rarely filled with hyphae; sporocarps containing 1—10 spores, diameter of subtending hypha at widest part 18—50 μm; outer surface of inner wall not ornamented; thin hyaline outer wall may not be obvious .... *Glomus mosseae* (Nicol. & Gerd.) Gerdemann & Trappe[5,24]
13.	(10) Sporocarp covering containing small scattered chlamydospores-... *Sclerocystis coccogena* (Pat.) von Höhn[8]
	Sporocarp covering not as above...14
14.	(13) Pore at point of attachment of spore and attached hypha usually with a distinct septum .... *Sclerocystis coremoides* Berk. & Broome[5]
	Pore at point of attachment of spore and attached hypha rarely with a distinct septum 15
15.	(14) True peridium a single layer of sinuous hyphae completely covering sporocarp.. *Sclerocystis sinuosa* Gerdemann & Bakshi[23]
	True peridium partially covering sporocarp .... *Sclerocystis rubiformis* Gerdemann & Trappe[5]
16.	(9) One subtending hypha or subtending hypha not visible 17
	Two subtending hyphae .... *Endogone crassa* Tandy[32]
17.	(16) Spores in sporocarps arranged randomly 18
	Spores in sporocarps arranged in radial rows .... *Glomus radiatum* (Thaxter) Gerdemann & Trappe[5]
18.	(17) Spore attachment obscured by mantle; spores bright yellow containing a yellow oil... *Glomus convolutum* Gerdemann & Trappe[5]
	Spores not as above, attachment visible..19
19.	(18) Pore at point of attachment of spore and subtending hypha usually with a distinct septum 20
	Pore at point of attachment of spore and subtending hypha never or rarely with a distinct septum..25
20.	(19) Number of spores in a sporocarp generally >12.............................21
	Number of spores in a sporocarp <12 Red-brown laminate ....(*Glomus* sp.)[77]
21.	(20) Spores often longer than broad (ellipsoid, ovoid, obovoid, or clavate)...22
	Spores not as above, generally globose occasionally longer than broad..24
22.	(21) Subtending hyphae always hyaline.......................... *Glomus fulvum* (Berk. & Broome) Trappe & Gerdemann[5,8]
	Subtending hyphae colored, though may be hyaline in immature spores 23
23.	(22) Sporocarp surface covered by a peridium; maximum breadth of spores >50 μm, but < 70 μm .... *Glomus canadense* (Thaxter) Trappe & Gerdemann[5,8]
	Sporocarp surface naked; maximum breadth of spores >100 μm but <125 μm .... *Glomus boreale* (Thaxter) Trappe & Gerdemann[5,8]

24.	(21)	Pore partially occluded and always with a distinct septum; spores 48—100 × 45—105 μm; wall not becoming laminated with age .... *Glomus pulvinatum* (Hennings) Trappe & Gerdemann[5]
		Pore not occluded with a septum in mature spores; spores 32—78 × 28—68 μm; wall becoming laminated with age .... *Glomus pallidum* Hall[24]
25.	(19)	Spores enclosed in a tightly appressed hyphal mantle; spores bright yellow .... *Glomus convolutium* Gerdemann & Trappe[5]
		Spores not enclosed in a hyphal mantle 26
26.	(25)	Number of spores in a sporocarp generally >12 27
		Number of spores in a sporocarp <12 30
27.	(26)	Spores interspersed with trumpet-shaped hyphae;[32] sporocarps whitish to light gray .... *Glomus tubiforme* Tandy[32]
		Spores not interspersed with trumpet-shaped hyphae; sporocarps colored but may be pale when young 28
28.	(26)	Width of spores 25—55 μm but generally <50 μm; spore wall up to 7 μm thick .... *Glomus microcarpum* Tul. & Tul.[5]
		Width of spores 35—400 μm but generally >50 μm; spore wall up to 17-μm thick 29
29.	(28)	Width of spores 35—105 μm .... *Glomus fasciculatum* (Thaxter sensu Gerd.) Gerdemann & Trappe[5,70]
		Width of spores 95—400 μm 30
30.	(26,29)	Spores reddish-brown; spore wall always laminated; thin outer wall never present; sporocarps containing 1—2 spores Red-brown laminate .... (*Glomus* sp.)[77]
		Spores hyaline, yellow or brown; spore wall may be laminate: spores with thick-walled subtending hyphae may have a thin outer wall; sporocarps generally containing many spores .... *Glomus macrocarpum* Tul. & Tul. var *macrocarpum*[5]
31.	(1)	One subtending hypha, or subtending hypha not observable 32
		Two subtending hyphae 55
32.	(31)	Spore wall two-layered 33
		Spore wall of five mostly separable layers .... *Glomus gerdemanni* Rose et al.[28]
33.	(32)	Sporocarps with a covering of large thin-walled vesicles or spores interspersed with large thin-walled vesicles .... *Glomus vesiculiferum* (Thaxter) Gerdemann & Trappe[5]
		Sporocarps not as above 34
34.	(33)	Sporocarp exuding a sticky latex when cut 35
		Sporocarp not exuding a sticky latex when cut 36
35.	(34)	Outer wall layer much thinner than inner; spores arranged randomly in sporocarps; spores enclosed in tightly appressed hyphal mantles .... *Endogone flammicorona* Trappe & Gerdemann[5]
		Outer wall layer not as above; spores in sporocarps arranged in clusters up to 2 mm broad separated by bands of soil or white hyphae; spores not enclosed in hyphal mantles .... *Endogone oregonensis* Gerdemann & Trappe[5]
36.	(34)	Subtending hyphae not observable or spores sessile or almost sessile .. 37
		Subtending hyphae observable... 47
37.	(36)	Maximum dimension of sporocarp <1 mm 38
		Maximum dimension of sporocarp >1 mm 40

38. (37) Spores sessile or almost sessile, borne in a cluster on the end of a swollen hyphal tip; outer wall layer much thicker than inner .... *Glomus fuegianum* (Spegazzini) Trappe & Gerdemann[8,24]
Spores in sporocarps arranged randomly; outer wall layer much thinner than inner 39
39. (38) Spores often filled with hyphae; sporocarp containing 1—3 spores; diameter of subtending hypha at widest part 8—26 µm; outer surface of inner wall layer of mature spores ornamented with minute echinulate projections; projections and thin outer wall layer may not be obvious .... *Glomus monosporum* Gerdemann & Trappe[5,24,25,85]
Spores rarely filled with hyphae; sporocarps containing 1—10 spores; diameter of subtending hypha at widest part 18—50 µm; outer surface of inner wall layer not ornamented; thin hyaline outer wall layer may not be obvious on all spores .... *Glomus mosseae* (Nicol. & Gerd.) Gerdemann & Trappe[5,24]
40. (37) Sporocarp central cavity present (cavity not caused by nematode activity) 41
Sporocarp central cavity absent 42
41. (40) Outer wall layer much thinner than inner; sporocarp lobed, irregular or variable; sporocarp central cavity opening to exterior.. *Endogone pisiformis* Link ex Fries[5,89]
Outer wall layer not as above; sporocarp globose to subglobose to ovoid; sporocarp central cavity not opening to exterior.. *Endogone incrassata* Thaxter[5]
42. (40) Spores arranged randomly; sporocarp globose to subglobose to ovoid 43
Sporocarps formed of aggregated clusters of zygospores, or small sporocarps aggregated into masses, sporocarps of convoluted strata or zygospores formed in radial rows 44
43. (42) Sporocarps whitish to light gray; odor of sporocarps not distinctive; sporocarps <1 mm diameter (but may be fused into mats) .... *Endogone alba* (Petch) Gerdmann & Trappe[5]
Sporocarps colored but may be pale if immature; sporocarps smelling distinctly of onions particularly if kept moist in a closed container; sporocarps >1.5 mm diameter......................... *Endogone incrassata* Thaxter[5]
44. (42) Spores in discrete clusters...45
Spores in radiate rows; mature spores often filled with hyphae similar to the glebal hyphae .... *Endogone acrogena* Gerdemann et al.[5]
45. (44) Sporocarps whitish to light gray; epigeous................... *Endogone alba* (Petch) Gerdemann & Trappe[5]
Sporocarps colored but may be pale if immature; hypogeous............46
46. (45) Outer wall layer (about 1-µm thick) much thinner than inner .... *Endogone verrucosa* Gerdemann & Trappe[5]
Outer wall layer thicker than inner .... *Endogone tuberculosa* Lloyd[5,8]

47.	(36) Outer wall layer much thinner than inner	48
	Outer wall layer thicker than inner or subequal	66
48.	(47) Maximum dimension of sporocarp <1 mm; number of spores in a sporocarp <12	49
	Maximum dimension of sporocarp >1 mm; number of spores in a sporocarp generally >12	50

49. (48) Spores often filled with hyphae; sporocarps containing 1—3 spores; diameter of subtending hypha at widest part 8—26 μm; outer surface of inner wall layer of mature spores ornamented with minute echinulate projections; projections and thin outer wall layer may not be obvious................................ *Glomus monosporum* Gerdemann & Trappe[5,24,25,85]

Spores rarely filled with hyphae; sporocarp containing 1—10 spores, diameter of subtending hypha at widest part 18—50 μm; outer surface of inner wall not ornamented; thin hyaline outer wall not always obvious .... *Glomus mosseae* (Nicol. & Gerd.) Gerdemann & Trappe[5,24]

50. (48) Spores hyaline to white at maturity; sporocarps segmented, whitish to light gray .... *Glomus segmentatum* Trappe[34]

Spores and sporocarps colored, though may be pale if immature.......51

51. (50) Pore at point of attachment of spore and attached hypha usually with a distinct septum; sporocarp always globose to subglobose to ovoid....52

Pore at point of attachment of spore and attached hypha never with a distinct septum; sporocarp much lobed, irregular, or variable 53

52. (51) Spores >100 μm diameter; wall layers readily separable; subtending hyphal wall two-layered, the outer readily separating from the inner; subtending hyphae colored, though may be hyaline in immature spores .... *Glomus caledonicum* (Nicol. & Gerd.) Gerdemann & Trappe[5,76]

Spores <75 μm diameter; wall layers not readily separable; subtending hyphal wall with a thin outer layer <1-μm thick not separating from inner; subtending hypha hyaline and fragile .... *Glomus fragile* (Berk. & Broome) Trappe & Gerdemann[5]

53. (51) Subtending hypha constricted at point of attachment; innermost wall layer hyaline .... *Glomus tenerum* Tandy[32]

Subtending hyphae not as above; innermost wall layer colored, but may be hyaline in immature spores................................54

54. (53) Spores yellow to bright brownish yellow, 95—140 μm in width; diameter of subtending hypha at widest part usually <10 μm; subtending hypha often inserted into the spore wall; pore at point of attachment of subtending hypha occluded by a septum-like plug; spores borne singly, in loose clusters, or in epigeous sporocarps .... *Glomus epigaeum* Daniels & Trappe[22]

Spores yellowish brown to dark brown, 95—400 μm in width; diameter of subtending hypha at widest part >10 μm; subtending hypha not inserted; pore without a plug but may be occluded by lateral wall thickenings; spores borne singly, in loose clusters, or in generally hypogeous sporocarps................................*Glomus macrocarpum* Tul. & Tul. var *macrocarpum*[5]

55.	(31) Sporocarp often exuding a sticky latex when cut	56
	Sporocarp not exuding a stick latex when cut	58
56.	(55) Outer wall layer much thinner than inner; spores in sporocarps arranged randomly; spores enclosed in a tightly appressed hyphal mantles	57
	Outer wall layer not as above; spores in sporocarps arranged in clusters up to 2 mm in width, separated by bands of soil or white hyphae, spores not enclosed in hyphal mantles ..... *Endogone oregonensis* Gerdemann & Trappe[5]	
57.	(56) Subtending hyphae persistent and conspicuous, at widest part >60 μm ..... *Endogone lactiflua* Berk. & Broome[5]	
	Subtending hyphae inconspicuous, at widest part >25 μm but <60 μm; mantle a flammenkrone ..... *Endogone flammicorona* Trappe & Gerdemann[5]	
58.	(55) Outer wall much thinner than inner wall	59
	Outer wall not as above	61
59.	(58) Sporocarp central cavity present (cavity not caused by nematode activity) ..... *Endogone pisiformis* Link ex Fries[5,89]	
	Sporocarp central cavity absent	60
60.	(59) Spores in sporocarps arranged randomly enclosed in tightly appressed hyphal mantles; subtending hyphae persistent and conspicuous ..... *Endogone reticulata* Tandy[32]	
	Spores in sporocarps arranged in radiate rows with the oldest toward center of sporocarp; spores not enclosed in a hyphal mantle; subtending hyphae inconspicuous or absent ..... *Endogone acrogena* Gerdemann et al.[5]	
61.	(58) Spores in sporocarps arranged randomly; sporocarp always globose to subglobose to ovoid; sporocarps smell distinctly of onions particularly if kept moist in closed container ..... *Endogone incrassata* Thaxter[5]	
	Spores in sporocarps arranged nonrandomly (see 10); sporocarp much lobed irregular or variable; odor of sporocarps not distinctive	62
62.	(61) Spores in sporocarps arranged in discrete clusters	63
	Spores in sporocarps not arranged as above	65
63.	(62) Spores enclosed in tightly appressed hyphal mantles; sporocarps whitish to light gray ..... *Endogone multiplex* Thaxter[8,89]	
	Spores not enclosed in hyphal mantles; sporocarps colored but may be pale if immature	64
64.	(63) Outer wall much thicker than inner wall ..... *Endogone tuberculosa* Lloyd[5,8]	
	Walls of approximately the same thickness ..... *Endogone aggregata* Tandy[32]	
65.	(62) Spores enclosed in tightly appressed hyphal mantles; maximum diameter of subtending hyphae at widest part >25 μm but <60 μm, generally persistent and conspicuous ..... *Endogone stratosa* Trappe et al.[5]	
	Spores not enclosed in hyphal mantles; subtending hyphae <25 μm wide often disappearing by maturity ..... *Endogone acrogena* Gerdemann et al.[5]	

66. (47) Spore wall 5—20 μm thick; outer layer much thicker than inner, spores sessile or almost sessile borne in a cluster on the end of a swollen hyphal tip .... *Glomus fuegianum* (Spegazzini) Trappe & Gerdemann[8,24]
Spore wall 1.2—2.4 μm thick with outer layer slightly thicker than inner; spores formed on the tips of simple subtending hyphae i.e., not as above .... *Glomus aggregatum* Schenck & Smith[31]

B. Key to Nonsporocarpic Species — Subtending Hyphae Observable

1. (0) Maximum breath of spores <15 μm, associated with VAM with hyphae rarely exceeding 2 μm in diameter; maximum diameter of subtending hyphae at widest part <3 μm.................................... *Glomus tenue* (Greenall) Hall[24]
Maximum breadth of spores >15 μm, and if associated with VAM hyphae regularly >2 μm; maximum diameter of subtending hyphae at widest part >3 μm 2
2. (1) Surface of spores ornamented with minute spines or mature spores surrounded by mantles; spores may have multiple subtending hyphae... 4
Spores not as above... 3
3. (2) Spore wall of one layer or if two-layered, outer thin and difficult to discern; if wall one-layered then it may be striated........................ 5
Spore wall of some spores at least, with two or more distinct layers ...21
4. (2) Surface of spores ornamented with rounded projections 1.2—3.7 μm high, regularly distributed over spore surface; usually more than one subtending hypha.. *Glomus multicaule* Gerdemann & Bakshi[23]
Surface of spores not as above 73
5. (4) Pore at point of attachment of spore and attached hypha usually with a distinct septum... 6
Pore at point of attachment of spore and attached hypha never or rarely with a distinct septum...10
6. (5) Spore attachment simple 7
Spore attachment funnel-shaped 9
7. (6) Spores 32—78 μm × 28—68 μm, white to pale yellow .. *Glomus pallidum* Hall[24]
Spores >68 μm wide, yellow, brown, or reddish-brown 8
8. (7) Thin outer wall layer never present; spores reddish-brown Red-brown laminate ....(*Glomus* sp.)[77]
Thin outer wall layer often visible; spores yellow to light brown......... 9
9. (8,6) Spores often filled with hyphae; sporocarp if present containing 1—3 spores; diameter of subtending hypha at widest part 8—26 μm; outer surface of inner wall layer of mature spores ornamented with minute echinulate projections; projections and thin outer wall layer may not be obvious on all spores .... *Glomus monosporum* (Gerdemann & Trappe)[5,24,25,85]
Spores rarely filled with hyphae; sporocarps if present containing 1—10 spores; diameter of subtending hypha at widest part 18—50 μm; outer surface of inner wall layer not ornamented; thin hyaline outer wall may not be obvious on all spores.......................... *Glomus mosseae* (Nicol. & Gerd.) Gerdemann & Trappe[5,24]

10. (5) Subtending hypha constricted at point of attachment......................11
    Subtending hypha not constricted 12
11. (10) Spore attachment funnel-shaped; pore at point of attachment of spore and parent hypha not occluded .... *Glomus merredum* Porter & Hall[27]
    Spore attachment simple, constricted at point of attachment .... *Glomus constrictum* Trappe[33]
12. (10) Subtending hyphae and spores hyaline 13
    Subtending hyphae and spores colored, though may be hyaline in immature spores 14
13. (12) Pore at point of attachment of spore and parent hypha occluded or partially occluded by lateral wall thickening; spores <50 μm in diameter .... *Glomus microcarpum* Tul. & Tul.[5]
    Pore at point of attachment of spore and parent hypha not occluded; spores 150—220 μm in diameter .... *Glomus lacteum* Rose & Trappe[29]
14. (12) Spore attachment simple 15
    Spore attachment not simple 20
15. (14) Spore surface ornamented with minute echinulations, a thin outer wall layer should be present on some spores; spores often filled with hyphae at maturity..................... *Glomus monosporum* Gerdemann & Trappe[5,24,25,85]
    Spore surface not ornamented; spores rarely filled with hyphae 16
16. (15) Width of spores 25—55 μm but generally <50 μm; spore wall up to 7 μm thick .... *Glomus microcarpum* Tul. & Tul.[5]
    Width of spores 35—400 μm but generally >50 μm; spore wall up to 17 μm thick 17
17. (16) Spores dark brown to black; subtending hypha yellow to brown with wall thickenings extending 50—150 μm along hypha; spores always ectocarpic .... *Glomus macrocarpum* Tul. & Tul. var *geosporum*[5]
    Spores hyaline, yellow, brown, or reddish-brown; subtending hypha not as above; spores ectocarpic or sporocarpic.................................18
18. (17) Width of spores 35—105 μm .... *Glomus fasciculatum* (Thaxter sensu Gerd.) Gerdemann & Trappe[5]
    Width of spores 95—400 μm 19
19. (18) Spores reddish-brown; spore wall always laminate; thin outer wall never present; sporocarps if present containing 1—2 spores; red-brown laminate ....(*Glomus* sp.)[77]
    Spores hyaline, yellow, or brown; spore wall may be striated; spores with thick-walled subtending hyphae may have a thin outer wall layer; sporocarps if present containing many spores.... *Glomus macrocarpum* Tul. & Tul. var *macrocarpum*[5]
20. (14) Spores often filled with hyphae; maximum diameter of subtending hyphae at widest part 8—26 μm (also see 9)....... *Glomus monosporum* Gerdemann & Trappe[5,24,25,85]
    Maximum diameter of subtending hyphae at widest part 25—60 μm... *Glomus merredum* Porter & Hall[27]

21.	(3) Spore wall two-layered but an extra very thin outer wall layer may be present	22
	Spore wall with three or more distinct layers	50
22.	(21) Spore attachment bulbous with one or rarely more lateral projections	23
	Spore attachment not as above	28
23.	(22) Surface of outermost wall ornamented	24
	Surface of outermost wall not ornamented	25
24.	(23) Subtending hypha with many attached fine hyphae; soil-borne vesicles echinulate, knobby or warty; soil-borne vesicles formed singly on a coiled hypha .... *Gigaspora coralloidea* Trappe et al.[5]	
	Subtending hypha without many attached fine hyphae; soil-borne vesicles smooth and round, or smooth but irregular in outline (tending polymorphic); soil-borne vesicles formed in clusters of 1—20 .... *Gigaspora heterogama* (Nicol. & Gerd.) Gerdemann & Trappe[5,76]	
25.	(23) Outermost wall always hyaline; spores hyaline to white at maturity	68
	Outermost wall colored, but may be hyaline in immature spores; spores colored though may be hyaline when young	26
26.	(25) Outer wall layer much thinner than inner; soil-borne vesicles formed in clusters of 1—20 .... *Gigaspora gigantea* (Nicol. & Gerd.) Gerdemann & Trappe[5,76]	
	Outer wall layer generally not as above though a thin outer wall layer may be present; soil-borne vesicles formed singly on a coiled hypha	27
27.	(26) Mature spores yellow to pale greenish yellow; soil borne vesicles 23—33 μm in diameter, smooth to knobby, borne singly.. *Gigaspora calospora* (Nicol. & Gerd.) Gerdemann & Trappe[5,25,76]	
	Mature spores bright golden yellow, becoming dull yellow with age and brown when moribund; soil-borne vesicles up to 100 μm in diameter, echinulate to knobby to polymorphic, borne in loose clusters .... *Gigaspora aurigloba* Hall[24]	
28.	(22) Ornamentation of outer surface of an inner wall layer generally present	29
	Ornamentation of outer surface of an inner wall layer not occurring	32
29.	(28) Wall layers readily separable, or some readily separable	30
	Wall layers not readily separable	31
30.	(29) Spore surface ornamented with minute echinulations, a thin outer wall layer should be present on some spores; spores often filled with hyphae at maturity .... *Glomus monosporum* Gerdemann & Trappe[5,24,25,85]	
	Spore outer wall layer thicker than inner; outer surface of inner wall layer ornamented with polygonal projections; spore develops in a hyaline hypha which subtends a hyaline vesicle; spore thus with a smooth hyaline outer wall layer and a brown ornamented inner; a stalk (termed a subtending hypha by Hall (1976)) continuous with the inner spore wall layer, joins the contents of the spore and the vesicle .... *Entrophospora infrequens* (Hall) Ames & Schneider[16,24]	
31.	(29) Spore wall may appear to be two-layered with inner and outer layers readily separable and with pits on outer surface of inner layer but,	

careful examination shows outer layer has a very thin layer appressed to its outer surface and a membrane is present on the inner surface of wall contiguous with septum in the subtending hypha.... *Glomus velum* Porter & Hall[27]

Outer wall layer hyaline 8—12 (−20) μm thick, amorphous; inner 10—15 μm thick, brown, ornamented with crowded spines 0.5 × 0.2 μm that extend into the outer layer .... *Glomus halonatum* Rose & Trappe[29]

32. (28) Pore at point of attachment of spore and attached hypha usually with a distinct septum 33

Pore at point of attachment of spore and attached hypha never or rarely with a distinct septum 41

33. (32) Outer wall layer much thinner than inner 34

Outer wall layer not as above 36

34. (33) Subtending hyphal wall two-layered, the outer readily separating from the inner; outer spore wall layer, when present, continuous with loose outer wall layer of subtending hypha .... *Glomus caledonicum* (Nicol. & Gerd.) Gerdemann & Trappe[5,76]

Subtending hyphal wall homogenous, or if two-layered, outer not readily separating from inner; outer spore wall layer when present not as above 35

35. (34) Spores >100 μm; subtending hypha generally funnel-shaped and colored though may be hyaline in immature spores .... *Glomus mosseae* (Nicol. & Gerd.) Gerdemann & Trappe[5,24]

Spores 53—73 × 49—62 μm in diameter; subtending hypha simple, always hyaline... *Glomus fragile* (Berk. & Broome) Trappe & Gerdemann[5]

36. (33) Outermost wall layer always hyaline 37

Outermost wall layer colored though may be hyaline in immature spores 40

37. (36) Outer layer of spore wall much thicker than inner; spore attachment often absent but when present one layered, an outer layer having been lost .... *Glomus albidum* Walker & Rhodes[25,36]

Spore wall layers of approximately the same thickness, outer layer when present obviously extending down subtending hypha as a tightly fitting sleeve 38

38. (37) Spore wall may appear to be two-layered with inner and outer layers readily separable and with pits on outer surface of inner layer, but careful examination shows outer layer has a very thin layer appressed to its outer surface and a membrane on the inner surface of wall continuous with septum in subtending hypha, i.e., wall of four layers .... *Glomus velum* Porter & Hall[27]

Spore wall not as above 39

39. (38) Spores hyaline to white at maturity; subtending hypha always hyaline; walls of subtending hypha thin and fragile............... *Glomus albidum* Walker & Rhodes[25,36]

Spores colored though may be hyaline when young; subtending hypha colored; walls of subtending hypha thick............. *Glomus etunicatum* Becker & Gerdemann[18]

40. (36) Surface of spore and subtending hypha covered in projections 4—30 ×

	2—5 μm, often with the tips reflexed .... *Complexipes moniliformis* Walker[35]
	Surface of spore not ornamented 70
41.	(32) Spore attachment simple 42
	Spore attachment on at least some spores distinctly funnel-shaped 48
42.	(41) Spores and subtending hyphae hyaline to white at maturity ...... WUM 5 (*Glomus* sp.)[25]
	Spores and subtending hyphae colored though may be hyaline when young 43
43.	(42) Spores sessile or almost sessile, borne in a cluster on the end of a swollen hyphal tip .... *Glomus fuegianum* (Spegazzini) Trappe & Gerdemann[8,24]
	Spores not as above 44
44.	(43) Subtending hyphae constricted at point of attachment 45
	Subtending hyphae not as above 46
45.	(44) Spores 50—75 μm in diameter, outermost wall layer always hyaline extending down subtending hypha as a tightly fitting sleeve .... *Glomus invermaium* Hall[24]
	Spores 150—330 μm in diameter, outermost wall layer colored, but may be hyaline in immature spores; outer spore wall layer not extending down subtending hypha as a sleeve .... *Glomus constrictum* Trappe[33]
46.	(44) Outer spore wall layer on some spores extending down subtending hypha as a sleeve 71
	Not as above 47
47.	(44) Spores colorless to pale yellow .... *Glomus claroideum* Schenck & Smith[31]
	Spores bright yellow to brownish yellow, brown, or dark brown 75
48.	(41) Subtending hyphal wall two-layered, the much thinner outer layer readily separating from the inner .... *Glomus merredum* Porter & Hall[27]
	Subtending hyphal wall homogenous, or if two-layered outer not as above 49
49.	(48) Outer spore wall layer much thicker than inner; pore at point of attachment of spore and parent hypha occluded or partially occluded by a plug and lateral wall thickening; outer spore wall layer not extending down subtending hypha .... *Glomus magnicaule* Hall[24]
	Spore wall layers of approximately the same thickness; pore at point of attachment of spore and parent hypha not occluded; outer spore wall obviously extending down subtending hypha 72
50.	(21) Spore attachment bulbous regularly with one or more lateral projections 51
	Spore attachment not as above 60
51.	(50) Outermost wall layer always hyaline; spores hyaline to white at maturity 52
	Outermost wall layer colored, but may be hyaline in immature spores; spores colored though may be hyaline when young 57
52.	(51) Mature spores generally <200 μm in diameter; spore wall of 3—5 layers of unequal thickness; outer two brittle and readily separable from 1—3 flexible inner layers .... *Gigaspora pellucida* Nicolson & Schenck[26]

	Mature spores >200 μm in diameter; wall not as above....................53
53.	(52) Wall of three layers (but look for older spores with many wall layers), outer two approximately equal and each much more than 2-μm thick, innermost membranous, outermost layer brittle, inner two flexible; soil-borne vesicles distally covered in straight or more commonly forked spines up to 10-μm high and about 2-μm wide, formed singly or in tight clusters of up to 20, each attached by a helically coiled hypha... *Gigaspora decipiens* Hall & Abbott[25]
	Wall of 3—14 layers; if three layered then wall not arranged as above...54
54.	(53) Spore wall 2.4—7.5 μm thick consisting of 2—5 layers each layer 1—2 μm thick; portion of spore wall surrounding subtending hypha with a distinct to barely detectable rose-pink tint................. *Gigaspora rosea* Nicolson & Schenck[26]
	Spores and spore wall not as above; spore wall usually more than 10-μm thick 55
55.	(54) Spore wall consisting of 3—5 layers of unequal thickness; outer two layers brittle; readily separable from 1—3 flexible inner layers .... *Gigaspora gilmorei* Trappe & Gerdemann[5]
	Spore wall of 2—14 layers; if 3—5 layers then wall not arranged as above...56
56.	(55) Young spore wall 20—35 μm thick of three subequal layers, innermost very thin outermost brittle and separable with difficulty from flexible inner layers; wall of old spores 34—47 μm thick; juvenile outer wall replaced by three inseparable subequal layers; soil-borne vesicles 33—54 μm in diameter with straight or more commonly forked spines up to 10 μm high and 2 μm wide................................ *Gigaspora decipiens* Hall & Abbott[25]
	Spore wall 4—24 μm thick of 3—10 layers each 1.5—4 μm thick; soil-borne vesicles 22—35 μm in diameter with warty projections up to 4-μm high and 5-μm wide 69
57.	(51) Spore surface ornamented with pores or irregular projections...........58
	Spore surface not ornamented..59
58.	(57) Spore surface ornamented with large pores 7—10 μm in diameter with smaller pores within them .... *Gigaspora nigra* Redhead[26]
	Spore surface ornamented with irregular projections 1—7 × 3—12 μm.. *Gigaspora gregaria* Schenck & Nicolson[26]
59.	(57) Spore wall of three layers, outer up to 15-μm thick, inner up to 20-μm thick, middle up to 2-μm thick .... Type 4 Sward et al. (*Gigaspora* sp.)[38]
	Spore wall of 2—4 layers; outer layer 6—16 μm thick, 1—3 hyaline inner layers each approximately 1-μm thick......... *Gigaspora aurigloba* Hall[24]
60.	(50) Pore at point of attachment of spore and attached hypha usually with a distinct septum 61
	Pore at point of attachment of spore and attached hypha never or rarely with a distinct septum..64
61.	(60) Outer surface of middle wall layer ornamented with crowded spines 0.2

	× 0.5 μm that extend into the outermost wall layer; subtending hyphae constricted at point of attachment to spore .... *Glomus halonatus* Rose & Trappe[29]
	No ornamentation of an outer surface of an inner wall layer; subtending hyphae not constricted 62
62.	(61) Spores light to dark brown; surface covered in projections 4—30 × 2—5 μm often with the tips reflexed; subtending hyphae with 2—5 distinct septa.. *Complexipes moniliformis* (Hall) Ames & Schneider[35]
	Spores and subtending hyphae not as above 63
63.	(62) Old spore wall yellow to light brown, up to 25-μm thick of three layers with outermost closely appressed to middle; young spore wall colorless up to 10-μm thick of three layers with innermost readily separable from middle; pits on outer surface of inner wall........... *Glomus velum* Porter & Hall[27]
	Old spore walls 7—31 μm thick with an inner portion 2—9 μm thick of 2—5 layers and an outer portion 5—20 μm thick; spores often with a hyaline mucilaginous coat 0.5—2 μm thick which with age beomes verrucose or rugose with folds up to 5 μm high .... *Glomus clarum* Nicolson & Schenck[26]
64.	(60) Surface of spore covered in hyaline knobs 1—3 × 0.4—1.2 μm... *Glomus scintillans* Rose & Trappe[29]
	Surface of spore not as above...65
65.	(64) Subtending hypha funnel shaped with a two-layered wall .... *Glomus merredum* Porter & Hall[27]
	Subtending hypha simple with an homogenous wall 66
66.	(65) Spore wall 5—16 μm thick consisting of many layers each approximately 1-μm thick; subtending hypha colored though may be hyaline in immature spores............................ *Glomus intraradices* Schenck & Smith[31]
	Spore walls not as above; subtending hypha always hyaline..............67
67.	(66) Spore wall (25-) 45—55 μm thick; innermost wall layer always hyaline .... Type 7 Sward et al. (*Glomus* sp.)[38]
	Spore wall 5—13 μm thick; innermost wall layer colored, but may be hyaline in immature spores............................ *Glomus gerdemanni* Rose et al.[28]
68.	(25) Outer layer of spore wall up to 10-μm thick; inner laminated up to 14-μm thick .... Type 3 Sward et al. (*Gigaspora* sp.)[38]
	Outer layer of spore wall 1-μm thick and striated; inner up to 6-μm thick .... *Gigaspora candida* Bhattacharjee et al.[21]
69.	(56) Spore wall 5—24 μm thick of 4—10 layers each 1.5—4-μm thick; soil-borne vesicles with warty projections up to 4-μm high and 5-μm wide.. *Gigaspora margarita* Becker & Hall[17]
	Spore wall 4—12 μm thick with up to 6 layers (only 1 in young spores), outer wall 1—2 μm thick readily cracking under light pressure, 2—5 inner layers inseparable and of varying thickness; soil-borne vesicles

	with spines 2.5—10 μm long with the spines frequently septate and occasionally bifurcate .... *Gigaspora albida* Schenck & Smith[31]
70.	(40) Spores up to 90-μm wide; subtending hypha simple but may be reflexed .... *Glomus aggregatum* Schenck & Smith[31]
	Spores 140—300 μm wide; subtending hypha funnel-shaped.. Type 8 Sward et al. (*Glomus* sp.)[38]
71.	(46) Spores 50—75 μm wide always with two-layered walls .... *Glomus invermaium* Hall[24]
	Spores 40—190 μm wide with 1, 2, or occasionally up to 4 striated walls; spores often formed within roots.............. *Glomus intraradices* Schenck & Smith[31]
72.	(49) Spores colored 210—250 μm wide; thin colorless outer wall layer present in immature spores but lost by maturity; inner wall can be of two layers and may be striated; septum can be present in subtending hypha.. *Glomus merredum* Porter & Hall[27]
	Spores hyaline to dirty white, up to 155-μm wide; spore wall of two subequal layers; septum absent .... *Glomus fecundisporum* Schenck & Smith[31]
73.	(4) Mature spores covered with a hyaline to brown mantle arising from the subtending hypha... *Glomus tortuosum* Schenck & Smith[31]
	Spores not as above 74
74.	(73) Spores ornamented with minute spines approximately 1-μm high .... WUM 4 (*Glomus* sp.)[25]
	Spores ornamented with an indistinct reticulum of shallow ridges 0.5—1 μm wide, most apparent on young spores devoid of contents.. *Glomus leptotichum* Schenck & Smith[31]
75.	(47) Spores bright yellow to bright brownish yellow, 95—140 μm wide; diameter of subtending hypha at widest part usually <10 μm; subtending hypha inserted into the spore wall; pore at point of attachment of subtending hypha occluded by a septum-like plug; spores borne singly, in loose clusters, or in epigeous sporocarps .... *Glomus epigaeum* Daniels & Trappe[22]
	Spores yellowish brown to dark brown, 95—400 μm wide; diameter of subtending hypha at widest part usually >10 μm; subtending hypha not inserted; pore without a plug but may be occluded by walls of subtending hypha; spores borne singly, in loose clusters, or in generally hypogeous sporocarps .... *Glomus macrocarpum* Tul. & Tul. var *macrocarpum*[5]

C. Key to Nonsporocarpic Species — Subtending Hyphae not Observable

1.	(0) Spores sessile, borne laterally on thin-walled hyphae ending in a thin-walled mother spore 2
	Spores not as above 11

2.	(1)	Outer surface of outermost wall layer ornamented with sparse warts, a reticulum, regular pits or pores, or regular projections 3
		Outer surface of outermost wall layer not ornamented but surface of an inner wall layer may be ornamented 7
3.	(2)	Surface of spore ornamented with a distinct reticulum with small spines in the spaces within the reticulum 4
		Surface of spore not as above.. 5
4.	(3)	Outermost wall layer thicker than any of the inner, i.e., wall layers not similar in thickness.. *Acaulospora elegans* Trappe & Gerdemann[5]
		Wall layers of similar thickness..................... *Acaulospora bireticulata* Rothwell & Trappe[30]
5.	(3)	Surface of outermost wall layer ornamented with spines... *Acaulospora spinosa* Walker & Trappe[37]
		Surface of outermost wall layer not as above 6
6.	(5)	Surface of outermost wall layer ornamented with pores .... *Acaulospora scrobiculata* Trappe[33]
		Surface of outermost wall layer ornamented with cerebriform folds up to 12-μm high.. *Acaulospora gerdemannii* Schenck & Nicolson[26]
7.	(2)	Spores <85 μm in diameter, minutely roughened; spore wall homogenous .... *Acaulospora trappei* Ames & Linderman[15]
		Spores >100 μm in diameter; spore wall two or more layered............. 8
8.	(7)	Outer surface of second wall layer ornamented with spines or an alveolate reticulum.. 9
		Outer surface of an inner wall layer not ornamented 10
9.	(8)	Outer surface of an inner wall layer ornamented with spines... *Acaulospora spinosa* Walker & Trappe[37]
		Outer surface of an inner wall layer ornamented with an alveolate reticulum .... *Acaulospora gerdemannii* Schenck & Nicolson[26]
10.	(8)	Maximum breadth of spores generally <250 μm; spores generally honey colored... *Acaulospora laevis* Gerdemann & Trappe[5]
		Maximum breadth of spores generally >250 μm; spores deep red .... *Acaulospora* large red[24]
11.	(1)	Surface of spores and sometimes subtending hyphae covered in minute spines and ridges about 1-μm high .... WUM 4 (*Glomus* sp.)[25]
		Surface of spores not ornamented though an outer surface of an inner wall layer may be ornamented..12
12.	(11)	Spore develops in a hyaline hypha which subtends a hyaline vesicle; spores thus with a smooth hyaline outer wall layer and a brown inner layer with polygonal projections on the outer surface .... *Entrophospora infrequens* (Hall) Ames & Schneider[16,24]
		Spore does not develop as above; ornamentation of outer surface of an

	inner wall layer not occurring; outer wall layer much thicker than inner wall..13
13.	(12) Spores hyaline to creamy white at maturity; wall layers separable......14
	Spores colored though may be hyaline when young formed in a cluster on the end of a swollen hyphal tip; wall layers not readily separable .... *Glomus fuegianum* (Spegazzini) Trappe & Gerdemann[8,24]
14.	(13) Spores ornamented with projections 12—30 μm high... White reticulate Hayman (*Glomus* sp.?)[90]
	Spores not as above 15
15.	(14) Spores 230—300 μm in diameter; wall 45—55 μm thick .... Type 7 Sward et al. (*Glomus* sp.)[38]
	Spores (85-) 95—168 (-198) μm in diameter; wall 1—10 μm thick .... *Glomus albidum* Walker & Rhodes[36]

D. Key to the Genera

1.	(0) Spores (zygospores, chlamydospores, sporangiospores) formed in compact sporocarps (Figures 6—13), sporocarps generally containing many spores but if few spores then peridium well developed (Figure 10)..2
	Spores not formed in sporocarps but may occur in loose clusters........ 6
2.	(1) Spores (sporangiospores) with thin walls and always containing few to many sporangiospores (Figure 1 (V); Figure 4)... *Modicella* Kanouse[5,71]
	Spores not containing sporangiospores though some spores may contain hyphae or spores of hyperparasites 3
3.	(2) Zygospores present formed after the union of two gametangia; gametangia attached to the same part of the spore (Figure 1 (i); Figure 2) or occasionally wide apart; gametangia often not visible on mature zygospores; zygospore contents regularly sized oil globules; zygospore wall nearly always of two distinct layers.......................... *Endogone* Link ex Fries[5]
	Spores (chlamydospores) not as above usually with only one subtending hypha..4
4.	(3) Spores in spherical clusters arranged around a central plexus of sterile hyphae with the sporocarps consisting of one to many such clusters (Figures 8 and 9) which may be interspersed with thin-walled vesicles (Figure 8).. *Sclerocystis* Berk. & Broome[5]
	Spores arranged randomly within the sporocarp or if with a distinct arrangement then not as above... 5
5.	(4) Sporocarps large (10—60 mm in diameter), hollow and orange, containing chlamydospores in their walls......................... *Glaziella* Berk.[5,8]
	Sporocarps solid containing one to many chlamydospores .... *Glomus* Tul. & Tul.[5]
6.	(1) Spores (chlamydospores) <15 μm in diameter associated with VAM with hyphae <1.5 μm wide (Figure 25) .... *Glomus* Tul. & Tul.[5]

	Spores (chalmydospores, azygospores) >25-μm in diameter and if associated with VAM then some hyphae >4-μm wide 7
7.	(6) Spores (azygospores) formed on bulbous subtending hyphae often bearing one to few short lateral projections (Figure 1 (ii)); soil-borne vesicles may be present (Figures 26 and 30) .... *Gigaspora* Gerdemann & Trappe[5]
	Spores (azygospores, chlamydospores) not formed on bulbous subtending hyphae as above; soil-borne vesicles absent 8
8.	(7) Spores (azygospores) sessile formed on the side of a wide thin-walled hypha that terminates in a large thin-walled vesicle (Figure 1 (iii); Figure 3) .... *Acaulospora* Gerdemann & Trappe[5]
	Spores not formed as above 9
9.	(8) Spores (azygospores) formed in a hypha which subtends a large thin-walled vesicle (Figure 1 (iv); Figure 22) .... *Entrophospora* Ames & Schneider[16]
	Spores (chlamydospores) not formed as above 10
10.	(9) Spores and subtending hyphae covered in club-shaped or reflexed projections (Figure 5 and 20) .... *Complexipes* Walker[35]
	Spores not as above .... *Glomus* Tul. & Tul.[5]

VII. GLOSSARY*

In order to make these keys and others[5,17,26,31,37,39] easier to use an illustrated glossary has been assembled. Other terms not listed below may be found in a good English Dictionary[91] or in *Ainsworth and Bisby's Dictionary of the Fungi.*[92] Further illustrations can be found in Hall and Abbott's[93] photographic slide collection which illustrates features of the Endogonaceae.

Azygospore — Zygospore originating from only one gametangium-like structure, e.g., *Gigaspora* (see also the section on *Gigaspora*) and *Acaulospora* (Figures 1 (ii,iii) and 3).

Bulbous subtending hypha — One which is markedly swollen below the spore and constricted at the point of attachment to the spore[76] (Figure 1 (ii)).

Chlamydospore — Terminal (rarely intercalary) thick-walled spore formed asexually and generally (c.f., *Glomus multicaule*) on a single subtending hypha (Figures 4, 15, 23, 25, and 29).

Clavate spore — Club-shaped spore, narrower at the base than at the tip, much longer than broad.

Concolorous — The same color throughout, e.g., subtending hypha concolorous with spore wall.

Constricted subtending hypha — Markedly pinched-in at the point of attachment to the spore (Figures 1 (viii) and 14).

---

* Since this chapter was written some of the confusing terminology used in descriptions of Endogonaceae has been revised by Trappe and Schenck (Taxonomy of the fungi forming endomycorrhizae. A. Vesicular-arbuscular mycorrhizal fungi (Endogonales), in *Methods and Principles of Mycorrhizal Research*, Schenck, N. C., Ed., American Phytopathological Society, St. Paul, Minn., 1982, chap. 1).

Echinulate — Covered in small spines (Figure 30).

Ectocarpic spores — Formed loose in the soil, root or rhizosphere; not formed within a sporocarp.

Flammenkrone — A tightly adherent mantle encircling each spore in a spiral sinuous manner with the lateral walls pressed together; in cross-section, hyphal walls adjacent to the spore and lateral walls very thick, the thickening becoming thinner away from the spore surface; thickened lateral walls of two adjacent hyphae form pointed flame-shaped projections[5] (Figure 11).

Funnel-shaped subtending hypha — Hypha gradually widening-out toward the point of attachment to the spore[76] (Figures 1 (vi) and 15).

Gametangium — A fertile hyphal tip separated from the suspensor by a septum (Figures 1 (i) and 2).

Germination compartment — A chamber formed between two of the spore wall layers prior to the formation of germ tubes in *Acaulospora laevis* and some species of *Gigaspora* (Figure 27).

Germination warts — Structures formed on the inner surface of the spore wall prior to the formation of germ tubes in those *Gigaspora* sp. which do not produce germination compartments (Figure 28).

Gleba — The hyphal net which gives rise to spores in a sporocarp (Figure 9).

Hyaline — Glass-like, colorless.

Hypogeous — Formed beneath the soil.

Intercalary — Formed between two "cells", e.g., a spore apparently formed between two hyphae (Figures 19 and 29).

Lactophenol — By weight: 1 part phenol, 1 part lactic acid, 1 part glycerol, 1 part water.

Laminated — Composed of many distinct layers (c.f., striated) (Figure 17).

Lateral projection — Short, thin hyphal projections formed on the side of the bulbous subtending hypha of, c.f., *Gigaspora* spp., often ending in very thin branched hyphae (Figure 1 (ii)).

Mantle — An individual envelope of hyphae surrounding a spore in a sporocarp (Figure 13).

Mature spore — No longer juvenile or immature; fully developed.

Melzer's reagent — Chloral hydrate 100 g, distilled water 100 mℓ, iodine 1.5 g, potassium iodide 5.0 g.

Membrane — A thin skin.

"Miracle" Mounting Fluid[60]

Gum arabic (acacia gum)	15 g
Chloral hydrate	100 g
Glycerol	10 g
Water	25 mℓ

Add to the water 1 crystal of chloral hydrate to prevent bacterial growth on gum. Soak gum in water for 24 hr. Add chloral hydrate and let stand until dissolved (several days). Add glycerol and strain through muslin before using.

Mother hypha — One that gives rise to a spore[39] (Figures 1 and 3).

Mother spore — The generally thin-walled bladder formed on the end of a hypha prior to the formation of the resting spore in *Acaulospora* sp. (Figures 1 and 3).

Mother vesicle — Mother spore (q.v.).

Parent hypha — One that gives rise to a spore in *Acaulospora* and *Entrophospora*[39] (Figures 1 (iii,iv) and 3).

Peg — q.v. Lateral projection.

Peridium — The infertile outer wall of sporocarp (Figures 10 and 12).

Plug — Ill-defined occlusion of the pore connecting the lumen of the subtending hypha to the spore contents, c.f., septum (Figures 7 and 24).

Polyvinyl lactic acid[61]

Polyvinyl alcohol	1.66 g
Water	10.0 mℓ
Lactic acid	10.0 mℓ
Glycerol	1.0 mℓ

PVA is dissolved in water and the lactic acid followed by the glycerine is added while stirring.

Polyvinyl lactophenol[60]

Stock solution: Dissolve 15 parts polyvinyl alcohol (50—75% hydrolyside: 20—25 centipoise viscosity when in a 4% aqueous solution at 20°C) in 100 mℓ distilled water over a water bath at 80°C. Store in dark bottle.

Working solution: 22 parts lactic acid
22 parts liquified phenol
56 parts stock solution

Pore
- The channel connecting a spore to its subtending hypha which may be occluded by the thickening of the spore wall/subtending hypha.
- Small holes in spore surface or inner wall layer.

Reticulate spore — One with the contents formed of small even-sized oil globules making the contents refractile, c.f., vacuolate.[77]

Septum — A thin, flat, or cup-shaped wall across the subtending hypha at or just below the point where the subtending hypha is attached to the spore (Figures 1, 2, 15, and 23).

Simple subtending hypha — One with parallel side walls, not increasing markedly in width at the point of attachment to spore (Figures 1 (vii), 19, 23, and 29).

Sporangium — A sac-like structure formed terminally on a hypha, the contents of which become converted into an indefinite number of asexual sporangiospores (Figures 1 (iv) and 4).

Sporangiospores — q.v. Sporangium.

Spore — Used loosely to refer to zygospores, azygospores, sporangia, and chlamydospores. The length of a spore is measured from the point of attachment of the subtending hypha to a point diagonally opposite (Figure 1).

Spore attachment — q.v. Subtending hypha.

Sporocarp — A tightly formed aggregation of glebal hyphae, spores, and possibly peridial hyphae (Figures 6 to 13).

Striated spore wall — Marked with delicate lines. The lines not running the full circumference of the spore wall, c.f., laminated (Figure 24).

Subtending hypha — The suspensor, gametangium, sporophore, or hypha which subtends a spore. In shape it may be simple, bulbous, funnel-shaped, or constricted (q.v.).

Trumpet-shaped hyphae — Ridged, swollen, hollow hyphae possibly formed by the elongation of a spore. Formed in sporocarps of *Glomus tubiforme*.[32]

Vacuolate spore — One with the contents formed of a few uneven-sized oil globules, i.e., reticulate[77] (Figure 29).

Vesicle — (1) A bladder formed intracellularly on the end of a hypha within a root (Figure 25); (2) A spherical, polymorphic, knobby or echinulate thin-walled bladder formed singly or in tight clusters of up to 20 in the soil by species of *Gigaspora* (Figures 26 and 30). (3) Thin-walled terminal bladders which are formed along with the thicker-walled spores in sporocarps of for example *Glomus vesiculiferum* and *Sclerocystis dussii* (Figure 8). (4) q.v. Mother spore.

Zygospore — The spore resulting from the union of two gametangia, for example, in *Endogone* (Figures 1 (i) and 2).

REFERENCES

1. Ham, A. M., Studies on vesicular-arbuscular mycorrhizas. IV. Inoculation of species of *Allium* and *Lactuca sativa* with *Pythium* isolates, *Trans. Br. Mycol. Soc.*, 45, 179, 1962.
2. Hawker, L. E., Studies on vesicular-arbuscular mycorrhizas. V. A review of the evidence relating to the identity of the causal fungi, *Trans. Br. Mycol. Soc.*, 45, 190, 1962.
3. Hawker, L. E., Harrison, R. W., Nicholls, V. O., and Ham, A. M., Studies on vesicular-arbuscular endophytes. I. A strain of *Phythium ultimum* Trow. in roots of *Allium ursinum* L. and other plants, *Trans. Br. Mycol. Soc.*, 40, 375, 1957.
4. Hawker, L. E. and Ham, A. M., Vesicular-arbuscular mycorrhizas in apple seedlings, *Nature (London)*, 180, 998, 1957.
5. Gerdemann, J. W. and Trappe, J. M., The Endogonaceae in the Pacific Northwest, *Mycol. Mem.*, 5, 1, 1974.
6. Butler, E. J., The occurrences and systematic position of the vesicular-arbuscular type of mycorrhizal fungi, *Trans. Br. Mycol. Soc.*, 22, 274, 1939.
7. Rosendahl, C. O, Some fossil fungi from Minnesota, *Bull. Torrey Bot. Club*, 70, 126, 1943.
8. Thaxter, R., A revision of the Endogoneae, *Proc. Am. Acad. Arts Sci.*, 57, 291, 1922.
9. Pirozynski, K. A. and Malloch, D. W., The origin of land plants: a matter of mycotrophism, *Biosystems*, 6, 153, 1975.
10. Weijman, A. C. M. and Meuzelaar, H. L. C., Biochemical contributions to the taxonomic status of the Endogonaceae, *Can. J. Bot.*, 57, 284, 1979.
11. Bucholtz, F., Beitrage zur Kenntnis der Gattung *Endogone* Link, *Beih. Bot. Zentralbl. Abstr.*, 2, 29, 147, 1912.
12. Benjamin, R. K., The merosporangiferous mucorales, *Aliso*, 4, 321, 1959.
13. Benjamin, R. K., Zygomycetes and their spores, in *The Whole Fungus*, Kendrick, B., Ed., National Museum of Natural Sciences and National Museums of Canada, Ottawa, 1979, chap. 23.
14. Zycha, H., Siepmann, R., and Linnermann, G., *Mucorales*, J. Cramer, Lehre, West Germany, 1969, 1.
15. Ames, R. N. and Linderman, R. G., *Acaulospora trappei* sp. nov., *Mycotaxon*, 3, 565, 1976.
16. Ames, R. N. and Schneider, R. W., *Entrophospora*: a new genus in the Endogonaceae, *Mycotaxon*, 8, 347, 1979.
17. Becker, W. N. and Hall, I. R., *Gigaspora margarita*, a new species in the Endogonaceae, *Mycotaxon*, 4, 155, 1976.
18. Becker, W. N. and Gerdemann, J. W., *Glomus etunicatus* sp. nov., *Mycotaxon*, 6, 29, 1977.
19. Bhattacharjee, M. and Mukerji, K. G., Studies on Indian Endogonaceae. II. The genus *Glomus*, *Sydowia*, 33, 14, 1980.
20. Bhattarcharjee, M., Mukerji, K. G., and Misra, S., Studies on Indian Endogonaceae. III. Further records, *Acta Bot. Indica*, 8, 99, 1980.
21. Bhattacharjee, M., Mukerji, K. G., Tewari, J. P., and Skoropad, W. P., Structure and hyperparasitism of a new species of *Gigaspora*, *Trans. Br. Mycol. Soc.*, 78, 184, 1982.
22. Daniels, B. A. and Trappe, J. M., *Glomus epigaeus* sp. nov., a useful fungus for vesicular-arbuscular mycorrhizal research, *Can. J. Bot.*, 57, 539, 1979.
23. Gerdemann, J. W. and Bakshi, B. K., Endogonaceae of India. Two new species, *Trans. Br. Mycol. Soc.*, 66, 340, 1976.

24. Hall, I. R., Species and mycorrhizal infections of New Zealand Endogonaceae, *Trans. Br. Mycol. Soc.*, 68, 341, 1977.
25. Hall, I. R. and Abbott, L. K., Some endomycorrhizal fungi from South Western Australia, *Trans. Br. Mycol. Soc.*, in press.
26. Nicolson, T. H. and Schenck, N. C., Endogonaceous mycorrhizal endophytes in Florida, *Mycologia*, 71, 178, 1979.
27. Porter, W. M. and Hall, I. R., *Glomus velum* and *Glomus merredum*, two new species in the Endogonaceae, in preparation.
28. Rose, S., Daniels, B. A., and Trappe, J. M., *Glomus gerdemannii* sp. nov., *Mycotaxon*, 8, 297, 1979.
29. Rose, S. and Trappe, J. M., Three new endomycorrhizal *Glomus* spp. associated with actinorrhizal shrubs, *Mycotaxon*, 10, 413, 1980.
30. Rothwell, F. M. and Trappe, J. M., *Acaulospora bireticulata*, sp. nov., *Mycotaxon*, 8, 471, 1979.
31. Schenck, N. C. and Smith, G. S., Additional new and unreported species of mycorhizal fungi (Endogonaceae) from Florida, *Mycologia*, 74, 77, 1982.
32. Tandy, P., Sporocarpic species of Endogonaceae in Australia, *Aust. J. Bot.*, 23, 849, 1975.
33. Trappe, J. M., Three new Endogonaceae: *Glomus constrictus, Sclerocystis clavispora, Acaulospora scrobiculata*, *Mycotaxon*, 6, 359, 1977.
34. Trappe, J. M., *Glomus segmentatus* sp. nov., *Trans. Br. Mycol. Soc.*, 73, 362, 1979.
35. Walker, C., *Complexipes moniliformis:* a new genus and species tentatively placed in the Endogonaceae, *Mycotaxon*, 10, 99, 1979.
36. Walker, C. and Rhodes, L. H., *Glomus albidus:* a new species in the Endogonaceae, *Mycotaxon*, 12, 509, 1981.
37. Walker, C. and Trappe, J. M., *Acaulospora spinosa* sp. nov. with a key to the species of *Acaulospora*, *Mycotaxon*, 12, 515, 1981.
38. Sward, R. J., Hallam, N. D., and Holland, A. A., *Endogone* spores in a heathland area of South-Eastern Australia, *Aust. J. Bot.*, 26, 29, 1978.
39. Hall, I. R. and Fish, B. J., A key to the Endogonaceae, *Trans. Br. Mycol. Soc.*, 73, 261, 1979.
40. Hall, I. R., Unpublished data, 1980.
41. Hayman, D. S., *Endogone* spore numbers in soil and vesicular-arbuscular mycorrhiza in wheat as influenced by season and soil treatment, *Trans. Br. Mycol. Soc.*, 54, 53, 1970.
42. Herskowizz, J. M. and Estey, R. H., Endogonaceae from Quebec soils, *Can. J. Bot.*, 56, 1095, 1978.
43. Mason, D. T., A survey of numbers of *Endogone* spores in soils cropped with barley, raspberry, and strawberry, *Hortic. Res.*, 4, 98, 1964.
44. Smith, T. F., The effect of season and crop rotation on the abundance of spores of vesicular-arbuscular mycorrhizal endophytes, *Plant Soil*, 57, 475, 1980.
45. Sutton, J. C. and Barron, G. L., Population dynamics of *Endogone* spores in soil, *Can. J. Bot.*, 50, 1909, 1972.
46. Hayman, D. S., Johnson, A. M., and Ruddlesdin, I., The influence of phosphate and crop species on *Endogone* spores and vesicular-arbuscular mycorrhizae under field conditions, *Plant Soil*, 43, 489, 1975.
47. Smith, T. F., A note on the effect of soil tillage on the frequency and vertical distribution of spores of vesicular-arbuscular endophytes, *Aust. J. Soil Res.*, 16, 359, 1978.
48. Baylis, G. T. S., Host treatment and spore production by *Endogone*, *N. Z. J. Bot.*, 7, 173, 1969.
49. Johnson, P. N., Mycorrhizas in a Coniferous Broadleaf Forest, Ph.D. thesis, Otago University, Dunedin, New Zealand, 1973.
50. Hall, I. R., Endogonaceous Fungi Associated with Rata and Kamahi, Ph.D. thesis, Otago University, Dunedin, New Zealand, 1973.
51. St. John, T. V., Hays, R. I., and Reid, C. P. P., A new method for producing pure vesicular-arbuscular mycorrhizal-host cultures without specialised media, *New Phytol.*, 89, 81, 1981.
52. Menge, J. A., Lembright, H., and Johnson, E. L. V., Utilization of mycorrhizal fungi in *Citrus* nurseries, *Proc. Int. Soc. Citric.*, 1, 129, 1977.
53. Tommerup, I. C. and Carter, D. J., Dry separation of micro-organisms from soil, *Soil Biol. Biochem.*, 14, 69, 1982.
54. Smith, G. W. and Skipper, H. D., Comparison of methods to extract spores of vesicular-arbuscular mycorrhizal fungi, *Soil Sci. Soc. Am. J.*, 43, 722, 1979.
55. Gerdemann, J. W. and Nicolson, T. H., Spores of mycorrhizal *Endogone* species extracted by wet sieving and decanting, *Trans. Br. Mycol. Soc.*, 46, 235, 1963.
56. Furlan, V., Bartschi, H., and Fortin, J. A., Media for density gradient extraction of endomycorrhizal spores, *Trans. Br. Mycol. Soc.*, 75, 336, 1980.
57. Furlan, V. and Fortin, J. A., A flotation bubbling system for collecting endogonaceous spores from sieved soil, *Nat. Can. (Quebec)*, 102, 603, 1975.

58. Mosse, B. and Jones, G. W., Separation of *Endogone* spores from organic soil debris by differential sedimentation on gelatin columns, *Trans. Br. Mycol. Soc.*, 51, 604, 1968.
59. Tommerup, I. D. and Kidby, D. K., Preservation of spores of vesicular-arbscular endophytes by L-Drying, *Appl. Environ. Microbiol.*, 37, 831, 1979.
60. Abbott, L. K. and Robson, A. D., Personal communication, 1979.
61. Omar, M. B., Bolland, L., and Heather, W. A., A permament mounting medium for fungi, *Bull. Br. Mycol. Soc.*, 13, 31, 1979.
62. Feder, N. and O'Brien, T. P., Plant micro-technique: some principles and new methods, *Am. J. Bot.*, 55, 123, 1968.
63. Mosse, B., Honey-coloured sessile *Endogone* spores. I. Life history, *Arch. Mikrobiol.*, 70, 167, 1970.
64. Mosse, B., Honey-coloured sessile *Endogone* spores. II. Changes in fine structure during spore development, *Arch. Mikrobiol.*, 74, 129, 1970.
65. Mosse, B., Honey-coloured sessile *Endogone* spores. III. Wall structure, *Arch. Mikrobiol.*, 74, 146, 1970.
66. Sward, R. J., The structure of the spores of *Gigaspora margarita*. I. The dormant spore, *New Phytol.*, 87, 761, 1981.
67. Sward, R. J., The structure of spores of *Gigaspora margarita*. II. Changes accompanying germination, *New Phytol.*, 88, 661, 1981.
68. Sward, R. J., The structure of spores of *Gigaspora margarita*. III. Germ tube emergence and growth, *New Phytol.*, 88, 667, 1981.
69. Godfrey, R. M., Studies of British species of *Endogone*. I. Morphology and taxonomy, *Trans. Br. Mycol. Soc.*, 40, 117, 1957.
70. Gerdemann, J. W., Vesicular-arbuscular mycorrhiza formed on maize and tuliptree by *Endogone fasciculata*, *Mycologia*, 57, 562, 1965.
71. Kanouse, B. B., Studies of two species of *Endogone* in culture, *Mycologia*, 28, 47, 1936.
72. Fassi, B., Micorrize ectotrofiche *Pinus strobus* L. prodotte da un' *Endogone* (*Endogone lactiflua* Berk.), *Allionia*, 11, 7, 1965.
73. Fassi, B., Fontana, A., and Trappe, J. M., Ectomycorrhizae formed by *Endogone lactiflua* with species of *Pinus* and *Pseudotsuga*, *Mycologia*, 61, 412, 1969.
74. Fassi, B. and Palenzona, M., Sintesi micorrizica tra *Pinus strobus, Pseudotsuga douglasii* ed "*Endogone lactiflua,*" *Allionia*, 15, 105, 1969.
75. Warcup, J. H., Unpublished data, 1980.
76. Nicolson, T. H. and Gerdemann, J. W., Mycorrhizal *Endogone* species, *Mycologia*, 60, 313, 1968.
77. Mosse, B. and Bowen, G. D., A key to the recognition of some *Endogone* spore types, *Trans. Br. Mycol. Soc.*, 51, 469, 1968.
78. Boedijn, K. B., Die Gattung *Glaziella* Berkeley, *Bull. Jard. Bot. Buitenzorg Ser. 3*, 11, 57, 1930.
79. Rogers, D. P., Fungi of the Marshall Islands, Central Pacific Ocean, *Pac. Sci.*, 1, 92, 1947.
80. Walker, L. B., Some observations on the development of *Endogone malleola* Hark., *Mycologia*, 15, 245, 1923.
81. Wilcox, H. E., Ganmore-Neumann, R., and Wang, C. J. K., Characteristics of two fungi producing ectendomycorrhizae in *Pinus resinosa, Can. J. Bot.*, 52, 2279, 1974.
82. Mosse, B., Fructifications of an *Endogone* species causing endotrophic mycorrhiza in fruit plants, *Ann. Bot. (London)*, 20, 349, 1956.
83. Mosse, B., Growth and chemical composition of mycorrhizal and non-mycorrhizal apples, *Nature (London)*, 179, 922, 1957.
84. Abbott, L. K. and Robson, A. D., Growth of subterranean clover in relation to the formation of endomycorrhizas by introduced and indigenous fungi in a field soil, *New Phytol.*, 81, 575, 1978.
85. Abbott, L. K. and Robson, A. D., A quantitative study of the spores and anatomy of mycorrhizas formed by a species of *Glomus* with reference to its taxonomy, *Aust. J. Bot.*, 27, 363, 1979.
86. Abbott, L. K., Comparative anatomy of vesicular-arbuscular mycorrhizas formed on subterranean clover, *Aust. J. Bot.*, 30, 485, 1982.
87. Dallwitz, M. J., A flexible computer program for generating identification keys, *Syst. Zool.*, 23, 50, 1974.
88. Hall, I. R., A summary of the features of endogonaceous taxa, Tech. Rep. No. 8 Ed. 3, Invermay Agricultural Research Centre, Mosgiel, New Zealand, 1, 1983.
89. Grand, L. F. and Randall, B. L., *Endogone pisiformis* and *Endogone multiplex* in North Carolina, *Mycologia*, 73, 1092, 1981.
90. Hayman, D. S., Mycorrhizal populations of sown pastures and native vegetation in Otago, New Zealand, *N.Z. J. Agric. Res.*, 21, 271, 1978.
91. *The Concise Oxford Dictionary,* 6th ed., Sykes, J. B., Ed., Clarendon Press, Oxford, 1976.
92. Ainsworth, G., *Ainsworth and Bisby's Dictionary of the Fungi,* 6th ed., Kew Commonwealth Mycological Institute, Kew, England, 1971.
93. Hall, I. R. and Abbott, L. K., Photographic slide collection illustrating features of Endogonaceae, *Bull. Br. Mycol. Soc.*, 17, 54, 1983.

Chapter 5

ISOLATION AND CULTURE OF VA MYCORRHIZAL (VAM) FUNGI

Christine M. Hepper

TABLE OF CONTENTS

I.	Introduction	96
II.	Isolation of VAM Fungi from Soil	96
III.	Surface-Sterilization of Resting Spores	97
IV.	Axenic Mycorrhizal Plants	98
	A. Methods of Preparation	98
	B. Use of Axenic Mycorrhizal Plants	100
V.	Pure Culture of VAM Fungi	103
	A. Requirement for Pure Cultures	103
	B. Metabolic Capabilities of VAM Fungi	103
	C. Progress Toward Culture of VAM Fungi	104
	1. Introduction	104
	2. Growth From Resting Spores	104
	3. Reasons For Failure of Growth	106
	4. Other Possible Approaches to Culture	108
VI.	General Conclusions	109
References		110

I. INTRODUCTION

The growth of vesicular-arbuscular mycorrhizal (VAM) fungi in pure culture in the absence of living host roots is an important objective. Currently their taxonomic classification is based almost entirely on the morphology of the chlamydospores and azygospores and their method of germination. If these fungi could be grown in culture, many more characteristics might be considered and classification on biochemical bases used as it is with other microorganisms. Other fundamental studies on VAM fungi would also be greatly facilitated, in particular, their genetic modification and selection to obtain superior strains.

Probably the most important application of pure culture envisaged at the moment is the large-scale production of inoculum for field experiments. Current methods of producing inoculum involve growing plants in semi-sterile rooting media and then using the roots with or without the infested medium (see Menge, Chapter 9). This can be used satisfactorily for glasshouse experiments and field trials, but from a commercial point of view it is essential that the inoculum should be free from pathogens (both fungal and nematode) and that it should be of uniform quality. This might be difficult to monitor in practice and problems could arise if inoculum were not rigorously tested. For this reason, axenic culture of VAM fungi has now become a subject of interest to some commercial firms who forsee the possibilities of producing high-quality inoculum under controlled conditions.

II. ISOLATION OF VAM FUNGI FROM SOIL

One of the first problems associated with a project on VAM may be to isolate strains of fungi from field soils and establish them as individual pot cultures in soil mixes which have been sterilized to kill all other mycorrhiza-forming species. If the fungi of interest produce large resting spores these can be isolated by passing a suspension of the soil through a series of graded sieves (the wet-sieving method).[1,2] The sieve mesh sizes should be chosen so that the first removes pieces of root and large particles of organic matter, and the subsequent sieves (with mesh aperture diameters of 450, 250, 150, 100, and 50 μm) retain spores and sporocarps of different sizes. The material held on each sieve can be examined under a microscope and individual spores selected for identification or as initial inoculum for a stock plant. When a soil is very difficult to wet-sieve, for instance, if it has a high organic matter content, other approaches must be adopted. One method is to plant seedlings in the soil and when they become infected, transplant them into a soil which is easier to wet-sieve and which has been sterilized to kill its indigenous endophytes.[3] The same approach can be used by transplanting a naturally infected seedling.[3] Fungi which do not form large resting spores, e.g., the fine endophyte described by Greenall,[4] must be isolated by taking pieces of washed infected root and using these as inoculum for a new plant growing in sterilized soil. However, a culture established in this way could consist of several fungi, all of which would be classified as fine endophytes but which might in fact be separate strains or species. The isolation of nonsporing VAM fungi from a soil which also contained spore-bearing types would be difficult, since the pieces of infected root would probably contain both types and so both would be present in any culture obtained in this way.

The wet-sieving method is the basis of many other techniques designed to improve the recovery of spores and facilitate the collection of large numbers of them free from soil debris. One method is to subject the material retained on a sieve to sedimentation in a density gradient. Ohms[5] first used centrifugation in a discontinuous gradient of 50% sucrose and water, but other workers[6] found that this treatment damaged the spores and later, modifications were introduced to include lower concentrations of sucrose.[7-9] The disadvantage of all these methods is that they subject the spores to osmotic shock and so exposure to the sucrose solution should be minimal. Despite this, it has been reported that spores isolated in this way germinate well.[9] The problem of osmotic shock can be obviated by the use of low osmotic pressure solutions. Mosse and Jones[6] used gelatin solutions and recently other nonosmotic media have been used successfully including one, Percoll, which can give a continuous density gradient.[10]

Another method of isolating spores is the flotation-adhesion technique[11] which depends on their relatively low density and hydrophobic surface properties. Smith and Skipper[12] compared several of these methods and found that adhesion-flotation, sucrose density gradient, and gelatin sedimentation gave statistically comparable figures for recovery of spores, but these were all lower than a direct count of the number of spores in an aliquot of soil suspension. These authors[12] suggested that in most procedures a proportion of spores were lost because they had different densities or surface properties.

III. SURFACE-STERILIZATION OF RESTING SPORES

Before resting spores of VAM fungi are surface-sterilized, it may be beneficial to store them as a suspension at 2 to 6°C for a few weeks. In the case of *Glomus mosseae* the sporocarps can be stored on damp filter paper to induce more rapid germination.[13,14]

The most commonly used spore-surface sterilant is chloramine T. Mosse[15] first described the use of a 2% (w/v) solution of it together with streptomycin at 200 μg mℓ^{-1} and a trace of surfactant to surface-sterilized spores of *G. mosseae*. The spores were washed by transferring them with a micropipette to watch-glasses containing sterile water. This technique has been modified to facilitate the washing of large numbers of spores, particularly those which have a tendency to float, by allowing the sterilant to drain from the spores and so minimize their handling.[14,16,17] By using chloramine T plus streptomycin as a sterilant for *G. mosseae* and plating the spores onto a variety of media to detect bacteria, fungi, and algae, it was shown that only 4% of them were contaminated and that the only contaminants were fungi.[18]

When this method was used for *Gigaspora margarita* it gave only 32 to 68% axenic spores[9] and so it was modified by including a prewash with 0.05% Tween 80 and subjecting the spores to partial vacuum (up to 600 mg Hg) during sterilization in 2% chloramine T. The spores were then stored in a solution of streptomycin (200 μg mℓ^{-1}) and gentamycin (100 μg mℓ^{-1}) for a week at 4°C and given a final sterilization in chloramine T for 20 min before plating out.[9] This treatment gave complete surface-sterility and if the spores were in a dormant condition, it was not detrimental to their viability.

An alternative procedure for surface-sterilization is to expose the spores to a dilute solution of sodium hypochlorite (0.5%) for up to 3 min.[1] With spores of *Gigaspora coralloidea, Gigaspora heterogama,* and *Glomus mosseae* this method gave a high percentage of bacterial contamination particularly when the spores were incubated above ambient temperature.[19] Sward[20] noted better penetration of *Gigaspora margarita* spores by electron microscopy fixatives after they had been surface-sterilized by immersion in 0.5% sodium hypochlorite for 20 sec, suggesting that the permeability of the spore wall was changed due to oxidation by the hypochlorite.

Tommerup and Kidby[21] made a comprehensive study of chemical and physical methods of surface-sterilizing spores using *Acaulospora laevis, Glomus caledonicum,* and *Glomus monosporum* isolated from pot culture or from a natural clover/grass pasture. They used four concentrations of chloramine T (2 to 12.8% equivalent to 0.16 to 1.11% available chlorine) at temperatures fom 22 to 35°C, and two concentrations of sodium hypochlorite (1 to 2% equivalent to 0.29 and 0.59% available chlorine) at 30°C. All treatments included a prewash with a surfactant (Tween 80), and this was also incorporated into the sterilant solution and the 1% sodium chloride solution used to wash the spores after sterilizing. The sterilizing solution was replaced every 5 min, the spores being sedimented by centrifugation to facilitate this. After washing, the spores were plated out on 0.02% yeast extract agar.

Spores isolated from natural pasture soils were always more difficult to surface-sterilize chemically than those from pot cultures. This may be due to higher levels of fungal hyperparasitism on VAM spores in the field than in pot cultures (see Chapter 7). With *Glomus caledonicum* from pot cultures a 20 min treatment at 30° with 5% chloramine T gave completely uncontaminated spores whereas with spores from pasture soil only 90% sterility could be achieved after 40 min and this could not be improved on, without detriment to the spore germination, by increasing the concentration of the sterilant. Spores of *Acaulospora laevis* isolated from pot cultures could be completely surface-sterilized by a 40 min wash at 30°C with 5% chloramine T but with those isolated from pastures, 60 min treatment gave only 62% sterility. Surprisingly, increasing the concentration of the sterilant induced more spores of *A. laevis* to germinate if they had been isolated from a pot culture. Thoroughly agitating the sterilizing liquid by vortexing did not overcome the problem of surface contamination. Treatment with 1% sodium hypochlorite for 5 min at 30°C was effective at surface-sterilizing *G. caledonicum* spores from pot cultures without any loss of viability, whereas *A. laevis* required treatment with 1 or 2% hypochlorite for 20 min to ensure sterility and these conditions were detrimental to germination and subsequent hyphal growth.

Tommerup and Kidby[21] recommended the routine use of 5% chloramine T at 30°C for surface-sterilizing spores, because sodium hypochlorite solution gave much less reliable and uniform results. Using 5% chloramine T, 90% sterility could be achieved with pot culture spores of *G. caledonium* or *G. monosporum*; slightly longer treatment was needed if the spores had been isolated from pastures. *Acaulospora laevis* spores from pot culture could be adequately surface-sterilized after 40 min treatment with chloramine T, but a proportion of those isolated from pasture soils were never effectively decontaminated. Chloramphenicol could be used in the agar medium at 100 to 200 μg mℓ^{-1} to suppress growth of bacteria without any change in the ability of the spores to germinate and grow. Pimaricin, which might have helped to eliminate the growth of fungal contaminants, inhibited the germination of *G. caledonium* spores at 10 μg mℓ^{-1}.

Of the physical methods of spore sterilization which were tried, heat treatment (up to 60°) was not effective, but ultrasonic irradiation did decrease surface contaminants to some extent.[21] Ultraviolet irradiation for 7 hr very effectively surface-sterilized spores of both *G. caledonicum* and *A. laevis* without impairing their ability to germinate and grow, but this treatment was considered impracticable for routine work.[21]

The general conclusion drawn from this work is that the choice of sterilant will depend on which fungus is being used and where it originated. It is always advisable to use the lowest concentration and shortest exposure time concomitant with sterility, since the sterilization process may affect metabolic processes in the spores, which although not essential for germination, may influence subsequent hyphal development.

IV. AXENIC MYCORRHIZAL PLANTS

A. Methods of Preparation

Symbioses between a variety of plants and VAM fungi have now been established under axenic conditions. Species of *Trifolium* have generally been used as host plants,[16,17,22-24] but other genera have also been tested.[7,22,25-27] *Glomus* spp. have most

FIGURE 1. Seedling of *Trifolium repens*, growing on filter paper supported on a glass slide and inoculated with resting spores of *Glomus mosseae*.

commonly been used as inoculum, in particular *G. mosseae*,[16,22,23] *G. caledonicum*,[17,23] and *G. fasciculatum*,[25-27] but *Gigaspora margarita* has also infected successfully.[24] The inoculum is usually in the form of surface-sterilized spores, which may be pregerminated or ungerminated but pieces of infected root can also be used.[14,17] The external mycelium which remains when an infected plant has been removed from the growing medium has even been used.[22,23] External mycelium appears to have a greater potential than spore inoculum to invade roots. A seedling planted in agar containing the external mycelium from another mycorrhizal plant rapidly became infected in the main root and laterals, whereas when spores were used as inoculum, the infection was slower to establish and was generally restricted to the lateral roots.[23] If spores are being used as inoculum they should be placed in contact with a young, actively-growing lateral root, to encourage growth from the germ tubes.[17,22,23]

Most systems used for axenic work completely enclose the plant, for example, in glass tubes[16,17,22,23] or deep dishes,[25-27] however the apparatus can be designed so that the shoots are allowed a free circulation of air[7] in some cases by placing the plants in a sterile isolator.[24,28] The choice of support for the plant roots is quite wide, the most usual being agar (7.5 to 15 g ℓ^{-1}),[16,22,23,25-27] soil,[7] agar and soil together,[29] filter paper resting on a microscope slide (Figure 1),[23] or sand.[24] Liquid media, either stationary[23] or continuously circulating as in nutrient film culture[28] have also been used. Macdonald[17] used a combination of a thin layer of agar to support the plant and continuously circulating nutrient solution.

A variety of plant nutrient solutions can be utilized but one essential is that they should be low in soluble phosphate. This has often been achieved by the use of poorly available P sources, e.g., $CaHPO_4 \cdot 2H_2O$,[16] phytates,[16,25-27] bone meal,[23] or hydroxyapatite,[24] but low levels of soluble phosphate have also been used.[17,23,26] Nitrogen has generally been supplied as nitrate[16,17,23,25,30] but a 7:1 NO_3:NH_4 ratio has also been used.[24] Mosse[22] found that several nitrogen sources (nitrate, ammonium, urea, and asparagine) almost completely inhibited VAM infection but this has not been substantiated, and it may be that it is the ratio of nitrogen to phosphorus which is important in this respect. Levels of heavy metals should be kept to a minimum if ungerminated spores are used because of the known inhibitory effects of low concentrations of zinc and manganese on some species of mycorrhizal fungi.[13,31] In the absence of any additional information, the plant nutrient solution should be adjusted to the optimum pH found for fungal infection in soil.

Under axenic conditions, high levels of infection can be obtained with good development of external mycelium (Figure 2). Eighty-five percent (85%) root infection has been reported in *Bouteloua gracilis* with *Glomus fasciculatum* as endophyte[25] and up to 68% with *Trifolium parviflorum* and *Glomus mosseae*.[23] The optimum conditions for high levels of infection in one species may not suit another host:fungus combination. Hepper reported differences in infection levels when *G. mosseae* and *G. caledonicum* were used as inocula for *Trifolium parviflorum*,[23] and in a system in which *Gigaspora margarita* infected up to 10% of the root system of *Trifolium repens*, *Glomus mosseae* gave only 2% infection.[24]

B. Use of Axenic Mycorrhizal Plants

Mosse[22] was the first to establish that surface-sterilized resting spores of VAM fungi could infect plants under axenic conditions. This was the essential evidence that the spores found associated with infected roots under natural conditions could initiate new symbioses in the absence of any other microorganism. Since a single spore can successfully infect a plant,[22,23] this provides a way of starting a pure line of a VAM fungus, by transplanting the infected seedling to a soil or sand medium and allowing new spores to form. In this way the intraspecific variation in the fungus can be studied, and this would provide a valuable technique for taxonomic studies.

The use of axenic plants is essential for the investigation of changes in the host brought about by mycorrhizal infection which are not attributable to the presence of any other microorganism. Allen and co-workers[26] used these methods to study the difference in phytohormone levels, root surface phosphatases,[27] and chlorophyll content[27] of mycorrhizal and nonmycorrhizal *Bouteloua gracilis* plants.

Pearson and Tinker[29] used a divided petri dish in which an axenic plant was grown in one compartment and inoculated with surface-sterilized spores. As the infection developed, external hyphae grew over the division into the other compartment. The rate of movement of phosphorus via the external mycelium was studied by applying ^{32}P to the sector containing the hyphae alone and monitoring its appearance in the host shoot tissue. This technique was developed to investigate the effect of environmental factors on this rate[32,33] and the transport of zinc and sulfur by mycorrhizal hyphae.[32]

By using root organ cultures growing on agar or in liquid, it has been possible to show that, providing carbon is supplied to the root, the shoot is not required for the initiation or spread of VAM infection in the plant.[14,30,34] This technique also provides a way of growing a split root system so that the effect of localized feeding of one part of a root system on the infection in another part can be studied.[30]

Many of the axenic techniques developed have proved extremely useful for studying preinfection stages, since the plants can be examined frequently under low magnification and the growth of the fungus in the rhizosphere observed.[17,23,30,34] This has established that root exudates stimulate the growth of VAM fungi before they penetrate the

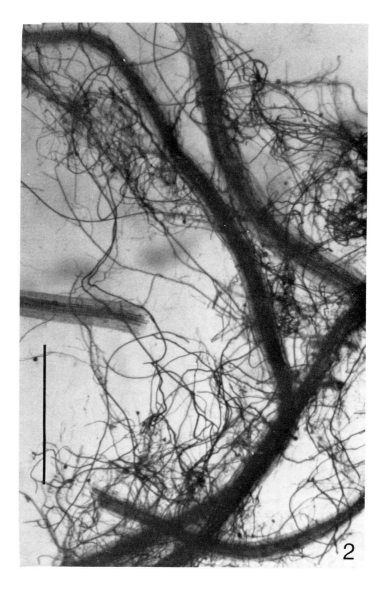

FIGURE 2. Roots of an axenic *Trifolium parviflorum* plant infected with *Glomus mosseae*, stained with trypan blue to show the internal infection and external mycelium. Scale line represents 1 mm.

root (Figures 3 and 4).[23,30,34] Formation of an appressorium provides an additional growth stimulus for the external hyphae before any intracellular phase has been established.[23] Material for a fine structure study on the early stages of infection could be obtained in this way since the time of formation of any particular entry point can be established fairly accurately.

Careful consideration should be made before plants, grown by the established axenic methods, are used for growth experiments in vitro. In the presence of soluble forms of phosphate, differing responses to mycorrhizal infection have been obtained: either an increase in the size of inoculated compared to uninoculated plants[27] or no difference between the two.[23] However, with sparingly soluble phosphorus sources no growth responses have been observed.[24,27]

Since it is possible to recover all the mycelium from agar media by melting, axeni-

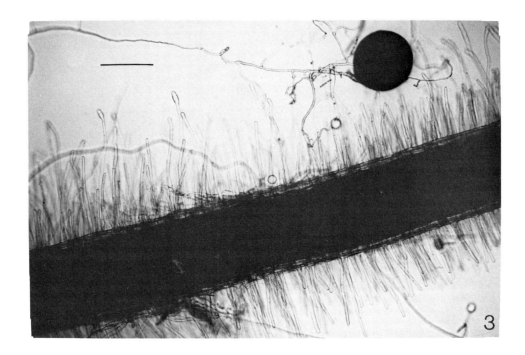

FIGURE 3. Spore of *Glomus caledonicum* germinating close to an axenic *Trifolium parviflorum* plant growing in liquid medium. Scale line represents 200 μm.

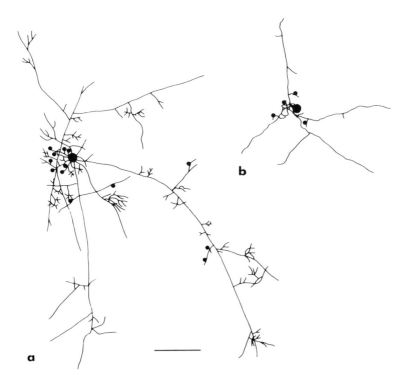

FIGURE 4. Spores of *Glomus mosseae* germinating in the presence (a) or absence (b) of root exudates of *Trifolium pratense*. Scale line represents 1 mm. (With permission from *Endomycorrhizas*, Sanders, F. E., Mosse, B., and Tinker, P. B., Eds., 1975. Copyright: Academic Press Inc. (London) Ltd.)

cally infected plants have been used to measure the ratio of external hyphae to internal infection.[16,23] This ratio was shown to approximate to 1.0 in *Trifolium parviflorum* seedlings infected with *Glomus mosseae* and growing in a medium with bone meal as a phosphorus source.[23] The ratio could be changed by incorporating organic compounds into the growing medium. In the presence of inositol, for example, the growth of the external mycelium appeared to be independent of the internal infection.[16]

In axenically maintained cultures many small vegetative spores (external vesicles) are found associated with the external mycelium but resting spores (chlamydospores or azygospores) have only occasionally been reported,[23,24] and so far no sporocarps of *G. mosseae* have been produced under these conditions.

V. PURE CULTURE OF VAM FUNGI

A. Requirement for Pure Cultures

The ability to culture VAM fungi in the absence of a host plant would simplify many of the procedures involved in setting up experiments and lead to more standardization.

It is possible, however, that if the fungi are grown in pure culture they will lose their ability to invade a root in the normal way. Initial difficulties were experienced when reinfecting plants with cultures of rust fungi unless the mesophyll cells of the leaves had been exposed by stripping off the epidermal layer.[35] Obligate biotrophic fungi may have a limited capacity to establish a parasitic phase after they have been grown in axenic culture and further studies might be necessary to find the mechanism whereby a VAM fungus could be induced to change from a saprophytic to a symbiotic mode. Despite this, the ability to culture VAM fungi would contribute greatly to our understanding of them and enable us to pursue many more fundamental studies.

B. Metabolic Capabilities of VAM Fungi

One approach to the problem of culturing VAM fungi is to see if there is any block in an essential metabolic process or a biochemical lesion which prevents their nonsymbiotic growth and necessitates a direct interaction with the metabolism of the host cells. So far the presence of some metabolic pathways in these fungi, when growing nonsymbiotically, have been illustrated or inferred. Ho and Trappe[36] demonstrated that spores of *G. mosseae* and *G. macrocarpum* var. *macrocarpum* were capable of converting nitrate to nitrite and implied the presence of nitrate reductase in both these fungi.

Macdonald and Lewis[37] tested the germ tubes emerging from *G. mosseae* spores for the presence of several enzymes using cytochemical techniques. They implied that three oxidative pathways were operative but confirmation that these were functional would require that the whole sequence of enzymes were shown to be present in the hyphae.

Another approach is to use compounds which inhibit specific metabolic pathways and this has allowed us to deduce which processes may take place during spore germination and subsequent hyphal development. The assumption is that, if an inhibitor of a specific pathway prevents spore germination or hyphal growth of the fungus, then the pathway is active during that process and is essential for it. Hepper[31] tested five inhibitors for their effect on germination and growth of *Glomus caledonicum*-cycloheximide which prevents cytoplasmic protein synthesis, actinomycin D which inhibits production of messenger RNA, proflavine hemisulfate and 5-fluorouracil which stop synthesis of other RNA species, and ethidium bromide which is thought to block the formation of mitochondrial DNA. From the results, it was inferred that the synthesis of cytoplasmic proteins, some forms of RNA, and mitochondrial DNA took place during germination and subsequent hyphal growth from these spores. From this work, it appeared that *de novo* synthesis of messenger RNA was not obligatory for germination but only for growth, suggesting that protein synthesis during germination was programmed by stored messenger RNA whereas new RNA was required for hyphal growth. The same situation probably occurs in other fungi, both saprophytes and obligate biotrophs, i.e., those fungi which appear to be totally dependent on a living host cell for growth under natural conditions. The response of G. caledonicum to these inhibitors closely resembles those of the saprophytic fungi which have been studied, except that some saprophytes do not synthesize mitochondrial DNA during germination. There is no evidence for a lesion which would prevent the nonsymbiotic growth of VAM fungi by limiting protein or nucleic acid synthesis. This is in contrast to other obligate biotrophic fungi, which are thought to have impaired ability to synthesize RNA and possibly also protein. The rate and extent of synthesis of these macromolecules by VAM fungi would, however, need to be determined by other methods before any firm conclusions could be made concerning the ability of the fungi to synthesize sufficient material for continued growth.

Beilby and Kidby[38] provided the most direct evidence of the metabolic capabilities of VAM fungi by showing that during spore germination and early germ tube growth of G. caledonicum, there was a net synthesis of lipid, with an increase in the concentration of free fatty acids and polar lipids in the spores, together with a decrease in neutral lipid. They also demonstrated the spores' ability to synthesize sterols.[39]

C. Progress Toward Culture of VAM Fungi

1. Introduction

There were many early attempts to culture VAM fungi using nonsterile roots as starting material and the results obtained have already been reviewed.[40] A variety of fungi grew from the roots, some of which could easily be grown in culture and identified, but these failed to form typical VAM infections in test plants and were probably surface contaminants. Other aseptate fungi, which were not capable of growing on synthetic media, but were entirely dependent on the host tissue were probably true VAM fungi regenerating from the roots.[41,42]

When mycorrhizal roots are chopped up and placed in water or on agar medium, fungal hyphae begin to regenerate from the cut ends. The new growth is from the intercellular hyphae and it occurs as readily from the proximal as from the distal end of the root.[43] By providing boiled hemp seeds as an intermediate substrate for growth, Barrett[44] was able to isolate a mucoraceous fungus from soil-grown roots which could be cultured on hemp seed extract agar with or without malt extract. Although not giving experimental details he stated that this fungus could reinvade a test plant giving typical VAM infections, which is the criterion of successful culture. Gerdemann[45] was unable to repeat this observation using the same isolates or others obtained in a similar way. Barrett's work therefore remains unconfirmed but of considerable interest. Now that axenically-infected plants can be obtained (see Section IV.B.) this work could be repeated without the possibility of contamination by other soil fungi associated with the roots.

2. Growth From Resting Spores

Surface-sterilized resting spores have generally been used to try to initiate cultures of VAM fungi. The approach has been to study the effect of various media or individual nutrients on the growth of hyphae from germinating spores. This presupposes that the nutrient requirements of the fungus when it grows nonsymbiotically without the spore will be the same as those of the germ tubes.

Gerdemann[1] looked at the effect of several media on a species of *Gigaspora* and found that only water containing pieces of hemp seed or nonsterile organic matter promoted hyphal extension. Typical fungal media such as potato dextrose agar, cornmeal agar, and malt extract agar did not support growth.

The effect of nutrients on the growth of hyphae from *G. mosseae* has also been examined.[15,46] Of the carbon sources tested, several organic acids promoted growth, in particular, oxaloacetic, pyruvic, acetic, citric, and tartaric but carbohydrates such as maltose, cellulose, sucrose, and glucose had an inhibitory effect on germ tube growth. Trehalose and mannitol, sugars which are commonly found in fungi, had no effect.[46] No nitrogen-containing compound was found which radically improved growth but sodium nitrate, casein, and yeast extract did change the rate of growth of the hyphae.[15] In another series of experiments using soil extract as a basic medium,[46] the effect of several sulfur-containing amino acids, amino acid mixtures, and proteins were tested. Of these only bovine serum albumin promoted hyphal growth.[46] Mosse[15] also found that dialysate from disintegrated cells of the green alga *Chlorella* or bean roots consistently improved the growth of hyphae from *G. mosseae* spores. Dialysate from disintegrated shoot tissue was not effective in improving hyphal growth.

Root exudates also promote growth from spores. This can be seen particularly well when roots and spores are maintained together in an axenic system, for instance on agar or in liquid medium (Figures 3 and 4). Much more fungal growth is obtained in the presence of roots even before appressorium formation and infection have taken place.[30,34]

With *G. caledonicum* it was found possible to substantially improve hyphal growth from spores by incorporating peptone, yeast extract, thiamin, and lima bean agar into a basic medium.[31] Peptone at 1 to 5 g ℓ^{-1} was the most stimulatory and this was shown to be due to the presence of three amino acids, cystine, glycine, and lysine.[47] Of these, lysine was the most effective giving a 3.4-fold increase at 825 mg ℓ^{-1}; cystine and glycine both gave a 2-fold increase, being optimal at 4.6 and 556 mg ℓ^{-1}, respectively. When all three amino acids were incorporated together in a medium they gave a fivefold increase in hyphal growth compared to that on water agar.[47] Yeast extract, thiamin, and lima bean agar were found to be optimal at 400, 5, and 500 mg ℓ^{-1}, respectively.[31] Another growth-promoting factor for *G. caledonicum* was found to be small pieces (1 mm^3) of boiled seeds, particularly from dicotyledonous plants.[31] The main storage material in the seed seemed to be unimportant, since both starchy and oily seeds had this effect but it is interesting that seeds such as hemp and soybean can provide the sterols needed by some fungi in culture.[48] When all stimulatory nutrients were combined in an agar medium and this was supplemented by placing pieces of *Phaseolus lunatus* seed on the surface, an eight-fold increase in growth compared to that on water agar could be obtained.[31] The mean length of hyphae from each spore on this medium was 535 mm and this was not increased by the addition of a range of sugars and organic acids.[31] Some carbohydrates, particularly glucose, have been found to be detrimental to the growth of hyphae from the resting spores of VAM fungi,[15,49] and to their germination.[49]

Even on a medium incorporating all the known stimulatory factors, hyphal growth of *G. caledonicum* ceased without any apparent utilization of the spore lipid reserves.[31] The reason for the cessation of growth is not understood. It is probably not due to exhaustion of an essential component in the parent spores, because if the spore is detached from these hyphae when growth stops on such a medium and is transferred to a fresh plate of the same medium, it will germinate again and the mycelium will regrow to the same extent for a second and third time.[50] This suggests that some form of auto-inhibition might be operating which limits the spread of mycelium beyond a certain point. Incorporation of activated charcoal or bentonite into the medium to adsorb a possible inhibitor did not overcome this effect,[50] but the growth of hyphae from *Gigaspora margarita* was improved by the presence of 0.01% activated charcoal in the medium, particularly when this was in contact with the parent spore.[51]

Since with mycorrhizal fungi the spore mass is large compared to that of the hyphae which grow from it, it has been a matter of speculation as to whether there is *de novo* growth, that is the incorporation of carbon from the medium into the hyphae, or whether new growth can be attributed to a redistribution of material within the spore under the influence of growth factors in the medium. Beilby and Kidby[38] have shown, however, that during the first 4 days on soil extract agar the dry weight of germinating *Glomus caledonicum* spores increased by 21%, demonstrating that they were synthesizing new material from the medium constituents.

Assessment of hyphal growth from individual resting spores to determine the effect of particular nutrients can be a problem. Some results have been based on direct observations,[1,15] drawings,[15] or grading on an arbitrary scale.[31] Dry weight measurement has been used[38] but it is necessary to pool a large number of replicates for each treatment, particularly if the parent spore and mycelium are not separated, because the mass of the spore is relatively large compared to the hyphae derived from it. Conventional methods of measuring fungal growth, such as colony diameter, have been used for *Gigaspora margarita*[51] which forms long hyphae, but this takes no account of the hyphal branching. Two grid intersect methods have been used for *Glomus caledonicum*. In one, the hyphae are stained and placed in a compartment with a 1 mm grid in the base[31] but in the other method, a grid is placed in the microscope eyepiece and the hyphae are stained *in situ* in the agar medium.[47,57] Both these methods can be used to calculate the length of hyphae growing from each spore or to make direct comparisons using the number of intersections that the hyphae make with the grid. These methods are difficult to use when the growth is very extensive. Estimation of fungal growth from individual spores has also been made using an image analyzing computer linked to a low power microscope.[31] This method has been used to make comparisons between treatments but if necessary it could be calibrated to give measurements of hyphal length. If a sensitive chemical method could be developed, based on the analysis of a component in the mycelium which was related to its mass, then this would be very useful. Estimation of biomass by measuring the glucosamine derived from hyphal wall polymers such as chitin and chitosan has been attempted but because of the limited sensitivity of the method, it was necessary to pool a large number of spores for each assay.[43]

Despite the large amount of hyphal extension which can occur on some media, growth has always been found to be dependent on the attachment of the hyphae to the parent spore, unless root infection has taken place and it has not been possible to obtain growth from mycelium which has been subcultured onto fresh medium.[15,31,52] This limitation has now been partly overcome, since if the parent spore is detached and removed at a fairly early stage (4 days), leaving the hyphae derived from it *in situ*, then they will continue to grow for up to 14 days (Figure 5).[53] The amount of new growth after spore detachment is generally 100 to 400% of that initially present and is dependent on nutrients in the medium, no growth of detached hyphae being obtained on water agar. It is likely therefore that this is not "residual" growth which has occurred because of the presence of some factors which have diffused from the spore into the germ tubes.

3. Reasons For Failure of Growth

There are many possible reasons why VAM fungi fail to make extensive and continuous growth unless they are part of a symbiotic partnership with a host root. They may have a simple nutritional requirement which, due to our lack of knowledge, has not yet been fulfilled or it may be necessary to supply some nutrients continuously at a low concentration.

At the other extreme, these fungi may have lost a considerable part of their genetic material and this necessitates their interaction with the host's metabolism. Brian[54] put forward a hypothesis that in obligate biotrophs, i.e., those organisms having no saprophytic phase under normal ecological conditions, some part of the genome is in the repressed condition and that the host supplies the inducer to allow the nucleic acid to be translated normally. If this is the case for VAM fungi, then the postulated inducer

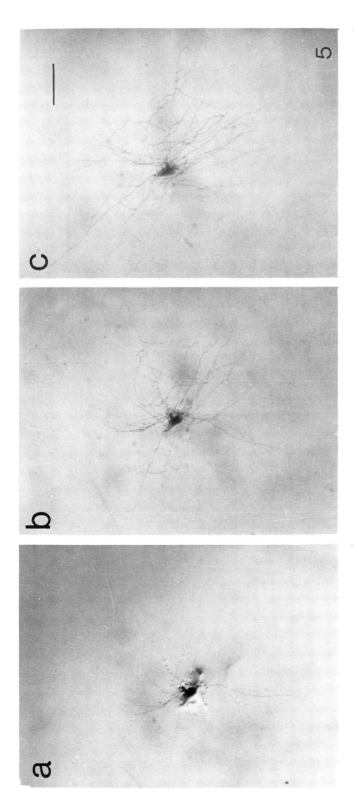

FIGURE 5. Hyphae of *Glomus caledonicum* growing on an agar medium after the parent spore has been removed (a) immediately after detachment of the parent spore, (b) after 4 days, and (c) after 11 days. Magnification is the same for all three; scale line represents 1 mm.

must be almost universally present in plant roots and might be extractable. Its capacity to promote growth of the fungi in vitro however, would be dependent on its ability to move to the appropriate site of action without the aid of a specific transfer mechanism.

One requirement for the culture of VAM fungi may be that they must first form a feeding structure which, perhaps because of its modified permeability properties, is able to take up nutrients which cannot be absorbed by the normal hyphae. If the arbuscule of VAM fungi is an absorbing structure analogous to the haustorium of some parasitic fungi, then it is interesting that arbuscule-like structures have been seen to form on external mycelium associated with roots.[30]

In a normal infection the arbuscule is positioned between the host cell wall and plasmalemma — it does not penetrate into the host cytoplasm.[55] Simulating this environment chemically and physically by careful control of pH, osmotic pressure, ionic concentration, etc., might possibly induce the fungus to grow in culture. So far, attempts to reproduce the chemical environment by providing a medium containing known components of plant roots and phloem exudates have not been successful.[56] Attempted simulation of the gaseous environment within the host tissue has also not improved fungal growth.[57] *Glomus mosseae* grows best in aerobic conditions, hyphal extension being restricted at atmospheric oxygen concentrations of less than 3%. High carbon dioxide concentrations (5%) inhibited growth at all oxygen tensions tested and appeared to permanently impair the spore's ability to develop.[57] In this part of the root cortex, plant growth substances might be at a critical level and play a vital role in the uptake of nutrients by the fungus. Very little work has been done on this aspect except some tests using ethylene which were negative.[46]

Some ectomycorrhizal fungi have two distinct types of mycelium in culture which have separate metabolic roles and which require different physical conditions. These capabilities and requirements might complement each other, the two types being analogous to the different phases of mycelium associated with a mycorrhizal root.[58] The type which is aerobic requires low pH and is capable of utilizing ammonia is like the external phase of the mycelium. The other type grows in submerged culture and has a greater capacity to hydrolyze starch than the aerobic phase; this is considered to represent the internal phase. If two distinct types of mycelium could be induced in VAM fungi then this might lead to their independent growth. It is already known that the hyphae from germinating spores of *Gigaspora margarita* have a tendency to form aerial and submerged phases, since the initial germ tube is negatively geotropic and the secondary branches deriving from it are positively geotropic.[59] Two growth phases have also been observed when *Glomus mosseae* spores are germinated on cellophane.[15]

Williams[60] has recently put forward a hypothesis that VAM endophytes are polyploid states of Basidiomycete fungi and that when a reduction in ploidy occurs they can be cultured fairly readily. As yet there is little evidence to support this idea, but it may lead to some new lines of research.

4. Other Possible Approaches to Culture

One of the obvious approaches is to test the effect of extracts of plant roots on the growth of mycorrhizal fungi. Except in the case of dialysate of disintegrated roots,[15] these extracts have generally been found to be inhibitory and this may be due to the release or production of toxic compounds during the extraction process. There is probably scope for the use of more sophisticated extraction techniques to overcome this problem. When preparing a whole root extract, it is likely that both stimulatory and inhibitory compounds would be released from the tissue and some form of fractionation would be necessary.

Recently a small number of fungi which were considered to be obligate biotrophs have been grown in culture. The techniques by which they have been induced to grow in vitro might be used to establish VAM fungi in culture but it will not necessarily follow that a method applicable to one species will be useful for another taxonomically and ecologically unrelated fungus.

Contact mucoraceous mycoparasites, which were considered to be obligate biotrophs can be grown if an extract of their fungal host is added to the medium. Haustorial mucoraceous mycoparasites have a different requirement in culture, since they have limited permeability to hexose sugars and will only grow if a surfactant is present in the medium. This approach has not been successful with a VAM fungus.[31]

A method used successfully for some rust fungi is to macerate the host tissue enzymatically with cellulase and pectinase, exposing the intact fungal colonies which can then be used to establish cultures.[61] This approach could be used with VAM fungi, although in general root tissue is much less susceptible to degradation by macerating enzymes than the leaf material.

With some fungi which are resistant to culture, the problem has been approached by starting a dual culture in vitro with a plant tissue culture in the hope that this will lead eventually to the establishment of a culture of the fungus alone. This has been used for some members of the Peronosporaceae with some limited success.[62,63] However, establishment of infection in plant tissue cultures (root organ cultures, see Section IV) did not lead to the independent growth of VAM fungi and experiments designed to study the possible interaction between plant calli from roots or stems and mycorrhizal fungi have yielded no promising results.[43,64]

VI. GENERAL CONCLUSIONS

There has been some progress toward culture of VAM fungi but the factor(s) on which the obligate relationship depends has not been elucidated. Further advances might be made if the nutritional requirements of a whole range of VAM fungi were studied. Work has generally been limited to only a very small number of species and it could be that others would be much easier to obtain in culture. For these studies it would be necessary to use species with spores which are capable of germination under axenic conditions. It is probable that not all VAM fungi will be found to have identical nutritional requirements but the independent growth of one species would provide valuable information to help culture the more refractory types.

The use of axenically infected roots as potential starting material for cultures should be extended, since putative VAM isolates have been obtained from roots in the past.[44] It is also essential to determine if the hyphae regenerating from cut roots are stimulated by those nutrients which have been found to improve the growth of hyphae arising from germinating spores.

Progress toward the culture of VAM fungi will be possible if systematic studies are made in which any small improvements in fungal growth are recorded by accurate hyphal length measurements. If these fungi have very complex nutritional requirements, then unlimited growth may be obtained when a range of compounds, each giving some stimulation, are supplied together. In this area of research on VAM fungi, more than any other, there is probably a pool of unpublished data relating to negative results or to the effect of nutrients or conditions which gave only marginal improvements in hyphal growth. All results of this type are valuable, even if they only prevent repetition of unprofitable lines of approach.

Another factor which may be important for the growth of these fungi is organic matter. It has been shown that VAM fungi are capable of sufficient growth in soil to establish an independent base from which they can infect a root[65] and that this saprophytic hyphal growth is dependent on the organic matter in the soil.[66] This evidence, together with the ability of *Glomus caledonicum* hyphae to make limited growth in the absence of a host plant or a resting spore,[53] is a promising basis for continuing work on axenic culture.

NOTE ADDED IN PROOF

It has recently been shown that inorganic sulfur-containing compounds, in particular potassium sulfite and potassium metabisulfite, can greatly increase the amount of hyphal growth from *Glomus caledonicum* spores on agar.[67]

REFERENCES

1. Gerdemann, J. W., Relation of a large soil-borne spore to phycomycetous mycorrhizal infections, *Mycologia*, 47, 619, 1955.
2. Gerdemann, J. W. and Nicolson, T. H., Spores of mycorrhizal *Endogone* species extracted from soil by wet-sieving and decanting, *Trans. Br. Mycol. Soc.*, 46, 235, 1963.
3. Mosse, B., Vesicular-Arbuscular Mycorrhiza Research for Tropical Agriculture, Res. Bull. No. 194, Hawaii Institute of Tropical Agriculture and Human Resources, University of Hawaii, Honolulu, 1981.
4. Greenall, J. M., The mycorrhizal endophytes of *Griselinia littoralis* (Cornaceae), *N.Z. J. Bot.*, 1, 389, 1963.
5. Ohms, R. E., A flotation method for collecting spores of a phycomycetous mycorrhizal parasite from soil, *Phytopathology*, 47, 751, 1957.
6. Mosse, B. and Jones, G. W., Separation of *Endogone* spores from organic soil debris by differential sedimentation on gelatin columns, *Trans. Br. Mycol. Soc.*, 51, 604, 1968.
7. Ross, J. P. and Harper, J. A., Effect of *Endogone* mycorrhiza on soybean yields, *Phytopathology*, 60, 1552, 1970.
8. Daniels, B. A. and Menge, J. A., Evaluation of the commercial potential of *Glomus epigaeus*, a vesicular-arbuscular mycorrhizal fungus, *New Phytol.*, 87, 345, 1981.
9. Mertz, S. M., Heithaus, J. J., and Bush, R. L., Mass production of axenic spores of the endomycorrhizal fungus *Gigaspora margarita*, *Trans. Br. Mycol. Soc.*, 72, 167, 1979.
10. Furlan, V., Bartschi, H., and Fortin, J. A., Media for density gradient extraction of mycorrhizal spores, *Trans. Br. Mycol. Soc.*, 75, 336, 1980.
11. Sutton, J. C. and Barron, G. L., Population dynamics of *Endogone* spores in soil, *Can. J. Bot.*, 50, 1909, 1972.
12. Smith, G. W. and Skipper, H. D., Comparison of methods to extract spores of vesicular-arbuscular mycorhizal fungi, *Soil Sci. Soc. Am. J.*, 43, 772, 1979.
13. Hepper, C. M. and Smith, G. A., Observations on the germination of *Endogone* spores, *Trans. Br. Mycol. Soc.*, 66, 189, 1976.
14. Hepper, C. M. and Mosse, B., Vesicular-arbuscular mycorrhiza in root organ cultures, in *Tissue Culture Methods for Plant Pathologists*, Ingram, D. S. and Helgeson, J. P., Eds., Blackwell Scientific, Oxford, 1980, 167.
15. Mosse, B., The regular germination of resting spores and some observations on the growth requirements of an *Endogone* sp. causing vesicular-arbuscular mycorrhiza, *Trans. Br. Mycol. Soc.*, 42, 273, 1959.
16. Mosse, B. and Phillips, J. M., The influence of phosphate and other nutrients on the development of vesicular-arbuscular mycorrhiza in culture, *J. G. Microbiol.*, 69, 157, 1971.
17. Macdonald, R. M., Routine production of axenic vesicular-arbuscular mycorrhizas, *New Phytol.*, 89, 87, 1981.
18. Maskall, C. S., Unpublished data, 1981.
19. Schenck, N. C., Graham, S. O., and Green, N. E., Temperature and light effects on contamination and spore germination of vesicular-arbuscular mycorrhizal fungi, *Mycologia*, 67, 1189, 1975.
20. Sward, R. J., The structure of the spores of *Gigaspora margarita*, *New Phytol.*, 87, 761, 1981.
21. Tommerup, I. C. and Kidby, D. K., Production of aseptic spores of vesicular-arbuscular endophytes and their viability after chemical and physical stress, *Appl. Environ. Microbiol.*, 39, 1111, 1980.
22. Mosse, B., The establishment of vesicular-arbuscular mycorrhiza under aseptic conditions, *J. G. Microbiol.*, 27, 509, 1962.
23. Hepper, C. M., Techniques for studying the infection of plants by vesicular-arbuscular mycorrhizal fungi under axenic conditions, *New Phytol.*, 88, 641, 1981.
24. St. John, T. V., Hays, R. I., and Reid, C. P. P., A new method for producing pure vesicular-arbuscular mycorrhiza-host cultures without specialised media, *New Phytol.*, 89, 81, 1981.

25. Allen, M. F., Moore, T. S., Christensen, M., and Stanton, N., Growth of vesicular-arbuscular mycorrhizal and non-mycorrhizal *Bouteloua gracilis* in a defined medium, *Mycologia*, 71, 666, 1979.
26. Allen, M. F., Moore, T. S., and Christensen, M., Phytohormone changes in *Bouteloua gracilis* infected by vesicular-arbuscular mycorrhizae. I. Cytokinin increases in the host plant, *Can. J. Bot.*, 58, 371, 1980.
27. Allen, M. F., Sexton, J. C., Moore, T. S., and Christensen, M., Influence of phosphate source on vesicular-arbuscular mycorrhizae of *Bouteloua gracilis*, *New Phytol.*, 87, 687, 1981.
28. Elmes, R., Hepper, C. M., and Maskall, C. S., Axenic NFT for mycorrhizal plants, *Rothamsted Exp. Stn. Annu. Rep.*, 1, 204, 1981.
29. Pearson, V. and Tinker, P. B., Measurement of phosphorus fluxes in the external hyphae of endomycorrhizas, in *Endomycorrhizas.*, Sanders, F. E., Mosse, B., and Tinker, P. B., Eds., Academic Press, London, 1975, 277.
30. Mosse, B. and Hepper, C., Vesicular-arbuscular mycorrhizal infections in root organ cultures, *Physiol. Plant Pathol.*, 5, 215, 1975.
31. Hepper, C. M., Germination and growth of *Glomus caledonius* spores: the effects of inhibitors and nutrients, *Soil Biol. Biochem.*, 11, 269, 1979.
32. Cooper, K. M. and Tinker, P. B., Translocation and transfer of nutrients in vesicular-arbuscular mycorrhizas. II. Uptake and translocation of phosphorus, zinc and sulphur, *New Phytol.*, 81, 43, 1978.
33. Cooper, K. M. and Tinker, P. B., Translocation and transfer of nutrients in vesicular-arbuscular mycorrhizas. IV. Effect of environmental variables on movement of phosphorus, *New Phytol.*, 88, 327, 1981.
34. Hepper, C. M. and Mosse, B., Techniques used to study the interaction between *Endogone* and plant roots, in *Endomycorrhizas*, Sanders, F. E., Mosse, B., and Tinker, P. B., Eds., Academic Press, London, 1975, 65.
35. Williams, P. G., Scott, K. J., Kuhl, J. L., and Maclean, D. J., Sporulation and pathogenicity of *Puccinia graminis* f. sp. *tritici* grown on an artificial medium, *Phytopathology*, 57, 326, 1967.
36. Ho, I. and Trappe, J. M., Nitrate reducing capacity of two vesicular-arbuscular mycorrhizal fungi, *Mycologia*, 67, 886, 1975.
37. Macdonald, R. M. and Lewis, M., The occurrence of some acid phosphatases and dehydrogenases in the vesicular-arbuscular mycorrhizal fungus *Glomus mosseae, New Phytol.*, 80, 135, 1978.
38. Beilby, J. P. and Kidby, D. K., Biochemistry of ungerminated and germinated spores of the vesicular-arbuscular mycorrhizal fungus *Glomus caledonius:* changes in neutral and polar lipids, *J. Lipid Res.*, 21, 739, 1980.
39. Beilby, J. P. and Kidby, D. K., Sterol composition of ungerminated and germinated spores of the vesicular-arbuscular mycorrhizal fungus *Glomus caledonius, Lipids,* 15, 375, 1980.
40. Harley, J. L., *The Biology of Mycorrhiza,* 2nd ed., Leonard Hill, London, 1969 263.
41. Magrou, M. J., Sur la culture de quelques champignons de mycorhizes a arbuscules et à vesicules, *Rev. Gen. Bot.*, 53, 49, 1946.
42. Jones, F. R., A mycorrhizal fungus in the roots of legumes and some other plants, *J. Agric. Res.*, 29, 459, 1924.
43. Hepper, C. M., Unpublished data, 1979.
44. Barrett, J. T., Isolation, culture and host relation of the phycomycetoid vesicular-arbuscular endophyte *Rhizophagus, Rec. Adv. Bot.*, 2, 1725, 1961.
45. Gerdemann, J. W., Vesicular-arbuscular mycorrhiza and plant growth, *Annu. Rev. Phytopathol.*, 6, 397, 1968.
46. Mosse, B., Culture of *Endogone, Rothamsted Exp. Stn. Annu. Rep.*, 1, 97, 1970.
47. Hepper, C. M. and Jakobsen, I., Hyphal growth from spores of the mycorrhizal fungus *Glomus caledonius:* effect of amino acids, *Soil Biol. Biochem.*, 15, 55, 1983.
48. Domnas, A. J., Srebro, J. P., and Hicks, B. F., Sterol requirement for zoospore formation in the mosquito-parasitizing fungus *Lagendium giganteum, Mycologia,* 69, 875, 1977.
49. Koske, R. E., Multiple germination by spores of *Gigaspora gigantea, Trans. Br. Mycol. Soc.*, 76, 328, 1981.
50. Hepper, C. M., Growth from spores, *Rothamsted Exp. Stn. Annu. Rep.*, 1, 189, 1980.
51. Watrud, L. S., Heithaus, J. J., and Jaworski, E. G., Evidence for production of inhibitor by the vesicular-arbuscular mycorrhizal fungus *Gigaspora margarita, Mycologia,* 70, 821, 1978.
52. Godfrey, R. M., Studies on British species of *Endogone:* III. Germination of spores, *Trans. Br. Mycol. Soc.*, 40, 203, 1957.
53. Hepper, C. M., Limited independent growth of a vesicular-arbuscular mycorrhizal fungus, *in vitro, New Phytol.*, 93, 537, 1983.
54. Brian, P. W., Obligate parasitism in fungi, *Proc. R. Soc. London Ser. B.*, 168, 101, 1967.
55. Cox, G. and Sanders, F., Ultrastructure of the host-fungus interface in a vesicular-arbuscular mycorrhiza, *New Phytol.*, 73, 901, 1974.

56. Macdonald, R. M. and Spokes, J. R., *Rothamsted Exp. Stn. Annu. Rep.*, 1, 234, 1979.
57. Le Tacon, F., Skinner, F. A., and Mosse, B., Spore germination and hyphal growth of a vesicular-arbuscular mycorrhizal fungus, *Glomus mosseae* (Gerdemann & Trappe), under decreased oxygen and increased carbon dioxide concentrations, *Can. J. Microbiol.*, 29, 1280, 1983.
58. Gyurko, P., Physiological investigations of ectomycorrhizal fungi, in *Mushroom Science X Part I* Proc. 10th Int. Congr. Sci. Cult. Edible Fungi, Delmas, J., International Society for Mushroom Science, Bordeaux, France, 1979, 919.
59. Watrud, L. S., Heithaus, J. J., and Jaworski, E. G., Geotropism in the endomycorrhizal fungus *Gigaspora margarita, Mycologia,* 70, 449, 1978.
60. Williams, P. G., Personal communication, 1981.
61. Lane, W. D. and Shaw, M., Axenic culture of flax rust isolated from cotyledons by cell wall digestion, *Can. J. Bot.*, 50, 2601, 1972.
62. Ingram, D. S. and Joachim, I., The growth of *Peronospora farinosa* f.sp. *betae* and sugar beet callus tissues in dual culture, *J. G. Microbiol.*, 69, 211, 1971.
63. Tiwari, M. M. and Arya, H. C., *Sclerospora graminicola* axenic culture, *Science,* 163, 291, 1969.
64. Mosse, B., Unpublished data.
65. Warner, A. and Mosse, B., Independent spread of vesicular-arbuscular mycorrhizal fungi in soil, *Trans. Br. Mycol. Soc.,* 74, 407, 1980.
66. Hepper, C. M. and Warner, A., Role of organic matter in the growth of a vesicular-arbuscular mycorrhizal fungus in soil, *Trans. Br. Mycol. Soc.* 81, 155, 1983.
67. Hepper, C. M., Inorganic sulphur nutrition of the vesicular-arbuscular mycorrhizal fungus *Glomus caledonicum, Soil Biol. Biochem.,* in press.

Chapter 6

THE EFFECT OF VA MYCORRHIZAE ON PLANT GROWTH

L. K. Abbott and A. D. Robson

TABLE OF CONTENTS

I. Introduction 114

II. Procedures for Studying the Effect of VAM Fungi on Plant Growth 114
    A. Suitable Controls 114
    B. The Need for Response Curves 116
    C. Sequential Harvesting and Assessing Infection 119

III. Effects of VAM on Plant Growth 120
    A. Enhanced Nutrient Uptake 120
        1. Phosphate 120
        2. Other Nutrients 121
    B. Non-Nutritional Effects 122
        1. Growth Depression 122
        2. Growth Stimulation 123

IV. Effect of Different Fungal Species and Strains on Plant Growth 123

V. Conclusions 125

References 126

I. INTRODUCTION

Vesicular-arbuscular mycorrhizal (VAM) fungi are ubiquitous in soils throughout the world. They form mycorrhizae with a majority of plant species[1,2] and show little host specificity.[3] However, the upsurge in interest in VAM during recent years[4,5] arose once they were shown to be involved in nutrient uptake by plants.[6-10]

Subsequent research has often been concentrated on the role of VAM in the nutrition of plants grown in infertile soils.[11] It has included studies seeking to exploit the symbiosis so that phosphate fertilizer (a nonrenewable resource) could be used more efficiently.[12] In particular, the role of VAM in the nutrition of agricultural and horticultural plants has received much attention.[13,14] Simultaneously, the importance of VAM in natural ecosystems[15-18] and in regeneration of mining sites[19,20] has been investigated.

In this chapter, we will consider the experimental approaches for examining the effects of VAM on plant growth, the effects of VAM on nutrient uptake specifically, and the effects of VAM on plant growth other than those related to increased nutrient uptake. It is intended as a review of approaches rather than as an encyclopedic examination of the effects of VAM on plant growth.

II. PROCEDURES FOR STUDYING THE EFFECT OF VAM FUNGI ON PLANT GROWTH

There are several procedures which should be employed in experiments which are designed to test the effect of VAM on plant growth. These are the careful selection of control treatments, the inclusion of plant growth response curves to added phosphorus (and other nutrients) in each experiment, and sequential harvesting of plants to allow the development of mycorrhizae to be studied in relation to plant growth. Without the information obtained from experiments designed to deal adequately with these areas, it is difficult to interpret data which illustrate changes in plant growth following inoculation with VAM. If we do not attempt to understand the reasons for effects of mycorrhizae on plant growth, we will be unable to predict those situations where plant growth will benefit from inoculation with introduced VAM fungi.

A. Suitable Controls

There are many problems associated with describing the effect of inoculation with VAM fungi on plant growth. Some of these have already been discussed.[2] They relate primarily to the natural occurrence of VAM fungi in most soils and to the difficulties associated with obtaining pure inoculum of these symbiotic fungi.

Various procedures have been used to remove the indigenous VAM fungi from soils prior to testing the effects of mycorrhizal inoculum on plant growth.[12] Some treatments, such as γ irradiation, may alter the nutritional status of the soil.[21] Furthermore, toxins may be formed after heating soils[22] which may be detrimental either to plant growth or to the development of mycorrhizae. These effects should not present a problem providing adequate control treatments are included in each experiment.[2,22]

The problem of impure inocula of VAM fungi has been recognized since the earliest inoculation experiments.[10] Because the fungi are biotrophic and have not yet been grown in pure culture, it is not possible to obtain large quantities of inoculum that are uncontaminated by other soil microorganisms. Inocula of various species of VAM fungi traditionally have been obtained from open pot cultures.[23,24] The plants are grown in a soil that has been pretreated (e.g., by steaming) to remove the indigenous VAM fungi. Inocula consisting of spores, fragments of infected root, or soil containing propagules of the fungi are added prior to sowing seed. The pots are usually kept in glasshouses, where they may remain for up to a year or more.[2] Consequently, inoculum from pot cultures will normally include a suite of associated microorganisms. Precautions must therefore be taken to ensure that contaminating microorganisms are not responsible for enhanced or retarded plant growth in experiments designed to assess the effect on plant growth of inoculation with VAM fungi.

The appropriate control treatment in an experiment testing the extent and mechanism of plant response to inoculation with VAM fungi will be dictated by the nature of the inoculum itself. Irrespective of the form of inoculum used, the control treatment should be matched so that it has the same population of associated microorganisms and the same quantities of nutrients as that containing the mycorrhizal fungi.

Clark[25] chose a well-matched control in an experiment using spores of *Endogone (Gigaspora) gigantea* as his mycorrhizal inoculum. Spores were pierced with a needle, then added to the control treatment. This ensured that the same microorganisms were added to all pots. Unfortunately, a tedious process such as this is not suitable for experiments on a large scale. An alternative method is to add surface disinfected spores[26] to the inoculated treatments and nothing to the control treatment. However, surface sterilization of spores may not remove all contaminating microorganisms.[26,27] Where infected roots (or roots and soil) are used as inoculum, similar material can be added to the control pots from uninoculated pot cultures prepared simultaneously.[28]

Filtered washings from the fungal inoculum (excluding mycorrhizal fungi) are often added to the control treatment[29,30] and vice versa[31] as a means of equalizing the microflora associated with the inoculum, over all treatments. This may be particularly important if toxins are formed following soil "sterilization".[22] However, it is not clear whether the microorganisms introduced with the inoculum are always matched quantitatively and qualitatively in all treatments.

Heat-treated inoculum has been used as a control treatment.[32] Care is needed in such cases to ensure that the microorganisms are the same in all pots, there are no persistent toxic effects of heating, and that the nutritional content of the inoculum is not altered by heating.

Another factor which cannot be ignored is the possibility of introducing pathogens with the inoculum. In this case it is not simply a matter of ensuring that the microorganisms associated with the inoculum are added to all pots. Inocula of VAM fungi can be contaminated by plant pathogens, for example *Fusarium*.[33] Some pathogens may be a problem for certain soils or plant species. Therefore, in these cases the soil and plant species used for pot cultures could be chosen to minimize the possibility of adding pathogens with the inoculum.[34]

An alternative approach to obtaining mycorrhizal and nonmycorrhizal plants for comparisons has been discussed by Smith and Smith.[35] In this case, inoculation and its associated complications are avoided in favor of soil sterilization, to eliminate the VAM fungi from the control treatment. This is a useful method for assessing the contribution which naturally occurring VAM fungi make to plant growth.[36] As for experiments where the fungi are added in inocula, the associated microorganisms in the untreated soil need to be re-introduced after soil sterilization. An extra precaution is needed to ensure that there have been no side effects of sterilization on plant growth. This can be achieved by growing a host which does not form VAM.[36,37] An additional check of this nature will only be of assistance if the growth of the nonmycorrhizal host plant responds to added nutrients or toxins in the same way as the mycorrhizal host (see next section).

Preinoculation of plants with VAM fungi and subsequent transplanting is one method of obtaining mycorrhizal plants.[38] However, this is not a suitable way of obtaining matched mycorrhizal and nonmycorrhizal plants for examining the effect of mycorrhizae on plant growth, because the two treatments will differ physiologically in addition to the presence or absence of mycorrhizas. Mycorrhizal plants prepared in this way may also have developed their infection under conditions which do not prevail in the test soil. Direct inoculation of plants in the test soil ensures that two important steps, the initiation of growth from a propagule and hyphal growth through soil to the root, are not bypassed.[12]

Until now we have dealt only with experiments in which the growth of mycorrhizal and nonmycorrhizal plants has been compared. However, the effect on plant growth of inoculating with VAM fungi in untreated soils containing indigenous VAM fungi is of major importance. This is so because most agricultural soils already contain indigenous populations of VAM fungi. Here, an additional control treatment may be necessary because the procedure of inoculation itself may interfere with the normal process of infection by the indigenous VAM fungi. To prevent this, a control for the inoculation procedure with no additions of any kind could be included as well as the control for the inoculum itself.

The difficulties in choosing a control treatment, which is identical to the inoculated treatment in all respects but does not include the VAM fungus, cannot be ignored. It is a topic which should not be treated casually whenever attempts are made to assess the effects of VAM on plant growth.

B. The Need For Response Curves

Effects of inoculation with VAM fungi in increasing plant growth can generally be overcome by increasing nutrient supply to a level where it is not limiting for the growth of uninoculated plants.[36,39-41] Therefore, the magnitude of the effect of VAM on plant growth will depend upon the nutritional status of the soil for plant growth. It is not possible to accurately define the nutritional status of a soil relative to potential plant growth either from chemical measurements on soil or from absolute levels of nutrient application. Plant species differ in the amount of a nutrient they require in the media to achieve their maximum growth (Table 1).[42] Furthermore, the amount of nutrient required for maximum growth of plants varies greatly with the soil[42] (see Table 1) and method of nutrient application.[43,44] The nutritional status of a soil for the growth of a particular plant is therefore best defined by quantitatively describing the nature of the response of plant growth to applied nutrients. The nutrient status of a soil at any level of nutrient application for the growth of a particular plant can thus be assessed by estimating the percentage of maximum growth achieved at that level. The actual level of applied nutrients has no relevance unless it is considered in relation to the amount which is necessary for the plant to achieve maximum yield in the soil in which it is grown.

There are at least two advantages in comparing the response curves of mycorrhizal and nonmycorrhizal plants to applied nutrients. First, it may allow for horizontal rather than vertical comparisons to be made. That is, rather than comparing the growth of mycorrhizal and nonmycorrhizal plants at a particular level of applied nutrient (vertical comparison), it may be possible to estimate how much nutrient is required for the same yield of mycorrhizal and nonmycorrhizal plants (horizontal comparison). This is illustrated using the data of Menge et al.[40] and Pairunan et al.[28]

In Figures 1 and 2 the response of both mycorrhizal and nonmycorrhizal plants to applied phosphate can be described by the Mitscherlich equation, i.e.,

$$Y = A - Be^{-cx} \qquad (1)$$

where Y is dry weight of shoots at any level of nutrient application (x), A is maximum dry weight of shoots, B is a constant for a particular soil-plant combination reflecting the responsiveness of the soil, and c is the curvature or Mitscherlich constant.

In the experiment of Pairunan et al.[28] (Figure 2), A and B were unaffected by inoc-

Table 1
AMOUNT OF PHOSPHORUS
REQUIRED FOR GROWTH OF
SUBTERRANEAN CLOVER AND
RYEGRASS ON 11 SOILS WITH
A WIDE RANGE OF ABILITY TO
ADSORB PHOSPHATE[42]

Soil	Phosphate adsorption (μg P/g soil)	Phosphate required for 90% of maximum growth (mg P/pot)[a]	
		Clover	Ryegrass
1	8	30	10
2	15	60	10
3	53	120	30
4	125	165	30
5	170	270	50
6	185	380	45
7	325	380	50
8	450	505	50
9	545	525	100
10	600	555	60
11	950	610	80

[a] Pots contained between 700 and 1200 g of soil depending on the bulk density of the soil. If a pot contained 1000 g soil a rate of phosphorus application of 1 mg/pot corresponds to 1 ppm or 1 μg g^{-1} soil.

ulation with a mycorrhizal fungus. Thus it was possible to fit a common Mitscherlich equation to the data,

$$Y = A - Be^{-c_1 x_1 - c_2 x_2}$$

where c_1 and x_1 are for mycorrhizal plants and c_2 and x_2 are for nonmycorrhizal plants. At any yield level the relative effectiveness of phosphorus for mycorrhizal compared with nonmycorrhizal plants is

$$\frac{c_1}{c_2} \text{ or } \frac{x_2}{x_1}$$

This estimate of relative effectiveness is independent of the level of nutrient application. When mycorrhizal and nonmycorrhizal plants are compared vertically (e.g., at a particular level of applied phosphorus), the effect of VAM is dependent upon the phosphorus level chosen. A similar analysis can be conducted using phosphorus content of tops (milligrams of P per plant) rather than dry weight of tops.[28,45]

Growth data may need to be transformed to logarithms before mathematical analysis if the variance is related to the mean or if nutrient supply influences shoot growth by affecting relative growth rate.[46] Furthermore, data may not always be adequately described by a Mitscherlich equation. For example, in some instances the response (even after log transformation) may be sigmoidal, i.e., hyperbolic with a threshold level[47] (Figure 3).

Sigmoidal response curves may be obtained for nonmycorrhizal plants even if the

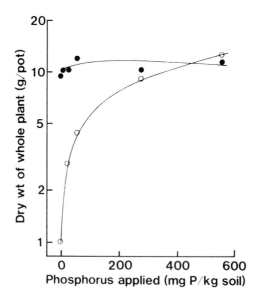

FIGURE 1. The effect of the application of superphosphate on the dry weight of mycorrhizal (•) and nonmycorrhizal (○) seedlings of Brazilian sour orange grown for 5 months on a loamy sand.[40]

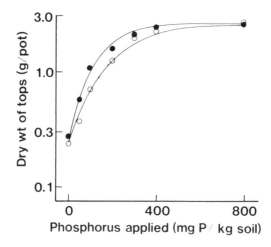

FIGURE 2. The effect of the application of superphosphate on the dry weight of shoots of mycorrhizal (•) and nonmycorrhizal (○) subterranean clover grown for 7 weeks on a lateritic podzol.[28]

response of mycorrhizal plants to applied nutrients can be described well by the Mitscherlich equation. If this is the case (see Figure 3 for an example) the extent to which mycorrhizae influence plant growth, as assessed by both a vertical and a horizontal comparison, is affected by the level of nutrient application.

The second advantage of having complete response curves is that it is only possible to claim that mycorrhizae increase or affect plant growth by processes other than by increasing nutrient uptake when either (1) the nutrient status is equivalent in mycorrhizal and nonmycorrhizal plants or (2) when nutrient supply does not limit the growth of either mycorrhizal or nonmycorrhizal plants. Accordingly, physiological comparisons of mycorrhizal and nonmycorrhizal plants must be conducted in plants achieving

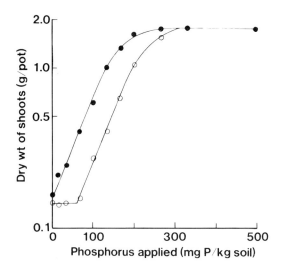

FIGURE 3. The effect of the application of KH_2PO_4 on the dry weight of shoots of mycorrhizal (●) and nonmycorrhizal (○) subterranean clover grown for 5 weeks on the subsoil from a lateritic podzol.(Unpublished data of N. S. Bolan.)

the same percentage of maximum growth and nutrient content. For example, in many instances mycorrhizal plants may have a lower root to shoot weight ratio than nonmycorrhizal plants at the same level of nutrient application.[48-52] However, at the same degree of nutritional limitation (same dry matter), mycorrhizal and nonmycorrhizal plants appear to have a similar distribution of dry weight beween roots and shoots.[51] That is, mycorrhizae influence dry matter distribution between roots and shoots by influencing nutrient status of the plant.

Similarly, mycorrhizal legumes may exhibit increased nodulation and nitrogen fixation compared to nonmycorrhizal plants when phosphorus supply limits the growth of nonmycorrhizal plants.[53,54] However, effects of mycorrhizae on nodulation and nitrogen fixation can be completely overcome by increasing phosphorus supply in the nonmycorrhizal treatments.[55,56]

A major difficulty with using complete response curves for mycorrhizal and nonmycorrhizal plants is that the size of experiments may become excessively large and unmanageable. A possible solution for some experimental questions is to have a complete response curve for the uninoculated or nonmycorrhizal control treatments, but to have only one or two levels of nutrient application for treatments which include mycorrhizal fungi.

C. Sequential Harvesting and Assessing Infection

Sequential harvests can be of great benefit in experiments designed to test the effect of VAM on plant growth. For example, they provide material for examining, in detail, the stages in the formation and development of VAM.[30,57,58] This can include the growth of hyphae in the soil surrounding infected roots. In addition, changes which occur in the host, such as nutrient concentration and growth can be examined in relation to the development of the fungus within the roots and in the soil.

Sequential harvests have been used to demonstrate the relationship between infection and phosphate inflow.[30] It has also been useful to determine the role of mycorrhizae in nodulation and nitrogen fixation by mycorrhizal subterranean clover plants.[54,56] By using sequential harvests it is possible to determine whether effects of inoculation with

VAM fungi are direct effects of the fungi or simply indirect effects of improved phosphorus status of the infected plants.

Experiments may become too large and unmanageable if several sequential harvests are included to study plant growth (tops and roots) as well as the development of mycorrhizal colonization. Valuable data can be gained if at least some mycorrhiza treatments are included with more than one harvest.

Growth responses to inoculation with VAM fungi have occasionally been reported with no corresponding details of the development of infection.[31,32,59] This practice is not wise because growth responses to inoculation with VAM fungi have sometimes been observed where there was little or no mycorrhiza formation.[60,61] These reports demonstrate the caution which must be taken before plant responses are attributed to the activity of VAM. They emphasize both the necessity for adequate controls, plant growth response curves, and observations on the development of mycorrhizal colonization during the experiment.

III. THE EFFECT OF VAM ON PLANT GROWTH

A. Enhanced Nutrient Uptake

VAM usually increase the growth of plants solely by enhancing nutrient uptake. There are three possible explanations for the greater uptake of mineral nutrients by mycorrhizal plants compared to nonmycorrhizal plants.[11,12,62] First, mycorrhizae may increase nutrient uptake by reducing the distance that nutrients must diffuse to plant roots.[63,64] Secondly mycorrhizal roots may differ from nonmycorrhizal roots in the relationship between rate of nutrient absorption and nutrient concentration at the absorbing surface.[65-67] Finally, mycorrhizal hyphae may chemically modify the availability of nutrients for uptake by plants (see Section III.A.1.). From a consideration of the published evidence it is likely that VAM increase nutrient uptake from soil primarily by shortening the distance that nutrients must diffuse through soil to the root.[12,62] It is likely, therefore, that effects of mycorrhizae in increasing nutrient uptake will be most marked for nutrients which move to roots principally by diffusion[68] and for plant species with coarse roots and sparse, short root hairs.[69-71] As a result, most studies of VAM have concentrated their attention upon phosphate nutrition of plants such as legumes and citrus.

1. Phosphate

The uptake of phosphate by plants from soil will usually be limited by the rate of movement of the phosphate to the plant root rather than by the rate of absorption at the root surface.[72] In soils with high capacities to adsorb phosphate, phosphate concentrations in soil solutions will be extremely low[73] and diffusion to plant roots will be extremely slow.[68,74] These are the soils in which it is likely that the greatest benefits will be gained from VAM.

Mycorrhizal and nonmycorrhizal roots appear to absorb phosphorus from the same labile pool of phosphorus in the soil. The specific activity of phosphorus in plants grown in ^{32}P-labeled soils is the same for mycorrhizal and nonmycorrhizal plants.[75-77] Moreover, mycorrhizal onions and subterranean clover were able to obtain firmly held phosphate no better than that applied freshly.[45]

There have been several claims that mycorrhizae have an ability to exploit nonlabile forms of soil phosphate such as tricalcium phosphate and rock phosphate[3,62,78,79] These claims are based on comparisons of the effects of mycorrhizae on growth of plants when supplied with a single level of phosphorus (applied either as rock phosphate or in a soluble form). These are typical examples of vertical comparison (see Section II.B.). A more valid comparison of the ability of mycorrhizal and nonmycorrhizal plants to obtain phosphorus from various sources can be made using the entire growth response curves (see Section II.B.). When horizontal comparisons were made from such curves for subterranean clover, the relative effectiveness of phosphorus from three sources differing widely in water and citrate solubilities was similar.[28] That is,

$$\frac{c_1}{c_2}$$

(see Section II.B.) was independent of the solubility of the phosphorus in these three sources.

Species and cultivars of plants may differ in their responses to mycorrhizal infection.[40,80-83] However, comparisons of this type have sometimes been conducted at a single level of nutrient supply and differences among plants in response may simply reflect differences in internal or external requirements[84] for phosphorus. Plant species may differ in their external requirements for phosphorus because of differences in their maximum growth rate, in their ability to take up phosphate, or in the utilization of phosphate within the plant.[85,86] The only direct effect of VAM on plant growth is probably through nutrient uptake. Therefore, assessments of the dependency of plant species on mycorrhizae should be made using horizontal comparisons and hence require response curves (see Section II.B.).

When comparisons of different hosts have been made using response curves it is apparent that plant species do differ in their response to mycorrhizal infection.[36,40] Indeed some species (for example *Manihot esculenta*, *Stylosanthes* spp.[36]) are so dependent on mycorrhizal infection that they have been described as being obligate mycorrhizal plants. The suggestions of Baylis,[69-71] that plants with fibrous root systems and long root hairs are probably less dependent on mycorrhizae than those with coarse roots and sparse root hairs, appear to be generally valid.

2. Other Nutrients

It is likely that VAM fungi will increase the uptake of any nutrients that move to plant roots primarily by diffusion. Indeed, inoculation with VAM fungi alleviated zinc[87,88] and copper deficiencies[89] in peach and citrus seedlings. Additionally, it has been suggested that rates of phosphorus application greater than those required to overcome phosphorus deficiency may induce deficiencies of micronutrients by decreasing the formation of VAM.[89,90]

To demonstrate unequivocally that mycorrhizae can alleviate a deficiency or a toxicity of a particular nutrient it is essential that the supply of all other nutrients not be limiting for growth of either mycorrhizal or nonmycorrhizal plants. Under these conditions (particularly in phosphate-adequate plants) the roots may be only sparsely infected.[39,75] Increasing phosphorus supply may decrease mycorrhizal infection to levels which are not sufficient to enhance the uptake of another nutrient which is deficient for plant growth. Similarly, lower concentrations of a toxic nutrient within a mycorrhizal plant may be due to increased plant growth (by relief of phosphorus deficiency) and thus dilution of the toxic nutrient.

Notwithstanding these difficulties, mycorrhizal roots can absorb zinc[65] and sulfur[91] from solution faster per gram of roots than can nonmycorrhizal roots. This may reflect differences in the relationship between the surface area and the weight of roots from mycorrhizal and nonmycorrhizal plants. Surface area of roots is an important factor in absorption of nutrients from well-stirred solutions. Additionally, these differences in rate of absorption by mycorrhizal and nonmycorrhizal roots may reflect differences in phosphorus status of the two types of plants.[91]

B. Non-Nutritional Effects
1. Growth Depression

Inoculation with VAM fungi has sometimes depressed plant growth, particularly where the phosphorus supply does not limit the growth of nonmycorrhizal plants.[3,69,92] The depression in plant growth may be transitory[54] or persistent.[80] Growth depressions may occur if a pathogen such as *Pythium* or *Fusarium* (which may decrease plant growth without obvious symptoms) is introduced with the mycorrhizal fungus.[33,93] This may be more of a problem in partially sterilized soils. Steamed soil, for example, may provide a suitable environment for the growth of contaminant microorganisms introduced with mycorrhizal fungi. Although growth depressions may be caused by the inadvertent inclusion of a pathogen with the inoculum, this possibility is rarely considered or mentioned as an explanation for reduced growth of inoculated plants. Growth depressions produced in this way would most likely be persistent if the pathogen became well established.

Although it has been suggested that some growth depressions may result from competition between the plant and the mycorrhizal fungus for phosphate,[94] this seems to be an unlikely explanation. Depressed growth of inoculated plants has mainly been shown at levels of phosphorus supply adequate for the growth of nonmycorrhizal plants.[3,69,92] Under these conditions there is probably little competition between the plant and the fungus for phosphate because the phosphate supply would not be likely to limit the growth of the plant or the fungus.

Competition for photosynthate between the host and the fungus may be the factor responsible for some growth depressions.[3] The biomass of VAM fungi within roots can represent up to 17% of the dry weight of roots.[95] Pang and Paul[96] demonstrated that mycorrhizal roots expend more energy than nonmycorrhizal roots. However, in their study, the greater expenditure of energy could be met from increased photosynthesis by plants.

Comparisons of the soluble carbohydrate concentrations in mycorrhizal and nonmycorrhizal plants may be relevant when assessing the degree of competition for photosynthate between mycorrhizal fungi and plants. Infection by plant roots by VAM fungi does not appear to decrease concentrations of carbohydrates within roots[97,98] but may decrease concentrations in shoots.[99] However, these results require careful interpretation because an increasing phosphorus supply which increases plant growth may decrease the concentration of soluble carbohydrates in both shoots and roots.[100,101] Concentrations of soluble carbohydrates within the roots of mycorrhizal plants may be less than those of nonmycorrhizal plants because the latter are more phosphorus-deficient. Indeed, concentrations of soluble carbohydrates were similar in roots of mycorrhizal onions and those of nonmycorrhizal onions supplied with phosphate.[102]

The development of VAM may normally be controlled to avoid competition for photosynthate between the plant and fungus. Treatments such as low light intensity, low root temperature, and high levels of phosphorus which are likely to decrease the supply of carbohydrates available for the mycorrhizal fungi, also decrease mycorrhizal infection.[101-104] Moreover, increasing the supply of phosphorus decreases mycorrhizal infection,[39,75] perhaps by decreasing the concentration of soluble carbohydrate in roots[44,101] or in root exudates.[98,105] When mycorrhizal plants are transferred or transplanted into another environment or soil, the roots may contain more mycorrhizal infection than would have developed if the plants had been grown, from germination, under the new environmental conditions. In situations where this is so, competition for photosynthate may occur until the level of infection adjusts to the lower level of photosynthate supply. The consequence may be depressed growth of the mycorrhizal plants. If the resulting adjustment in the development of infection is relatively minor, then the growth depression may be transitory.

Growth of plants inoculated with mycorrhizal fungi is sometimes less than that of uninoculated plants even when there is very little infection.[80] In situations such as this, competition for photosynthate between the plant and mycorrhizal fungus would be an unlikely explanation for depressed growth.

Phosphorus toxicity has been suggested as an explanation for certain growth depressions in mycorrhizal plants.[106] However, characteristic symptoms of phosphorus toxicity in crop and pasture species have not generally been observed where growth has been depressed by inoculation with mycorrhizal fungi. Furthermore, growth has been depressed when phosphorus concentrations within shoots and roots are much lower than those likely to be associated with phosphorus toxicity.[80]

Growth depressions have usually been reported with little attempt to identify the mechanism by which growth has been reduced. More attention should be paid to evaluating the possible alternative explanations for reduced growth of plants inoculated with VAM fungi.

2. Growth Stimulation

It has been suggested that VAM may stimulate plant growth by physiological effects other than by enhancement of nutrient uptake.[3] In one experiment that supports this view mycorrhizal seedlings of sweetgum *(Liquidambar styraciflua)* produced approximately 40 times more dry weight than did nonmycorrhizal seedlings.[31] Tenfold increases in the rate of application of nitrogen, phosphorus, and potassium fertilizers did not affect the dry weight of either mycorrhizal or nonmycorrhizal seedlings. It is possible that mycorrhizae increased plant growth in this study by increasing the uptake of micronutrients such as copper or zinc. However, severe deficiencies of these particular micronutrients generally induce characteristic symptoms of deficiency in plants and these were not reported. Schultz et al.[31] suggested that stimulation of hormonal activity or enhanced water uptake may be involved in the growth responses to inoculation.

Water uptake may be enhanced by mycorrhizal colonization.[107-109] However, in most cases the response is associated with differences in nutrient status between mycorrhizal and nonmycorrhizal plants.[108,109] Similarly, mycorrhizal plants appear to be more tolerant of some plant diseases than nonmycorrhizal plants because of differences in nutrient status.[110] Differences in hormone levels[111] between mycorrhizal and nonmycorrhizal plants may also reflect differences in nutrient status. Research into hormone production by VAM fungi and by mycorrhizae is just beginning.[111] All of these areas need further research to determine whether VAM play a role in improving plant growth other than through their effect on nutrient uptake.

IV. EFFECT OF DIFFERENT FUNGAL SPECIES AND STRAINS ON PLANT GROWTH

Species and strains of VAM fungi have been shown to differ in the extent to which they increase nutrient uptake and plant growth.[30,32,39,51,82,112-114]

These observations have led to a discussion of efficient or superior strains of mycorrhizal fungi.[115] Such terminology has often been used with the implication that any given species may have some kind of "innate effectiveness". We have used the term "effectiveness" to refer to the ability of the fungus to increase plant growth in a phosphate-deficient soil.[116] There are at least four factors which are related to effectiveness defined in this way: (1) the ability to form extensive and well-distributed hyphae in soil; (2) the ability to form extensive infections throughout the developing root system; (3) the ability of hyphae to absorb phosphorus from the soil solution; and (4) the longevity of the transport mechanism along hyphae and into the root.[12] These characteristics are all involved with the uptake and transport of nutrients and they may differ for different species or strains of fungi. The only one of these four characteristics

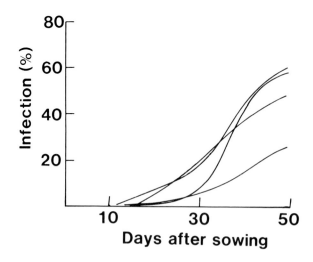

FIGURE 4. Percentage of total root length infected at different times for four species of VAM fungi.[30]

which has so far been shown to differ among species of VAM fungi is the ability to infect roots. There have been no reports which compare VAM fungi for their ability to form hyphae in soil, absorb phosphorus from the soil solution, or transport phosphorus along hyphae and into the root. The fungi may indeed differ in any of these three attributes. However, currently we only have data to show that they differ in their ability to infect roots rapidly and/or extensively.

Where the development of infection has been studied with time in relation to plant growth, those species which were effective at increasing plant growth were the ones which resulted in the most rapid and extensive infection of roots (Figure 4).[30] The rate at which different fungal species reach their plateau infection level will depend upon the number of infective propagules and their ability to spread within soil or along and within roots by secondary infections.[58,117-119] The effectiveness of a particular species of VAM fungus will also be related to its ability to infect roots of the host at a time most appropriate for increased uptake of a nutrient which is deficient for plant growth.

Frequently, at a final harvest, there is no difference in the percentage of roots infected by different species of fungi although they may have produced differences in plant growth.[120] This could arise if fungi differed in the rate at which they formed mycorrhizae (Figure 5).[51]

Species of VAM fungi may indeed differ in their ability to form hyphae in soil,[57] both in the distribution and quantity of hyphae. It has already been shown that the development of infection is correlated with the ability of a fungus to increase plant growth.[121] However, we do not know whether the growth of external hyphae is a characteristic which is independent of the development of infection within roots. Do species of VAM fungi have a constant and "innate" ability to take up nutrients and increase plant growth which is independent of the ability of the fungus to infect the host? So far there is no evidence to suggest this. However, experiments designed to compare species for their ability to take up nutrients and transport them to the plant are needed to test such an hypothesis.

The procedures which have sometimes been used to select "efficient" VAM fungi are not always satisfactory. This is because the inoculum potential for the fungi being compared is not usually standardized. Therefore, differences in the ability of species to increase plant growth may simply reflect differences in the infectivity or quantity of propagules in the original inoculum.[117,119,121]

The performance of particular fungi in tests of their ability to enhance plant growth

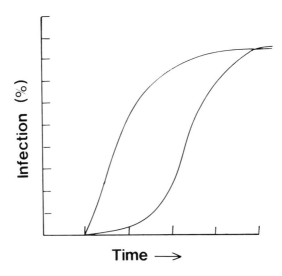

FIGURE 5. Possible patterns of development of mycorrhizal infection for two VAM fungi reaching the same plateau level of infection.

will depend upon their ability to absorb nutrients as well as their capacity to form mycorrhizas rapidly and extensively.[12] Therefore, further research to explain the factors which affect the formation of mycorrhizae by different species of fungi will be crucial for understanding differences among VAM fungi in their ability to affect plant growth.

Certainly, it has already been shown that some VAM fungi are more efficient than others at increasing the growth of plants. Unfortunately, comparisons of fungi have not always been made using the best inoculum of each fungus, thereby giving each its maximum opportunity.[121] More research is needed to define the characteristics which make fungi efficient at enhancing plant growth. Clearly, there is scope for selecting fungi for field inoculation which are more efficient at increasing plant growth than those already resident in many soils. To be successful however, they must also have characteristics which enable them to persist after inoculation.[12]

V. CONCLUSIONS

There have been many experiments reported in recent years which have been designed to examine the effect of VAM on plant growth and metabolism. Generally, the problems involved in choosing an adequate control treatment have not been considered. Furthermore, experiments have frequently been conducted without adequately defining the nutritional status of either mycorrhizal or nonmycorrhizal plants. Thus, in many of these experiments the data are difficult to interpret because of inadequate designs. In spite of this, it is clear that VAM fungi enhance plant growth primarily through their effects on nutrient uptake. It is likely that effects of mycorrhizae in increasing nutrient uptake will be most marked for nutrients which move to roots principally by diffusion and for host plants with coarse roots and few root hairs. Most attention has been given to the role of VAM fungi in increasing the uptake of phosphorus. Because of the generally low levels of infection in phosphorus-adequate plants, it is probable that VAM will seldom alleviate deficiencies of other nutrients. However, in some situations, the supply of more than one nutrient may be marginally deficient for plant growth. Here, mycorrhizae may increase plant growth by increasing the uptake of both nutrients (for example, phosphorus and another nutrient). The effect of mycorrhizae on nutrient uptake and plant growth has not been adequately examined when the supply of more than one nutrient is limiting growth. Similarly, non-nutritional effects of VAM on plant growth, such as growth depressions, or stimulation of growth by hormonal activity require further investigation.

REFERENCES

1. Nicolson, T. H., Vesicular-arbuscular mycorrhiza-a universal plant symbiosis, *Sci. Prog. (Oxford)*, 55, 561, 1967.
2. Gerdeman, J. W., Vesicular-arbuscular mycorrhiza and plant growth, *Annu. Rev. Phytopathol.*, 6, 397, 1968.
3. Smith, S. E., Mycorrhizas of autotrophic higher plants, *Biol. Rev. Cambridge Philos. Soc.*, 55, 475, 1980.
4. Mosse, B., Advances in the study of vesicular-arbuscular mycorrhiza, *Annu. Rev. Phytopathol.*, 11, 171, 1973.
5. Mosse, B., Stribley, D. P., and Le Tacon, F., Ecology of mycorrhizae and mycorrhizal fungi, *Adv. Microbiol. Ecol.*, 5, 137, 1981.
6. Mosse, B., Growth and chemical composition of mycorrhizal and non-mycorrhizal apples, *Nature (London)*, 179, 922, 1957.
7. Peuss, H., Untersuchungen zur Okologie und Bedeutung der Tabakmycorrhiza, *Arch. Mikrobiol.*, 29, 112, 1958.
8. Winter, A. G. and Meloh, K. A., Untersuchungen uber den Einfluss der endotrophen Mycorrhiza auf die Entwicklung von *Zea mays* L., *Naturwissenschaften*, 45, 319, 1958.
9. Baylis, G. T. S., Effect of vesicular-arbuscular mycorrhizas on growth of *Griselinia littoralis* (Cornaceae), *New Phytol.*, 58, 274, 1959.
10. Baylis, G. T. S., Rhizophagus. The catholic symbiont, *Aust. J. Sci.*, 25, 195, 1962.
11. Tinker, P. B. H., Effects of vesicular-arbuscular mycorrhizas on higher plants, in *Symbiosis*, Jennings, D. H. and Lee, D. L., Eds., Cambridge University Press, Cambridge, 1975, 325.
12. Abbott, L. K. and Robson, A. D., The role of vesicular-arbuscular mycorrhizal fungi in agriculture and the selection of fungi for inoculation, *Aust. J. Agric. Res.*, 33, 389, 1982.
13. Tinker, P. B., Effects of vesicular-arbuscular mycorrhizas on plant nutrition and plant growth, *Physiol. Veg.*, 16, 743, 1978.
14. Menge, J. A., Lembright, H., and Johnson, E. L. V., Utilization of mycorrhizal fungi in citrus nurseries, *Proc. Int. Soc. Citric.*, 1, 129, 1977.
15. Baylis, G. T. S., The significance of mycorrhizas and root nodules in New Zealand vegetation, *Proc. R. Soc. N.Z.*, 89, 45, 1961.
16. Nicolson, T. H., Mycorrhiza in the Gramineae. II. Development in different habitats, particularly sand dunes, *Trans. Br. Mycol. Soc.*, 43, 132, 1960.
17. Koske, R. E., A preliminary study of interactions between species of vesicular-arbuscular fungi in a sand dune, *Trans. Br. Mycol. Soc.*, 76, 411, 1981.
18. Koske, R. E., Sutton, J. C., and Sheppard, B. R., Ecology of *Endogone* in Lake Huron sand dunes, *Can. J. Bot.*, 53, 87, 1975.
19. Daft, M. J. and Nicolson, T. H., Arbuscular mycorrhizas in plants colonizing coal wastes in Scotland, *New Phytol.*, 73, 1129, 1974.
20. Reeves, F. B., Wagner, D., Moorman, T., and Kiel, J., The role of endomycorrhizae in revegetation practices in the semi-arid west. I. A comparison of incidence of mycorrhizae in severely disturbed and natural environments, *Am. J. Bot.*, 66, 6, 1979.
21. Jakobsen, I. and Andersen, A. J., Vesicular-arbuscular mycorrhiza and growth in barley: effects of irradiation and heating of soil, *Soil Biol. Biochem.*, 14, 171, 1982.
22. Rovira, A. and Bowen, G. D., The effects of microorganisms upon plant growth. II. Detoxication of heat sterilized soils by fungi and bacteria, *Plant Soil*, 25, 129, 1966.
23. Gerdemann, J. W., The effect of mycorrhiza on the growth of maize, *Mycologia*, 56, 342, 1964.
24. Gilmore, A. E., Phycomycetous mycorrhizal organisms collected by open-pot culture methods, *Hilgardia*, 39, 87, 1968.
25. Clark, F. B., Endotrophic mycorrhizae influence yellow poplar seedling growth, *Science*, 140, 1220, 1963.
26. Mosse, B., The establishment of vesicular-arbuscular mycorrhiza under aseptic conditions, *J. Gen. Microbiol.*, 27, 509, 1962.

27. Hepper, C. M., Techniques for studying the infection of plants by vesicular-arbuscular mycorrhizal fungi under axenic conditions, *New Phytol.*, 88, 641, 1981.
28. Pairunan, A., Robson, A. D., and Abbott, L. K., The effectiveness of vesicular-arbuscular mycorrhizas in increasing growth and phosphorus uptake of subterranean clover from phosphorus sources of different solubilities, *New Phytol.*, 84, 327, 1980.
29. Hall, I. R., Response of *Coprosma robusta* to different forms of endomycorrhizal inoculum, *Trans. Br. Mycol. Soc.*, 67, 409, 1976.
30. Sanders, F. E., Tinker, P. B., Black, R. L. B., and Palmerley, S. M., The development of endomycorrhizal root systems. I. Spread of infection and growth-promoting effects with four species of vesicular-arbuscular endophyte, *New Phytol.*, 78, 257, 1977.
31. Schultz, R. C., Kormanik, P. P., Bryan, W. C., and Brister, G. H., Vesicular-arbuscular mycorrhiza influence growth but not mineral concentrations in seedlings of eight sweetgum families, *Can. J. For. Res.*, 9, 218, 1979.
32. Carling, D. E. and Brown, M. F., Relative effect of vesicular-arbuscular mycorrhizal fungi on the growth and yield of soybeans, *J. Soil Sci. Soc. Am.*, 44, 528, 1980.
33. Hall, I. R. and Kelson, A., An improved technique for the production of endomycorrhizal infested soil pellets, *N.Z. J. Agric. Res.*, 24, 221, 1981.
34. Ferguson, J. J., Inoculum Production and Field Application of Vesicular-Arbuscular Mycorrhizal Fungi, Ph.D. thesis, University of California, Riverside, 1981.
35. Smith, F. A. and Smith, S. E., Mycorrhizal infection and growth of *Trifolium subterraneum*: comparison of natural and artificial inocula, *New Phytol.*, 88, 311, 1981.
36. Yost, R. S. and Fox, R. L., Contribution of mycorrhizae to P nutrition on an oxisol, *Agron. J.*, 71, 903, 1979.
37. Smith, F. A. and Smith, S. E., Mycorrhizal infection and growth of *Trifolium subterraneum*: use of sterilized soil as control treatment, *New Phytol.*, 88, 299, 1981.
38. Mosse, B., Plant growth responses to vesicular-arbuscular mycorrhiza. X. Responses of *Stylosanthes* and maize to inoculation in unsterile soils, *New Phytol.*, 78, 277, 1977.
39. Abbott, L. K. and Robson, A. D., Growth stimulation of subterranean clover with vesicular-arbuscular mycorrhizas, *Aust. J. Agric. Res.*, 28, 639, 1977.
40. Menge, J. A., Labanauskas, C. K., Johnson, E. L. V., and Platt, R. G., Partial substitution of mycorrhizal fungi for phosphorus fertilization in the greenhouse culture of citrus, *J. Soil Sci. Soc. Am.*, 42, 926, 1978.
41. Stribley, D. P., Tinker, P. B., and Snellgrove, R. C., Effect of vesicular-arbuscular mycorrhizal fungi on the relations of plant growth, internal phosphorus concentration and soil phosphate analysis, *J. Soil Sci.*, 31, 655, 1980.
42. Barrow, N. J., The response to phosphate of two annual pasture species. I. Effect of the soil's ability to adsorb phosphate on comparative phosphate requirement, *Aust. J. Agric. Res.*, 26, 137, 1975.
43. Parberry, N. H., Phosphate placement in a brown lateritic soil, *Agric. Gaz. N.S.W.*, 57, 405, 1946.
44. Jasper, D. A., Robson, A. D., and Abbott, L. K., Phosphorus and the formation of vesicular-arbuscular mycorrhizas, *Soil Biol. Biochem.*, 11, 501, 1979.
45. Barrow, N.J., Malajczuk, N., and Shaw, T. C., A direct test of the ability of vesicular-arbuscular mycorrhiza to help plants take up fixed soil phosphate, *New Phytol.*, 78, 269, 1977.
46. Barrow, N. J., The response to phosphate of two annual pasture species. II. The specific rate of uptake of phosphate, its distribution and use for growth, *Aust. J. Agric. Res.*, 26, 145, 1975.
47. Campbell, N. A. and Keay, J., Flexible techniques in describing mathematically a range of response curves of pasture species, *Proc. 11th Int. Grassl. Congr.*, University of Queensland Press, Brisbane, 1970.
48. Hayman, D. S. and Mosse, B., Plant growth responses to vesicular-arbuscular mycorrhiza. I. Growth of *Endogone*-inoculated plants in phosphate-deficient soils, *New Phytol.*, 71, 19, 1971.
49. Mosse, B. and Hayman, D. S., Plant growth responses to vesicular-arbuscular mycorrhiza. II. In unsterilized field soils, *New Phytol.*, 70, 29, 1971.
50. Becker, W. N. and Gerdemann, J. W., Colorimetric quantification of vesicular-arbuscular mycorrhizal infection in onion, *New Phytol.*, 78, 289, 1977.
51. Abbott, L. K. and Robson, A. D., Growth of subterranean clover in relation to the formation of endomycorrhizas by introduced and indigenous fungi in a field soil, *New Phytol.*, 81, 575, 1978.
52. Smith, S. E., Inflow of phosphate into mycorrhizal and non-mycorrhizal plants of *Trifolium subterraneum* at different levels of soil phosphate, *New Phytol.*, 90, 293, 1982.
53. Crush, J. R., Plant growth responses to vesicular-arbuscular mycorrhiza. VII. Growth and nodulation of some herbage legumes, *New Phytol.*, 73, 743, 1974.
54. Smith, S. E. and Daft, M. J., Interactions between growth, phosphate content, and N_2 fixation in mycorrhizal and non-mycorrhizal *Medicago sativa*, *Aust. J. Plant Physiol.*, 4, 403, 1977.
55. Carling, D. E., Riehle, W. G., Brown, M. F., and Johnson, D. R., Effects of a vesicular-arbuscular mycorrhizal fungus on nitrate reductase and nitrogenase activities in nodulating and non-nodulating soybeans, *Phytopathology*, 68, 1590, 1978.

56. Robson, A. D., O'Hara, G. W., and Abbott, L. K., The involvement of phosphorus in nitrogen fixation by subterranean clover (*Trifolium subterraneum* L.), *Aust. J. Plant Physiol.*, 8, 427, 1981.
57. Bevege, D. I. and Bowen, G. D., *Endogone* strain and host plant differences in development of vesicular-arbuscular mycorrhizas, in *Endomycorrhizas,* Sanders, F. E., Mosse, B., and Tinker, P. B., Eds., Academic Press, London, 1975,77.
58. Smith, S. E and Walker, N. A., A quantitative study of mycorrhizal infection in *Trifolium:* separate determination of the rates of infection and of mycelial growth, *New Phytol.*, 89, 225, 1981.
59. Azcon-G. de Aquilar, C. and Barea, J. M., Field inoculation of *Medicago* with V-A mycorrhiza and *Rhizobium* in phosphate-fixing agricultural soil, *Soil Biol. Biochem.*, 13, 19, 1981.
60. Csinos, A. S., *Gigaspora margarita* inoculation of *Nicotiana tabacum:* responses to fertilization, *Can. J. Bot.*, 59, 101, 1981.
61. Jensen, A., Influence of four vesicular-arbuscular mycorrhizal fungi on nutrient uptake and growth in barley *(Hordum vulgare)*, *New Phytol.*, 90, 45, 1982.
62. Rhodes, L. H. and Gerdemann, J. W., Nutrient translocation in vesicular-arbuscular mycorrhizae, in *Cellular Interactions in Symbiosis and Parasitism,* Cook, C. B., Pappas, P. W., and Rudolph, E. D., Eds., The Ohio State University Press, Columbus, 1980, 173.
63. Hattingh, M. J., Gray, L. E., and Gerdemann, J. W., Uptake and translocation of ^{32}P-labelled phosphate to onion roots by endomycorrhizal fungi, *Soil. Sci.*, 116, 383, 1973.
64. Rhodes, L. H. and Gerdemann, J. W., Phosphate uptake zones of mycorrhizal and nonmycorrhizal onions, *New Phytol.*, 75, 555, 1975.
65. Bowen, G. D., Skinner, M. F., and Bevege, D. I., Zinc uptake by mycorrhizal and uninfected roots of *Pinus radiata* and *Araucaria cunninghamii, Soil Biol. Biochem.*, 6, 141, 1974.
66. Bowen, G. D., Bevege, D. I., and Mosse, B., The phosphate physiology of vesicular arbuscular mycorrhizas, in *Endomycorrhizas,* Sanders, F. E., Mosse, B., and Tinker, P. B., Eds., Academic Press, London, 1975, 241.
67. Cress, W. A., Throneberry, G. O., and Lindsey, D. L., Kinetics of phosphorus absorption by mycorrhizal and nonmycorrhizal tomato roots, *Plant Physiol.*, 64, 484, 1979.
68. Nye, P. B. and Tinker, P., *Solute Movement in the Soil-Root System,* Blackwell Scientific, Oxford, 1977, 342.
69. Baylis, G. T. S., Experiments on the ecological significance of phycomycetous mycorrhizas, *New Phytol.*, 66, 231, 1967.
70. Baylis, G. T. S, Root hairs and phycomycetous mycorrhizas in phosphate deficient soil, *Plant Soil*, 33, 713, 1970.
71. Baylis, G. T. S., The magnolioid mycorrhiza and mycotrophy in root systems derived from it, in *Endomycorrhizas,* Sanders, F. E., Mosse, B., and Tinker, P. B., Eds., Academic Press, London, 1975, 375.
72. Nye, P. H., The rate-limiting step in plant nutrient absorption from soil, *Soil Sci.*, 123, 292, 1977.
73. Fox, R. L., Studies on phosphorus nutrition in the tropics, in *Mineral Nutrition of Legumes in Tropical and Subtropical soils,* Andrew, C. S. and Kamprath, E. J., Eds., Commonwealth Scientific and Industrial Research Organization, Melbourne, 1978, 169.
74. Lewis, D. G. and Quirk, J. P., Phosphate diffusion in soil and uptake by plants. III. ^{31}P movement and uptake by plants as indicated by ^{32}P autoradiography, *Plant Soil*, 26, 445, 1967.
75. Sanders, F. E. and Tinker, P. B., Phosphate flow into mycorrhizal roots, *Pestic. Sci.*, 4, 385, 1973.
76. Hayman, D. S. and Mosse, B., Plant growth responses to vesicular-arbuscular mycorrhiza. III. Increased uptake of labile P from soil, *New Phytol.*, 71, 41, 1972.
77. Powell, C. L., Plant growth responses to vesicular-arbuscular mycorrhiza, VIII. Uptake of P by onion and clover infected with different *Endogone* spore types in ^{32}P labelled soils, *New Phytol.*, 75, 563, 1975.
78. Murdoch, C. L., Jackobs, J. A., and Gerdemann, J. W., Utilization of phosphorus sources of different availability by mycorrhizal and non-mycorrhizal maize, *Plant Soil*, 27, 329, 1967.
79. Powell, C. L. and Daniel, J., Mycorrhizal fungi stimulate uptake of soluble and insoluble fertilizer from a phosphate deficient soil, *New Phytol.*, 80, 351, 1978.
80. Hall, I. R., Scott, R. S., and Johnstone, P. B., Effect of vesicular-arbuscular mycorrhizas on response of "Grassland Huia" and "Tamar" white clovers to phosphorus, *N.Z. J. Agric. Res.*, 20, 349, 1977.
81. Hall, I. R., Effect of vesicular-arbuscular mycorrhizas on two varieties of maize and one of sweet corn, *N.Z. J. Agric. Res.*, 21, 517, 1978.
82. Nemec, S., Response of 6 Citrus rootstocks to 3 species of *Glomus, Proc. Fla. State Hortic. Soc.*, 91, 10, 1978.
83. Azcon, R. and Ocampo, J. A., Factors affecting the vesicular-arbuscular infection and mycorrhizal dependency of thirteen wheat cultivars, *New Phytol.*, 87, 677, 1981.
84. Loneragan, J. F., Nutrient requirements of plants, *Nature (London)*, 220, 1307, 1968.

85. Barrow, N J., Problems of efficient fertilizer use, *Proc. 8th Int. Colloq. Plant Anal. Prob.*, Ferguson, A. D., Bieleski, R. L., and Ferguson, I. B., Eds., New Zealand Department Scientific and Industrial Research Information Ser. No. 134, Government Printer, Wellington, 1978.
86. Barrow, N. J., Phosphorus uptake and utilization by tree seedlings, *Aust. J. Bot.*, 25, 571, 1977.
87. Gilmore, A. E., The influence of endotrophic mycorrhizae on the growth of peach seedlings, *J. Am. Soc. Hortic. Sci.*, 96, 35, 1971.
88. La Rue, J. H., McClellen, W. D., and Peacock, W. L., Mycorrhizal fungi and peach nursery nutrition, *Calif. Agric.*, May, 6, 1975.
89. Timmer, L. W. and Leyden, R. F., Stunting of citrus seedlings in fumigated soils in Texas and its correction by phosphorus fertilization and inoculation with mycorrhizal fungi, *J. Am. Soc. Hortic. Sci.*, 103, 533, 1978.
90. Lambert, D. H., Baker, D. E., and Cole, H., The role of mycorrhizae in the interactions of phosphorus with zinc, copper and other elements, *J. Soil Sci. Soc. Am.*, 43, 976, 1979.
91. Rhodes, L. H. and Gerdemann, J. W., Influence of phosphorus nutrition on sulphur uptake by vesicular-arbuscular mycorrhize of onion, *Soil Biol. Biochem.*, 10, 361, 1978.
92. Cooper, K. M., Growth responses to the formation of endotrophic mycorrhizas in *Solanum*, *Leptospermum*, and New Zealand ferns, in *Endomycorrhizas*, Sanders, F. E., Mosse, B., and Tinker, P. B., Eds., Academic Press, London, 1975, 291.
93. Bolan, N. S., Unpublished data.
94. Crush, J. R., The effect of *Rhizophagus tenuis* mycorrhizas on ryegrass, cocksfoot and sweet vernal, *New Phytol.*, 72, 965, 1973.
95. Hepper, C. M., A colorimetric method for estimating vesicular-arbuscular mycorrhizal infection in roots, *Soil Biol. Biochem.*, 9, 15, 1977.
96. Pang, P. C. and Paul, E. A., Effects of vesicular-arbuscular mycorrhiza on ^{14}C and ^{15}N distribution in nodulated fababeans, *Can. J. Soil Sci.*, 60, 241, 1980.
97. Nemec, S. and Guy, G., Carbohydrate status of mycorrhizal and nonmycorrhizal citrus rootstocks, *J. Am. Soc. Hortic. Sci.*, 107, 177, 1982.
98. Graham, J. H., Leonard, R. T., and Menge, J. A., Membrane-mediated decrease in root exudation responsible for phosphorus inhibition of vesicular-arbuscular mycorrhiza formation, *Plant Physiol.*, 68, 548, 1981.
99. Buwalda, J. G. and Goh, K. M., Host-fungus competition for carbon as a cause of growth depressions in vesicular-arbuscular mycorrhizal ryegrass, *Soil Biol. Biochem.*, 14, 103, 1982.
100. Rossiter, R. C. and Beck, A. B., Physiological ecological studies on the oestrogenic isoflavones of subterranean clover (*T. subterraneum* L.). II. Effects of phosphorus supply, *Aust. J. Agric. Res.*, 17, 447, 1966.
101. Same, B. I., Robson, A. D., and Abbott, L. K., Phosphorus, soluble carbohydrates and endomycorrhizal infection, *Soil Biol. Biochem.*, 15, 593, 1983.
102. Hayman, D. S., Plant growth responses to vesicular-arbuscular mycorrhiza, VI. Effect of light and temperature, *New Phytol.*, 73, 71, 1974.
103. Daft, M. J. and El-Giahmi, A. A., Effect of arbuscular mycorrhiza on plant growth. VIII. Effects of defoliation and light on selected hosts, *New Phytol.*, 80, 359, 1978.
104. Furlan, V. and Fortin, J. A., Formation of endomycorrhiza by *Endogone calospora* on *Allium cepa* under three temperature regimes, *Nat. Can. (Quebec)*, 100, 467, 1973.
105. Ratnayake, M., Leonard, R. T., and Menge, J. A., Root exudation in relation to supply of phosphorus and its possible relevance to mycorrhizal formation, *New Phytol.*, 81, 543, 1978.
106. Mosse, B., Plant growth responses to vesicular-arbuscular mycorrhiza. IV. In soil given additional phosphate, *New Phytol.*, 72, 127, 1973.
107. Safir, G. R., Boyer, J. S., and Gerdemann, J. W., Mycorrhizal enhancement of water transport in soybean, *Science*, 172, 581, 1971.
108. Safir, G. R., Boyer, J. S., and Gerdemann, J. W., Nutrient status and mycorrhizal enhancement of water transport in soybean, *Plant Physiol.*, 43, 700, 1972.
109. Nelsen, C. E. and Safir, G. R., The water relations of well-watered, mycorrhizal and non-mycorrhizal onion plants, *J. Am. Soc. Hortic. Sci.*, 107, 271, 1982.
110. Graham, J. H. and Menge, J. A., Influence of vesicular-arbuscular mycorrhizae and soil phosphorus on take-all disease of wheat, *Phytopathology*, 72, 95, 1982.
111. Allen, M. F., Moore, T. S., and Christensen, M., Phytohormone changes in *Bouteloua gracilis* infected by vesicular-arbuscular mycorrhizae. II. Altered levels of gibberellin-like substances and abscisic acid in the host plant, *Can. J. Bot.*, 60, 468, 1982.
112. Mosse, B., The influence of soil type and *Endogone* strain on the growth of mycorrhizal plants in phosphate deficient soils, *Rev. Ecol. Biol. Soil*, 9, 529, 1972.
113. Powell, C. L., Metcalfe, D. M., Buwalda, J. G., and Waller, J. E., Phosphate response curves of mycorrhizal and non-mycorrhizal plants. II. Responses to rock phosphates, *N.Z. J. Agric. Res.*, 23, 477, 1980.

114. Daniels, B. A. and Menge, J. A., Evaluation of the commercial potential of the vesicular-arbuscular mycorrhizal fungus, *Glomus epigaeus*, *New Phytol.*, 87, 345, 1981.
115. Powell, C. L., Mycorrhizal fungi stimulate clover growth in New Zealand hill country soils, *Nature (London)*, 264, 436, 1976.
116. Abbott, L. K. and Robson, A. D., Infectivity and effectiveness of five endomycorrhizal fungi: competition with indigenous fungi in field soils, *Aust. J. Agric. Res.*, 32, 621, 1981.
117. Daniels, B. A., McCool, P. M., and Menge, J. A., Comparative inoculum potentials of spores of six vesicular-arbuscular mycorrhizal fungi, *New Phytol.*, 89, 385, 1981.
118. Powell, C. L., Effect of inoculum rate on mycorrhizal growth responses in pot-grown onion and clover, *Plant Soil*, 62, 231, 1981.
119. Wilson J. M. and Trinick, M., Factors affecting the estimation of numbers of infective propagules of vesicular-arbuscular mycorrhizal fungi by the most probable number method, *Aust. J. Soil Res.*, 22, 73, 1983.
120. Owusu-Bennoah, E. and Mosse, B., Plant growth responses to vesicular-arbuscular mycorrhiza. XI. Field inoculation responses in barley, lucerne and onion, *New Phytol.*, 83, 671, 1979.
121. Abbott, L. K. and Robson, A. D., Infectivity and effectiveness of vesicular-arbuscular mycorrhizal fungi: effect of inoculum type, *Aust. J. Agric. Res.*, 32, 631, 1981.

Chapter 7

BIOLOGICAL INTERACTIONS WITH VA MYCORRHIZAL FUNGI

D. Joseph Bagyaraj

TABLE OF CONTENTS

I.	VA Mycorrhizal (VAM) Fungi as Components of the Rhizosphere 132	
II.	Interactions Between VAM and Beneficial Soil Organisms 132	
	A. Legume Bacteria 132	
	B. Free-Living Nitrogen-Fixing Bacteria 135	
	C. Phosphate-Solubilizing Organisms 135	
	D. Beneficial Actinomycetes 136	
III.	Interactions Between VAM and Plant Pathogenic Organisms 137	
	A. Fungi 137	
	1. Root Pathogens 137	
	a. VAM Supporting the Severity of Disease 137	
	b. VAM With No Effect on Disease 138	
	c. VAM Decreasing the Severity of Disease 138	
	2. Foliar Pathogens 139	
	B. Bacteria 139	
	C. Viruses 139	
	D. Nematodes 139	
	1. Sedentary Endoparasites 140	
	2. Migratory Endoparasites 141	
	E. Mechanisms of Suppression of Pathogens by VAM 142	
IV.	Microorganisms Intimately Associated with VAM 143	
	A. Companion Fungi 143	
	B. Bacterial Associates 144	
	C. Bacterium-Like Organelles 144	
V.	Organisms Parasitic and Predatory on VAM 144	
	A. Parasitic Microflora 144	
	B. Predatory Microfauna 145	
VI.	Conclusions 146	
References 146		

I. VA MYCORRHIZAL (VAM) FUNGI AS COMPONENTS OF THE RHIZOSPHERE

In 1904, Hiltner[1] coined the term rhizosphere to denote the region of the soil subject to the influence of plant roots and characterized by intense microbial activity. Many workers have subsequently shown that the microflora of the rhizosphere differs both quantitatively and qualitatively from that in the soil beyond the influence of the root.[2-5] The rhizosphere effect was greatest with bacteria followed by actinomycetes and fungi. The increased microbial activity in the rhizosphere has been attributed to the extra nutrients available in the region. The extra nutrient supply comes from root exudates, sloughed-off parts of the root cap, mucigel, root hair residues, and abraded epidermal cells.[6,7] The microflora of the rhizosphere could influence plant growth in many ways as all plant nutrients pass through this region.[8,9] Attempts have also been made to alter the rhizosphere microflora through soil amendments, foliar sprays, and microbial seed inoculation for the benefit of plant growth,[10-12] some of which have found a place in practical agriculture such as seed or root bacterization with efficient strains of rhizobia to increase nitrogen fixation[13] and with the bacteriocinogenic strain of *Agrobacterium radiobacter* var. *radiobacter* to biologically control the crown gall pathogen *A. radiobacter* var. *tumefaciens*.[14]

Concepts of rhizosphere microbiology are presently undergoing considerable change.[15] They are becoming more dynamic and concerned with colonization patterns and with interactions in the rhizosphere. Also, the physiology and distribution of roots are now often included as emphasis shifts from the mere identification and enumeration of organisms that can be isolated on synthetic media. Therefore, the important role of VAM fungi (neglected earlier as not culturable on laboratory media) as components of the rhizosphere population is receiving more recognition.[16] Of the various microorganisms colonizing the rhizosphere, VAM fungi occupy a unique ecological position as they are partly inside and partly outside the host. The internal phase of a mycorrhizal fungus does not encounter competition and antagonism from other soil microorganisms, and has an ensured source of nutrients from the host. This advantage enables them to achieve a large and more functional biomass in more intimate contact with the root and thus increases their chances of exerting a greater effect on plants than other microbial species restricted only to the rhizosphere. Tinker,[17] reviewing the present position of VAM research pointed out that root surface ecology is a subject of great potential promise, and more investigations are needed on the biological interactions between VAM fungi and beneficial and pathogenic soil organisms, and their effects on plant growth. However, it has been stated recently[16,17] that the topic of interaction of VAM with other organisms is in a greatly confused state, with a variety of isolated observations from which no useful generalizations could be made. In this chapter the different aspects of biological interactions with VAM fungi are reviewed, attempts made to draw generalizations wherever possible, and suggestions given for future lines of work.

II. INTERACTIONS BETWEEN VAM AND BENEFICIAL SOIL ORGANISMS

A. Legume Bacteria

The tripartite association of flowering plants, mycorrhizal fungi, and nitrogen-fixing bacteria has been a subject of interest since 1896 when Janse[18] first described such a symbiotic association in the legume *Pithecolobium montanum*. Jones[19] examined 18 species of nodulated legumes and found that 15 were colonized by VAM fungi. Asai[20] first suggested that mycorrhizae were a necessary precondition for effective nodulation in many legumes. He showed that several legumes grew poorly and failed to nodulate in autoclaved soil unless they were mycorrhizal. The possibility of mycorrhizae increasing the effectiveness of *Rhizobium* was suggested by Ross and Harper[21] when they found that soybean plants with mycorrhiza contained higher concentrations of nitrogen than plants without. Further indirect evidence that VAM may enhance nitrogen fixation by *Rhizobium* in soybean roots was provided by Schenck and Hinson.[22]

Direct evidence of mycorrhizae increasing nodulation in legumes was obtained by Crush[23] in 1974. He found that VAM strongly stimulated the growth and nodulation of *Centrosema pubescens, Stylosanthes guyanensis,* and *Trifolium repens.* The tropical legumes were much more dependent on mycorrhiza for growth than the temperate species. The dependence of the temperate species on mycorrhiza was increased when the available P was reduced by diluting the experimental soil with sand. Mosse et al.[24] showed that effective nodulation of legumes in a very P-deficient Brazilian Cerrado soil could be achieved only by introducing both mycorrhiza and extra P, albeit the P being in the form of sparingly soluble rock phosphate. Here, the effect of the mycorrhiza and rock phosphate was synergistic and increased the dry weight of *Centrosema, Stylosanthes,* and *Trifolium* plants 7 to 22 times. Increased nodulation, nitrogen fixation and plant growth in *Stylosanthes* because of mycorrhiza and rock phosphate was also observed in unsterile soils from Nigeria.[25] Plants were nodulated and grew best in acidic soils when both inoculated with VA endophytes and given rock phosphate. In neutral and alkaline soils rock phosphate remained essentially unavailable to both mycorrhizal and nonmycorrhizal plants. Since then, several workers from different parts of the world have reported increased nodulation, nitrogen fixation, plant growth, plant N and P content in *Vigna unguiculata*,[26,27] *Medicago sativa*,[28] *Peuraria phaseoloides* and *Stylosanthes guyanensis*,[29] and *Leucaena leucocephala*[30] when the plants were inoculated with VAM fungi and given rock phosphate. Similar observations were made on French bean[31] and peanuts[32] inoculated with *Rhizobium* and mycorrhizal fungi in soil fertilized with bonemeal, and also in soybean,[33] lucerne,[34] peanuts,[32] cowpea,[35] and pigeonpea[36] given superphosphate. Mycorrhizal fungi and rhizobia were also found to be synergistic with respect to nitrogen fixation rates and mycorrhizal infection levels of sweetvetch in a very P-deficient soil with no added P.[37] Dual inoculation increased the number of leaves, plant height, and above- and below-ground biomass. The authors stressed the importance of this study as sweetvetch is widely used in revegetating disturbed land and deteriorated rangelands in the U.S.

Abbott and Robson[38] compared the growth of subterranean clover inoculated with *Rhizobium* and mycorrhiza at different levels of superphosphate in a high P-fixing soil and found that the dual inoculation markedly stimulated nodulation, mycorrhizal infection, growth, and P uptake only at the intermediate superphosphate rate. In most of the *Rhizobium*-VAM interaction studies, measurement of nodulation and nitrogenase activity have usually been made relatively late in ontogeny, when mycorrhizal plants were already larger than nonmycorrhizal controls. A detailed investigation conducted by Smith and Daft[39] with *Medicago sativa* revealed that mycorrhiza-induced increases in nodulation, nitrogen fixation, and phosphate concentration were apparent at 2 weeks after inoculation and preceded any effect on plant growth, which appeared 10 weeks after inoculation. A similar observation was also made in soybean by Asimi et al.[40] Their work further recorded that adding soluble P at a low level (0.25 g KH_2PO_4/kg soil) eliminated the mycorrhizal effect on plant growth, but still significantly stimulated nodule formation and nitrogenase activity. Mycorrhizal effects on nodulation and nitrogenase were successively eliminated by P fertilizer at 0.5 and 1.0 g KH_2PO_4/kg soil, respectively. Smith and Bowen[41] studied the effect of soil temperature on VAM-*Rhizobium* interactions and reported that 16°C was optimum for infection of *Trifolium subterraneum* roots by both the organisms, and that no competition between the two organisms for infection sites occurred. The symbioses between *T.*

subterraneum, mycorrhizal fungi, and *Rhizobium* were affected by two nitrification inhibitors 2-chloro-6 (trichloromethyl) pyridine (N-serve) and 2-trichloromethyl pyridine (2 TMP).[42] Both the chemicals reduced mycorrhizal infection and nitrogen fixation as measured by acetylene reduction. Addition of ammonium sulfate also had an inhibitory effect on nodulation and the development of mycorrhizal entry points.

Most of the studies on VAM-*Rhizobium* interaction suggest that infection with efficient endophytes significantly improve P nutrition, and consequently nodulation and nitrogen fixation. Legumes have repeatedly been shown to require high levels of phosphate for growth and effective nodulation.[43,44] It is also known that nitrogen fixation has a high phosphate requirement.[45,46] While the principal effect of mycorrhiza on nodulation is undoubtedly phosphate-mediated, mycorrhiza may have other secondary effects. Comparison of plants with same dry weight and root:shoot ratios indicated that mycorrhizal plants had higher root phosphate contents, increased nodulation, and higher rates of nitrogenase activity than equivalent nonmycorrhizal plants.[35] Mycorrhizal inoculation of white clover more than doubled nodulation even though the degrees of mycorrhizal infection in inoculated and uninoculated plants were similar and P concentrations in both were equally large.[47,48] It appears, therefore, that in addition to effects on phosphorus nutrition, mycorrhizae aid other processes involved in nodulation and nitrogen fixation.[30,49-51] Such potentially limiting factors may include supply of photosynthate,[52] trace elements,[53,54] and plant hormones.[55,56] These parameters could be subject to mycorrhizal influence as indicated by effects of shading and daylength on mycorrhizal *Medicago,*[57] by mycorrhizal effects on the uptake of copper and zinc,[21,58-60] and by production of plant hormones directly by VAM[61,62] (or indirectly by bacteria in the rhizosphere[63-65] stimulated by mycorrhizae).

Most of the experiments discussed so far on VAM-*Rhizobium* interactions have been conducted in the glass house or under controlled environments. Very few field studies have been made. Ross and Harper[21] recorded a 29% increased yield in soybean grown in fumigated field plots with double inoculation of VAM and *Rhizobium* over single inoculation with *Rhizobium.* A similar observation was made in a later study under normal (unsterilized) field conditions with soybean[66] (19% increased yield) and cowpea[67] (38% increased yield). Coming to forage legumes, inoculation of white clover plants in the field more than doubled nodulation.[47,48] In a grass-legume mixture study, inoculation of the legume with *Rhizobium* was found to stimulate mycorrhizal fungi in the rhizosphere of the associated grass and improved the grass P content.[68] Azcon et al.[69] recorded 66% increased yield of *Medicago sativa* from dual inoculation with VAM and *Rhizobium* compared with inoculation by either organism.

Legumes play a very important role in many agricultural systems, providing high protein grain and herbage as well as maintaining and improving soil fertility. This has extra importance in the tropics because of the grain legume programs introduced to increase protein content of the diet, and the fact that tropical soils are very deficient in phosphorus.[70] It was said[71] a few years back that if the beneficial effects of dual inoculation with rhizobia and VAM fungi observed under pot culture studies could be emulated under field conditions, it would have a large economic impact on agriculture. Recent field studies have shown great advantages of dual inoculation by the two symbionts. The practical difficulties in extending this technology to the farmers should not be forgotten because of the inability to obtain large quantities of VAM inocula, since the fungi cannot be grown on culture media. The method of application of inoculum would also pose a problem as most of the grain and forage legumes are directly sown in the field. However, recent work with onion and maize in New Zealand suggests that it is possible to machine-drill mycorrhizal inoculum just below the seed at the time of sowing, resulting in an increased yield of 15 to 17%.[72,73] Though the rate of inoculum used in the onion trial was 1900 kg/ha, it was suggested that it could be reduced to 500 kg/ha; in the maize trial, the inoculum rate used was 400 kg/ha. As pointed out by

Smith and Daft[39] pelleting of legume seeds with not only appropriate rhizobia but also with the propagules of mycorrhizal fungi (as has been successfully done with citrus[74]) could be used. Hayman et al.[75] recently studied different methods for inoculating field crops with mycorrhizal fungi and suggested the possibility of fluid drilling a slurry in methyl cellulose base consisting of concentrated inocula of mycorrhiza and *Rhizobium* along with the seed. These recent studies and speculations encourage intensifying field studies on VAM-*Rhizobium* dual inoculations with different rates and methods of inoculation which would make the dual inoculation a practicable proposition in agriculture. Perhaps an immediate application could be in forestry as some leguminous tree species like *Acacia*,[76] *Robinia*,[77] *Leucaena*[30] (a fodder tree legume also used in land reclamation, erosion control, reforestation, as a shade crop, and for hedges in many parts of the world[78]), and many others have VAM, and hence could be preinoculated with selected rhizobia and mycorrhiza in the nursery to produce nodulated mycorrhizal seedlings before planting out in the field during the afforestation programs.

B. Free-Living Nitrogen-Fixing Bacteria

Nearly 25 genera of free-living bacteria can fix atmospheric nitrogen.[79] Species of *Azotobacter*, *Beijerinkia*, *Derxia*, *Clostridium*, and *Azospirillum* are well known among these. Several experiments in temperate regions showed that nitrogen fixation in *Azotobacter* inoculated soil is not more than 10 to 15 kg N/ha/year, depending on the availability of carbon sources. Up to 20% increased yield of vegetable crops have been reported after inoculation with *Azotobacter* under field conditions.[11] During the last decade specific and abundant colonization of the rhizosphere of the *Paspalum* grass with *Azotobacter paspali* has been observed.[80] More recently, *Azospirillum lipoferum* has been found in the cortical cells of the roots of maize where it fixed large quantities of atmospheric nitrogen.[81]

Bagyaraj and Menge[63] studied the interaction between *Azotobacter chroococcum* and the VAM fungus *Glomus fasciculatum* in tomato and found a synergistic effect on plant growth. Mycorrhizal infection increased the *A. chroococcum* population in the rhizosphere which was maintained at a high level for a longer time and *A. chroococcum* enhanced infection and spore production by the mycorrhizal fungus. Similar interactions have also been observed between *A. paspali* and VAM fungi in *Paspalum*,[82] and between *A. chroococcum* and *G. fasciculatum* in tall fescue *(Festuca arundinacea)*[83] and lettuce.[84] Manjunath et al.[85] recently conducted a triple interaction study between the free-living nitrogen-fixing bacterium *Beijerinkia mobilis*, phosphate solubilizing fungus *Aspergillus niger*, and the mycorrhizal fungus *G. fasciculatum* and found a synergistic beneficial effect on the growth of onions with all three organisms. Sometimes the beneficial effect on plant growth from free-living nitrogen-fixing organisms was attributed to hormone production[63,65,86,87] rather than, or in addition to, nitrogen fixation. As pointed out by Mosse et al.,[16] one might speculate that the hormones produced by bacteria could exert some synergistic effect on plant growth or mycorrhizal efficiency. The studies conducted so far reveal a definite synergistic interaction between free-living nitrogen-fixing bacteria and VAM in the rhizosphere with consequential improvement on plant growth. However, there is no information as to how *Azospirillum*, the bacterium of more recent interest, and VAM fungi would interact with each other and in turn affect plant growth. This warrants investigation.

C. Phosphate-Solubilizing Organisms

VAM fungi are well known for their ability to improve P nutrition of plants;[17,71] the source from which they take P and the ability of mycorrhizal plants to use poorly soluble forms of P was a popular topic a few years ago.[7,71] Many soil microorganisms solubilize unavailable forms of P and these bacteria, called "phosphobacteria", have been used as "bacterial fertilizers" with considerable success in the U.S.S.R.,[88] especially on vegetable crops.[11]

Some workers, principally in Spain and India, have studied the interactions between VAM fungi and phosphate-solubilizing microorganisms and their effect on plant growth. Barea et al.[89] found that phosphate-solubilizing bacteria (*Pseudomonas* sp. and *Agrobacterium* sp.) inoculated onto seeds or seedlings maintained high populations longer in the rhizospheres of mycorrhizal than nonmycorrhizal roots of lavender and maize. Lavender plants treated with the VAM fungus and the phosphate-solubilizing bacteria also recorded increased plant dry matter and P uptake in soils amended with rock phosphate.[90] They later found that their phosphate-solubilizing bacteria produced growth hormones which also increased plant growth.[65] Cell-free extracts from a phosphate-solubilizing bacterium (*Pseudomanas* sp.), and also from *Rhizobium meliloti,* increased growth of plants comparable to a mixture of plant hormones containing auxins, gibberellins, and cytokinins.[65] They pointed out that as auxins control root formation, gibberellins increase leaf area and root growth, and cytokinins regulate many basic processes of plant growth, all of which in turn could affect synthesis of mycorrhiza.

Raj et al.[91] studied the effect of *Glomus fasciculatum* and a nonphytohormone-producing strain of the phosphate-dissolving bacterium *Bacillus circulans* on phosphate solubilization, growth of finger millet, and phosphorus uptake from ^{32}P-labeled tricalcium phosphate and superphosphate. Their results clearly revealed that VAM fungi do not solubilize unavailable forms of P but still enhanced the P uptake, which was attributed to a better exploration of soil. Recently, Tinker[17] also suggested that mycorrhizal plants can use rock phosphate more efficiently by lessening the distance between any rock phosphate particle and the nearest absorbing surface. Raj et al.[91] also observed that phosphate solubilizing bacteria survived for a longer period in the rhizosphere of mycorrhizal roots, thus confirming the earlier findings of Barea et al.[89] The phosphate-solubilizing bacteria rendered more P soluble, while mycorrhiza enhanced P uptake; thus with combined inoculation there was a synergistic effect on P supply and dry matter production. Such synergistic interactions have also been observed between *Glomus fasciculatum* and the phosphate-solubilizing fungi *Aspergillus niger* and *Penicillum funiculosum,* with each organism stimulating the growth of the other with consequential improvement in growth and P, K, Zn, and Mn nutrition of finger millet.[92] Recently Baya et al.[93] found that a large percentage of phosphate-solubilizing bacteria also produce vitamins such as B_{12}, riboflavin, niacin, pantothenate, and biotin. Ecto- and orchidaceous mycorrhizal fungi[94] have been shown to be dependent on and stimulated by certain vitamins. Although roots may provide an adequate supply of vitamins for mycorrhizal development, rhizosphere and rhizoplane bacteria could also be a source of vitamins for fungi. The high incidence of vitamin-producing bacteria[95] and actinomycetes[5] in the rhizosphere and the greater amount of vitamins they produce suggest that vitamins synthesized by these organisms may contribute significantly to mycorrhizal development. If so, inoculation with mycorrhizal fungi and vitamin-producing organisms could result in improved plant growth.

All the interaction studies between VAM fungi and phosphate-solubilizing organisms have been done in pots; there is need for extending these studies to the field using cheaper forms of phosphate. If the synergistic interaction observed under pot trials could be repeated under field conditions it would help to cut down the cost on P fertilizers in practical agriculture.

D. Beneficial Actinomycetes

Very few studies have been carried out on the interaction between VAM fungi and actinomycetes. An increase in the total actinomycete population in the rhizosphere of mycorrhizal tomato plants was first reported by Bagyaraj and Menge,[63] and this was subsequently confirmed by a similar observation in finger millet.[92] Krishna et al.[96] recently studied the interaction between the VAM fungus *Glomus fasciculatum* and the actinomycete *Streptomyces cinnamomeus* (known for its stimulatory effect on plant growth[97]) introduced into the rhizosphere of finger millet and its effect on plant growth. Individual inoculation of either organism resulted in enhanced plant growth compared to uninoculated control. Simultaneous inoculation with both the organisms had an antagonistic effect on each other, each suppressing the growth and multiplication of the other in the rhizosphere, resulting in decreased plant growth compared to single inoculation. It was suggested that the antibiotics produced by *Streptomyces cinnamomeus* (which are inhibitory to some fungi in the Mucorales and may be also *Glomus fasciculatum*) could be responsible for the suppression of *G. fasciculatum* in the rhizosphere.

The actinomycete provisionally accepted as the genus *Frankia* is known to produce nitrogen fixing nodules on the roots of nonlegumes.[98,99] The plants forming actinorhizal nodules are typically tripartite associations.[100-102] Twenty-five (25) species of these plants representing 11 genera were found to be colonized by either ectomycorrhizal fungi, VAM fungi, or both ecto- and VAM fungi.[103] The actinomycete is housed in nodules where atmospheric nitrogen is fixed and made available to the host plant; the mycorrhizal fungus is both inter- and intracellular within the root tissue and may be found within the nodules. In an interaction study, snowbrush plants dually inoculated with *Glomus gerdemanii* and an actinomycete showed increased total dry weight of shoots and roots; number of nodules; weight of nodular tissue; as well as levels of N, Ca, and P; and increased N fixation. While legumes are the primary means of adding fixed atmospheric nitrogen to agricultural soils, nonlegumes are the major means of adding nitrogen to forests, bogs, and arid areas, especially in temperate regions of the world. Economically important nonlegumes belonging to the genera *Alnus, Myrica, Casurina,* and *Ceanothus* are some typical examples.[103] The indication that VAM can improve the growth of nodulated nonlegumes through enhanced nitrogen fixation as well as phosphate uptake opens up the possibility of exploiting such tripartite associations in certain ecosystems where soils have less mycorrhizal fungi and/or nitrogen-fixing microflora and are low in nitrogen and phosphorus, such as mine spoils and eroded or disturbed soils.

III. INTERACTIONS BETWEEN VAM AND PLANT PATHOGENIC ORGANISMS

A. Fungi
1. Root Pathogens

Although the VAM fungi have long been found in association with diseased roots, the first reported attempt to specifically study the interaction of a plant pathogenic fungus and a species of VAM fungus was that of Safir[104] in 1968. Since then several studies have been made. Most of these studies suggest that mycorrhizae decrease the severity of disease caused by plant pathogenic fungi, while a few suggest that there is either no effect or there is an increase in severity of disease due to mycorrhizal infections.

a. VAM Supporting the Severity of Disease

Ross[105] was the first to report the increase in severity of a plant disease due to a VAM fungus, *Glomus macrocarpum* var. *geosporum*. Nearly 90% of the susceptible soybean plants with mycorrhizae showed internal stem discoloration symptoms of *Phytophthora* root rot, while less than 20% of the nonmycorrhizal plants developed these symptoms. It was pointed out that very large vesicles and chlamydospores produced by the mycorrhizal fungus might have damaged host tissue, thus creating an avenue for the penetration and development of the pathogen. The soil used was also rich in phosphorus, probably accounting for the lack of response to mycorrhizae. Davis et al.[106] noted that the incidence of *Verticillium dahliae* was greater in cotton plants infected with *G. fasciculatum* than those in nonmycorrhizal plants. In another study, mycorrhizal plants were not found to be resistant to *Phytophthora* rot when citrus, avocado, and lucerne seedlings were inoculated with such a heavy inoculum of *P. parasitica* that pathogen-treated plants grew very little.

b. VAM With No Effect on Disease

There are a number of reports in which VAM had no effect on disease severity. Ramirez,[107] studying the interaction between mycorrhizal fungi and *Phytophthora palmivora* in papaya, observed that the mycorrhizal fungi, *Glomus macrocarpum* var. *macrocarpum* and *Gigaspora heterogama*, did not reduce the infection of roots by the pathogen. Mycorrhizal citrus plants offered no protection against the pathogen, *Phytophthora parasitica*.[108] Similarly, there was no significant reduction in *Thielaviopsis* root rot of citrus[109] or take-all of wheat[110] because of mycorrhizal inoculation. Although Menge et al.[108] observed a preferential attraction of *P. parasitica* zoospores toward mycorrhizal citrus roots compared to nonmycorrhizal roots, Hall and Finch[111] did not observe a greater chemotactic response of the zoospores of the same pathogen to mycorrhizal roots of avocado.

c. VAM Decreasing the Severity of Disease

Most reports in the literature indicate that VAM fungi decreased disease severity. Schonbeck and Dehne[112] found that mycorrhizal cotton plants could withstand the stress of infection by the pathogen *Thielaviopsis basicola* better than nonmycorrhizal plants. Baltruschat and Schonbeck[113] obtained a 10-fold decrease in chlamydospore production of *T. basicola* on mycorrhizal tobacco plants at low inoculum levels (10,000 endoconidia per milliliter). Extracts from roots of mycorrhizal plants inhibited chlamydospore production of *T. basicola* on malt extract agar. In later studies, they reported that the chlamydospore production of *T. basicola* was negatively correlated with mycorrhizal colonization of the roots.[114] Chou and Schmitthenner[115] conducted a complex interaction study between a VAM fungus, root nodule bacterium, and two root pathogenic fungi, *Phythium ultimum* and *Phytophthora megasperma* var. *sojae*. They reported that the presence of the mycorrhizal fungus reduced the number of plants killed by *P. megasperma* var. *sojae*, which was confirmed by a later study.[116] Prior root colonization by the mycorrhizal fungi, *Gigaspora margarita* or *Glomus macrocarpum* reduced the damage caused by *Phytophthora parasitica* to two citrus root stocks, Carrizo citrange, and sour orange.[117] To ensure good mycorrhizal establishment on citrus roots, plants were exposed for 110 days to mycorrhizal fungi before challenging them with the pathogen.

Paget,[118] working with *Cylindrocarpon destructans,* a mild pathogen of strawberry, obtained less plant stunting and reduced root infection when roots were colonized by mycorrhizal fungi. Prior root colonization by the mycorrhizal fungus, *Glomus mosseae* reduced the damage to tomato by *Fusarium oxysporum* f. sp. *lycopersici.*[119] In poinsettia,[120] simultaneous inoculation with a mycorrhizal fungus and the pathogens (*Pythium ultimum* and *Rhizoctonia solani*) had no effect, while preinoculated mycorrhizal plants offered resistance to the infection by the pathogens added 20 days later. The presence of *G. mosseae* in the roots of onion increased resistance to *Pyrenochaeta terrestris*.[104,121] In a field survey, the incidence of mycorrhizae was higher in onion fields with low levels of *P. terrestris* than in fields with high levels of *P. terrestris*.

Becker[121] challenged individual mycorrhizal and nonmycorrhizal roots on the same plant with *P. terrestris* and found that the inward invasion of the pathogen was restricted more on mycorrhizal roots. He attributed this to the cell wall thickenings (callosities or lignitubers) of the host at the point of penetration by the pathogen. Mycorrhizae have also been reported to reduce the severity of disease caused by *Olpidium brassicae* on lettuce and tobacco,[122] *Rhizoctonia solani* on *Brassica naprus*,[123] and *Codinella fertilis* on clover.[124] Davis and Menge[125] found that mycorrhizal inoculation offered protection to citrus plants inoculated with *Phytophthora parasitica* at a low concentration (20 chlamydospores per gram of soil). An experiment with seven mycorrhizal fungi on two citrus root stocks showed that different VA infections might confer variable tolerance or resistance on *P. parasitica*. Reduction in the severity of take-all of wheat due to VAM was observed when a low inoculum level of the pathogen, *Gaeumannomyces graminis* var. *tritici* (0.1% W/W, gram of oat kernel inoculum per gram of soil) was used in a P-deficient soil.[126] Recently Krishna and Bagyaraj[127] found that the mycorrhizal fungus *Glomus fasciculatum* added with the root rot pathogen *Sclerotium rolfsii* reduced the severity of disease and also the number of sclerotia produced by the pathogen in peanut plants. In microplot nurseries in the field Barnard[128] observed fewer dead plants and 50% less root infection caused by *Cylindrocladium scoparium* in mycorrhizal yellow poplar plants compared with nonmycorrhizal plants.

2. Foliar Pathogens

There is only one report on the effect of VAM on the incidence of disease caused by fungi on the aerial parts of the plant.[122] Disease severity of *Helminthosporium sativum* and *Erysiphe graminis* on barley, of *Colletotrichum lindemuthianum* and *Uromyces phaseoli* on french bean, of *Erysiphe cichoracearum* on cucumber, and of *Botrytis cinerea* on lettuce was increased in mycorrhizal compared with nonmycorrhizal plants.

B. Bacteria

There are only two reports in which plant susceptibility to bacterial pathogens was studied in relation to the incidence of VAM fungi in roots. Weaver and Wehunt[129] found no relationship between the percentage of mycorrhizal roots of peach seedlings and their susceptibility to bacterial canker caused by *Pseudomonas syringae*. In another study[130] in the Philippines, mycorrhizal inoculation was found to reduce the severity of the bacterial wilt of tomatoes caused by *Pseudomonas solanacearum*, an important soil pathogen occurring throughout the world.

C. Viruses

There have been only a few reports on virus infections in mycorrhizal and nonmycorrhizal plants.[122,131,132] In all cases the resistance of the plants to viruses appeared to be reduced by the mycorrhizae. The numbers of local lesions caused by tobacco mosaic virus were increased in mycorrhizal plants. The virus titer of tomato mosaic virus and potato virus X in tomato, and arabis mosaic virus in petunia and strawberry leaves and roots increased because of mycorrhizal inoculation. Tobacco plants were grown with a split root system, with one half being mycorrhizal. After virus inoculation of leaves, a higher virus content was found in the mycorrhizal than in nonmycorrhizal roots.[122] Individual root cells invaded by the mycorrhizal fungus could still act as host cells for tobacco mosaic virus. The virus multiplication increased remarkably in root cells containing the fungus in its arbuscular stage.[133]

D. Nematodes

The first interaction study between a nematode *(Heterodera solanacearum)* and the

VAM fungus *Gigaspora gigantea* reported that the mycorrhizal fungus increased the susceptibility of tobacco to the nematode.[134] Since then several studies have been made.[135-138] Diversity of interactions between VAM and plant parasitic nematodes have been observed, but in most cases mycorrhizae have reduced the severity of disease caused by plant parasitic nematodes. Based on the feeding habits plant parasitic nematodes can be placed into three groups: (1) sedentary endoparasites, (2) migratory endoparasites, and (3) ectoparasites. Sedentary endoparasites penetrate roots as vermiform juveniles; after feeding commences, the body swells and the nematode becomes immobile. Further development and reproduction by the nematode depends on the modified plant cells called giant cells or syncytia adjacent to its head providing nourishment. Migratory endoparasites remain vermiform throughout their life cycle. They penetrate and migrate throughout the root tissue, feeding on different cells without establishing a permanent feeding site. Ectoparasites remain outside the root, using their stylet to feed on internal cells. So far, no investigations involving ectoparasites and VAM fungi have been reported. The interactions between VAM and nematodes belonging to these two feeding groups are discussed below.

1. Sedentary Endoparasites

Most of VAM and nematode interaction studies have been made with sedentary endoparasites, especially with the root knot nematodes belonging to the genus *Meloidogyne*. Atilano et al.[139] observed that the presence of the nematode *Meloidogyne arenaria* completely negated the beneficial effect of the mycorrhizal fungus *Glomus fasciculatum* in grapevine. In studies involving inoculation of soybean with *M. incognita* and three VAM fungi, Schenck et al.[140] found that any given interaction was strongly influenced by nematode inoculum level, cultivar resistance to the nematode, and the species of fungal symbiont. The fungi used were *Gigaspora heterogama, Glomus macrocarpum,* and *Endogone (Gigaspora) calospora*. Although the endophytes tested usually stimulated plant growth and increased nematode densities, nematode reproduction was influenced differently by each symbiont. Peanut was recently[138] found to be a more suitable host for *M. arenaria* in the presence of the mycorrhizal fungi *Glomus etunicatum* or *Gigaspora margarita*. These symbionts caused an eightfold increase in number of eggs per gram of root and pods compared with reproduction in nonmycorrhizal plants. The mycorrhizal stimulation of plant growth was not affected by the nematodes, suggesting that the symbionts made the host plant tolerant to the parasite. Other tests using peanut plants with split root systems revealed that nematode reproduction was increased only when both organisms were present on the same side of the root system.

VAM can also suppress the activity of plant parasitic nematodes. Baltruschat et al.[141] reported that only 75% of the larvae of the root knot nematode *Meloidogyne incognita* developed into adults in mycorrhizal tobacco plants. Sikora and Schonbeck[142] also obtained a significant reduction in the initial number of *M. incognita* larvae that developed into adults on mycorrhizal carrot, tobacco, tomato, and oat plants. The population of *M. hapla* remained suppressed on mycorrhizal carrot roots for up to 18 weeks. After observing mycelia of the mycorrhizal fungi in the nematode galls, it was suggested that the presence of the mycorrhizal fungus plus physiological changes induced by the fungus on the host could be the cause for the depressed nematode population. In this experiment the plants were preinoculated with the mycorrhizal fungus and later with the nematodes.[142] The presence of *G. macrocarpum* significantly reduced the number of *M. incognita* galls in soybean.[143] Vesicles and arbuscules were observed in the nematode galls[143] but not in giant cells.[138]

When nematode-resistant and nematode-susceptible cotton cultivars were inoculated jointly with *Gigaspora margarita* and *Meloidogyne incognita* the beneficial effect of mycorrhiza offset the nematode damage to the susceptible cultivar.[144] This effect was attributed directly to host vigor rather than to any antagonistic physiological change induced by symbiosis. This conclusion was supported by similar numbers of eggs per gram of root on both mycorrhizal and nonmycorrhizal plants. Therefore, mycorrhizal plants, because of larger root systems supported a greater root knot nematode population than did nonmycorrhizal plants. The nematodes did not suppress mycorrhiza synthesis or fungal sporulation. The overall effect of mycorrhizal infection was simply to increase the tolerance of the susceptible cultivar to the nematode pathogen.[144] In our study,[145] inoculation of tomato with the mycorrhizal fungus *G. fasciculatum* significantly reduced the number of nematode galls produced by *M. incognita* and *M. javanica*. Plants inoculated with nematodes only developed many large-sized multiple galls, while those inoculated with mycorrhiza plus nematodes developed small primary galls. Cooper[124] has recently found that mycorrhizal inoculation reduced infection of tomato, clover, and lucerne plants by *M. hapla* and offset detrimental effects on plant growth caused by the nematode.

Hussey and Roncadori[138] found that *Glomus etunicatum* and *G. mosseae* differed from *Gigaspora margarita* in their effects on the reproduction of the root knot nematode *M. incognita* on a susceptible cotton cultivar. The total number of *M. incognita* (eggs per plant) was not increased on cotton colonized by *G. etunicatum* or *G. mosseae* over control plants even though the root systems of the mycorrhizal plant were considerably larger than those of the nonmycorrhizal plants. Therefore, when rated as eggs per gram of root, reproduction was lower on mycorrhizal plants than on controls, indicating that these symbionts make cotton resistant to *M. incognita*. In another study with soybean,[138] they found that mycorrhizal fungi stimulated plant growth, and that *M. incognita* reproduction (eggs per gram of root) was unaffected by *G. margarita* and suppressed by *G. etunicatum*. This indicated that the former makes the plant tolerant to the nematodes and the latter increases plant resistance. The work of O'Bannon et al.[146] with rough lemon provides another example of a nematode—symbiont (*Tylenchulus semipenetrans-Glomus mosseae*) interaction in which mycorrhizal infection offsets a loss in plant growth caused by nematodes by increasing the plant's nematode tolerance.

In a recent study on grapevine cuttings,[147] the nematode *M. arenaria* reduced root colonization by *G. fasciculatum* and also the growth of mycorrhizal and nonmycorrhizal cuttings. At low nematode inoculum levels (approximately 200 eggs per plant) the presence of mycorrhizae enhanced plant growth, but no significant benefit was achieved by mycorrhizae when high nematode inoculum (approximately 2000 eggs per plant) was used. This suggests that the number of nematodes in soil could play an important role in deciding the possible biological control by mycorrhizal fungi. Further, the sequence in which plants become colonized by the mycorrhizal symbiont and infected by nematodes may affect the interaction between these organisms.[148] In a recent study,[138] peach seedlings preinoculated with VAM were found to be more resistant to the root knot nematode than plants inoculated with both organisms concurrently. Application of mycorrhiza (*G. fasciculatum*) first, followed by nematodes (*M. incognita*) was also found to reduce nematode infection better than simultaneous application of mycorrhiza plus nematode or nematode first followed by mycorrhiza in tomato.[149]

2. Migratory Endoparasites

Less is known about interactions between VAM and migratory endoparasites. Cotton was a less suitable host for the migratory endoparasitic nematode, *Pratylenchus brachyurus*, when the roots were colonized by the mycorrhizal fungus *G. margarita*.[150] Nematode counts per gram of root were significantly less for mycorrhizal plants than for nonmycorrhizal plants. Decreased nematode reproduction was caused by the mycorrhizal fungus altering the cortex to make it an unfavorable food source for the nematode or by the fungus competing with the nematode for space in the cortex. Enhanced plant growth due to mycorrhizal inoculation was unaffected by *P. brachyurus*. The only other report of a migratory endoparasite interacting with a mycorrhizal fungus involved *Radophilus similis* and *G. etunicatum* on rough lemon.[151] Reproduction of the nematode was similar on mycorrhizal and nonmycorrhizal seedlings; however, seedlings inoculated first with the mycorrhizal fungus and then with nematodes were 28 times larger than seedlings inoculated with nematodes alone at the end of the experiment.

In the studies discussed above, the effects of the nematodes on sporulation by the VAM fungi were variable. In some studies fewer spores were formed, but in others no differences were found. In most studies, however, nematodes had little influence on root colonization by the mycorrhizal fungi.

E. Mechanisms of Suppression of Pathogens by VAM

It is apparent from the investigations on VAM-plant pathogen interactions that VAM can usually (though not always) deter or reduce the severity of disease caused by soil-borne plant pathogens. On the other hand, the few studies conducted have indicated that mycorrhizae make plants more susceptible to fungal[136] and viral[132,133] diseases on the aerial parts of the plant. More studies are needed to confirm this. Wilhelm[152] pointed out that interactions between VAM and plant pathogens in the rhizosphere have important implications in biological control. Like most instances of biological control, mycorrhizae can never confer complete immunity against any root disease. They could impart a degree of resistance or tolerance against soil-borne plant pathogenic fungi, bacteria, and nematodes. Most of the evidence is however from laboratory, greenhouse, or microplot studies, suggesting the need for extending these studies under actual field situations. Various mechanisms by which VAM could suppress root pathogens have been postulated and a few demonstrated.

Resistance in plants could be due to morphological alterations caused by mycorrhiza on the host plant. Thickening of the cell walls through lignification and production of other polysaccharides in mycorrhizal plants preventing the penetration and growth of the pathogens *Fusarium oxysporum*[153] and *Phoma terrestris*[121] have been demonstrated. A stronger vascular system observed in mycorrhizal plants[136] will increase the flow of nutrients, impart greater mechanical strength, and diminish the effect of vascular pathogens. High chitinase activity of the mycorrhizal tissue also may confine the growth of the pathogen in the host. Histopathological studies of nematode galls caused by *Meloidogyne incognita* showed that galls in mycorrhizal plants had fewer giant cells or syncytia, which are needed for the development of nematode larvae, compared to nonmycorrhizal plants.[149] These studies indicated that the nematodes in mycorrhizal plants were smaller and took a longer time to develop into adults.[149] Smaller syncytia with fewer cells have been reported to confer resistance on the host against nematodes.[154,155]

Alteration in the severity of plant disease in mycorrhizal plants may be due to physiological and biochemical changes in the host induced by the mycorrhizal fungi. It is very well established that mycorrhizal plants usually contain higher concentrations of phosphorus than nonmycorrhizal plants.[15,71] Addition of supplementary phosphorus to plants inoculated with nematodes offset the nematode symptoms and paralleled the influence of VAM.[138] In another recent study[126] with VAM and the take-all fungus, *Gaeumannomyces graminis* var. *tritici*, high phosphate status of the mycorrhizal roots was found to decrease root exudation in wheat plants which influenced the infection process of the roots of the pathogen. Ratnayake et al.[156] showed that high P status in the host plant increased the phospholipid content which in turn decreased the root membrane permeability and led to less root exudation. The phosphate-induced decrease in root exudation was correlated with a subsequent decrease in take-all disease severity. The influence of soil and mycorrhiza on take-all thus appeared to be the same.

At this stage it would be appropriate to mention that altered root exudation in mycorrhizal plants in turn could affect the microflora in the rhizosphere. Quantitative changes in the microbial population in the rhizosphere of tomato due to VAM inoculation have been reported.[63] Studies on the qualitative changes in the microbial population in the rhizosphere due to mycorrhizal inoculation are needed especially to know the selective stimulation, if any, of antagonists against root pathogens.

In virus-infected plants, the increase in susceptibility and in virus titer may be due to increased phosphate levels in the mycorrhizal plants. There are indications that successful infections by viruses and their multiplication are correlated with improved phosphate supplies to the plant.[132,157] This is possible considering the role of phosphate in nucleic acids, which form the important constituent of plant viruses.

Higher levels of amino acids, especially arginine, have been recorded in mycorrhizal roots of several plant species. Arginine and root extracts from mycorrhizal plants both reduced *Thielaviopsis basicola* chlamydospore production.[114] Blockage in the ornithine cycle by the mycorrhizal fungus has been proposed as the cause of increased arginine levels. Sikora[158] noted decreased penetration and slower development of root knot larvae in mycorrhizal roots and suggested alteration of the root physiology by the mycorrhizal fungus as the reason. Increased phenylalanine and serine in the tomato roots due to inoculation with *G. fasciculatum* has recently been observed.[149] These two main amino acids are known to be inhibitory to root knot nematodes.[159-161]

The larger amounts of reducing sugars found in onion roots may explain a lower incidence of pink root rot disease.[104] Krishna and Bagyaraj's[127] recent study revealed a high concentration of ortho-dihydroxy (O-D) phenols in mycorrhizal plants compared to nonmycorrhizal plants. Addition of O-D phenol at an equivalent concentration to that found in mycorrhizal roots was inhibitory to the growth of *S. rolfsii* in vitro.

This diversity of interactions between VAM and plant pathogenic microorganisms shows that each pathogen-mycorrhizal fungus-plant combination is unique and generalizations regarding such interactions are difficult to make. However, the possibility of biologically controlling the root pathogens looks promising. Many root pathogens can currently be controlled only with expensive physical or chemical soil treatments. As recently pointed out by Schenck[137] mycorrhizae offer an alternative approach, and we should pursue their potential as biological control agents despite the obstacles. The main obstacle is the inability to produce large quantities of the inoculum, as these fungi are obligate symbionts. Despite that limitation, VAM may be used in certain crops where seedlings are raised in a nursery and then planted in the field (e.g., finger millet, chillies, tobacco, tomato, asparagus, citrus, and other horticultural crops). Small quantities of mycorrhizal inoculum added to the nursery bed would result in healthier mycorrhizal seedlings which would not only be able to withstand transplant shock better, but also challenge the infection by root pathogens and grow better.

IV. MICROORGANISMS INTIMATELY ASSOCIATED WITH VAM

A. Companion Fungi

Williams[162] has recently observed certain "companion fungi" living in close association with VAM. He has isolated these "companion fungi" from 13 out of 14 lines of VAM fungi in pot culture. Isolates differ from one another in cultural and morphological features and in their effect on plant growth. In glasshouse tests, different isolates stimulated (by up to 95%), inhibited (by up to 50%) or had no effect on growth, depending on the host species and the soil. In a field trial with pasture in a low P soil, one isolate of the companion fungus gave a 3-fold increase in dry matter, equivalent to an application of 250 kg/ha superphosphate. These observations question whether the improved growth obtained by inoculating plants with roots and soil from a pot culture of VA endophyte is due to VAM fungus alone or a cumulative effect of the mycorrhizal fungus and the companion fungus. The overall significance of this discovery will now have to await much more detailed study of the companion fungi.

B. Bacterial Associates

Epiphytic associations of an *Azotobacter* sp. with spores of *Glomus fasciculatum* have been observed.[83,163] Mosse[164] isolated *Pseudomonas* sp. as a common associate with VAM spores, and found the bacterium helped the mycorrhizal fungi in infecting the roots. It was suggested that it could be due to either the production of enzymes or growth-promoting substances which increase the cell wall plasticity and thereby increase fungal infections.[165] The bacteria associated with different endophytes were even found to have a stimulatory effect on onion growth.[166] Compared with the effects of the endophytes themselves, these effects were small, but nevertheless statistically significant in increasing the plant dry matter production.

C. Bacterium-Like Organelles

The occurrence of apparently nonpathogenic bacterium-like organelles (BLOs) in the cytoplasm of fungi is a recently recognized phenomenon,[167] but was first noticed in the VAM fungus *Acaulospora laevis* by Mosse[168] in 1970. They are referred to as BLOs because they look like bacteria but they have not been cultured and thus their bacterial nature is debatable. However, on the basis of their morphology (i.e., size and apparent division patterns) and cytology (i.e., plasmalemma, Gram-positive type cell wall, lack of classical nucleus and nuclear membrane, and presence of ribosome-like structures), they could be tentatively regarded as bacteria.[169] BLOs have been observed in the cytoplasm of *Endogone flammicorona*,[170,171] *Glomus fasciculatum*,[172] *Glomus caledonicum*,[169] an unidentified *Glomus* sp.,[173] and two unidentified VAM fungi.[174,175] The BLOs described by these authors are either rod-shaped and enclosed in vacuoles,[172,173,175] or irregularly coccal and not found in fungal vacuoles and hence free in the cytoplasm.[168,169,171,174] The recent work on BLOs of *G. caledonicum* mycorrhizal with onions revealed that they are found in the cytoplasm of intercellular hyphae, arbuscules, and resting spores, and often lie close to the fungal nucleus.[169] Smith[176] regards prokaryotic cytobionts without host membranes (such as BLOs) as probably noncultivable. In the light of this opinion, success in culturing BLOs from VAM fungi is unlikely. With the current information it is difficult to interpret the role played by BLOs in VAM fungi.

V. ORGANISMS PARASITIC AND PREDATORY ON VAM

A. Parasitic Microflora

As early as 1922, Thaxter[177] observed that chlamydospores of the mycorrhizal fungus *Endogone* were filled with masses of hyphae. This was later confirmed and the parasite identified as *Cephalosporium*.[178] Mosse[179] observed fine hyphae within the chlamydospores of *Endogone (Glomus)* spp. associated with strawberry roots, and considered it to be an actinomycete parasite belonging to the genus *Micromonospora*. Godfrey,[180] in illustrating the parasitism of chlamydospores of mycorrhizal fungi, correlated the parasite with the presence of fine radial canals in the chlamydospore walls. She believed that the parasite penetrated the chlamydospore through these canals. Describing different spore types of mycorrhizal fungi, Gerdemann and Nicolson[181] and Mosse and Bowen[182] also reported the presence of internal spore-like structures within mycorrhizal spores. The occurrence of a parasite on the azygospores of *Gigaspora margarita* and other mycorrhizal fungi was reported by Schenck and Nicolson[183] and identified as the chytrid *Rhizidiomycopsis stomatosa*.[184] The parasite was not obligately parasitic but commonly occurred on azygospores collected from "pot cultures" of *G. margarita* maintained in the glasshouse or from poorly drained fields. Addition of 5% unsterilized field soil to sterilized or fumigated soil was found to reduce the number and size of chalmydospores formed by two VAM fungi.[185] The "suppressive factor" in unsterilized soil was nondiffusive and thermolabile, and speculated to be of biological nature. Transmission electron microscopy (TEM) of *Glomus caledonicum* spores revealed the presence of a variety of bacteria on the outer and middle layers of the three-layered spore wall. These bacteria eroded the spore wall suggesting them to be parasites.[169] The deep location of these bacteria could probably explain the difficulties encountered in sterilization of mycorrhizal spores in the artificial synthesis of axenic VAM. Recently, Bhattacharjee et al.[186] studied the fine structure of the azygospores of *Gigaspora candida* using scanning electron microscopy (SEM) and illustrated perforations in the spore wall caused by certain hyphae and sporulating structures of an actinomycete-like organism.

Ross and Ruttencutter[187] investigating population dynamics of two VAM fungi *Glomus macrocarpum* var. *geosporum* and *Gigaspora margarita* in soil, observed that chlamydospore production was reduced by two hyperparasites, a species of *Phylyctochytrium* and a *Pythium*-like fungus. *Glomus macrocarpum* was more susceptible to hyperparasitism than *Gigaspora margarita*. These hyperparasites were postulated to play an important role in limiting populations of VAM fungi. Hyperparasitization in pot cultures of *Glomus epigaeum* and *G. fasciculatum* by three fungi, *Anguillospora pseudolongissima*, *Humicola fuscoatra*, and *Phylyctochytrium* sp. was reported by Daniels and Menge.[188] They found that VAM fungi with light-colored spores are more susceptible to attack by parasitic fungi. Surface sterilization of the mycorrhizal spores with sodium hypochlorite (which is commonly used for sterilizing spores) increased the susceptibility suggesting the need for looking into an alternate method for surface sterilization of mycorrhizal spores. The fungicide Mancozeb® inhibited the growth of the hyperparasite without entirely inhibiting the VAM fungi.

The parasites of mycorrhizal fungi have been referred to as hyperparasites by some workers[180,186,188] (see above). Here I would like to point out that hyperparasite is a term usually used to define a parasite attacking another parasite. Since VAM fungi are usually beneficial symbionts and not parasites, it would be better to call these "organisms parasitic on VAM fungi" rather than "hyperparasites of VAM fungi". Whatever the terminology, the ecological significance of this parasitic association is just beginning to emerge.[186-188] The presence of parasitized "pot cultures" of VAM inoculum may explain some erratic results which occur in tests with VAM. Methods for the commercial production of VAM fungi have been designed.[189] Parasites of VAM fungi could be a serious problem in the commercial production of VAM and no doubt play a role in the variable results provided by different batches of mycorrhizal inoculum. Menge et al.[190] observed increased mycorrhizal infection and sporulation due to the application of two pesticides and speculated that this resulted from reduced population of VAM parasites.[188] Plants such as citrus and sweetgum are extremely dependent on mycorrhizal fungi for survival.[188] Reduction of mycorrhizal infection in these plants due to parasitism of the VAM fungi used, in effect, could reduce the growth of these plants. In this respect, parasites of VAM fungi could be considered as secondary plant pathogens, even though they are not primary pathogens of higher plants.[188] More studies on the presence and activity of these parasites are desirable in view of the importance of VAM in agriculture, forestry, and revegetation of disturbed areas.

B. Predatory Microfauna

The collembola, *Folsomia candida* has recently been observed eating the external hyphae of the mycorrhizal fungus *Glomus fasciculatum*, and thereby reducing the effectiveness of the mycorrhiza on leeks.[191] It was suggested that this feeding activity of the soil microfauna may be one of the reasons as to why it is much easier to demonstrate increases in yield due to mycorrhiza in pots containing sterilized soil than it is in the field. Mycophagous nematodes, *Aphelenchoides* spp. feeding on VAM fungi and thereby controlling the inoculum density of the VA propagules is currently being investigated.[192] More studies are needed on the interactions between VAM fungi and soil fauna with the ultimate objective of a better understanding of the significance of VAM in natural ecosystems.

VI. CONCLUSIONS

Certain generalizations and conclusions can now be drawn. VAM markedly improve nodulation and nitrogen fixation by legume bacteria mainly by providing the high phosphorus requirement for the fixation process. Mycorrhizal infection also allows introduced populations of beneficial soil organisms like *Azotobacter* and phosphate-solubilizing bacteria to maintain higher numbers than around nonmycorrhizal plants, and to exert synergistic effects on plant growth. In general, VAM decrease the severity of root diseases. Most of these studies have been conducted in pots and hence there is a need for exploiting this potential under field conditions. To begin with, experiments could be done with plants important in agriculture, horticulture, and forestry which are usually raised in nursery and then planted out in the field. As only small quantities of the mycorrhizal inoculum would be required, the method could be applied immediately to practical farming without much difficulty.

While the inability to culture VAM fungi in vitro remains the single most inhibitory factor in VAM research, the recent discovery of the culturable "companion fungi" associated with VAM warrants concentrated research to clarify the situation. Research on parasites and predators of VAM should be intensified as they are probably going to play an important role in the success or failure of VAM inoculum trials in the field and also on survival and persistence of indigenous and introduced endophytes. This brings out the need for an international culture collection center or world bank of mycorrhizal fungi, which would maintain "clean inoculum" available for experimentation in any part of the world.

Finally, it should not be forgotten that rhizosphere is a complex region in the soil-plant interface with high microbial activity. We are attempting to manipulate the organisms in the rhizosphere for the benefit of plant growth. This is not easy to achieve, as we are trying to change a population in which every member has its own niche. However, the results obtained so far with biological interaction studies between VAM fungi and other soil organisms are encouraging and vindicates the need for strengthening research in this area.

REFERENCES

1. Hiltner, L., Uber neure Erfahrungen und Probleme auf dem Gebiet der Bodenbakteriologie und unter besonderer Berucksichtigung der Grundingung und Brache, *Arb. Dtsch. Landwirtsch. Gesellschaftswiss*, 98, 59, 1904.
2. Parkinson, D., Soil microorganisms and plant roots, in *Soil Biology*, Burges, A. and Raw, F., Eds., Academic Press, London, 1967, 449.
3. Bowen, G. D. and Rovira, A. D., Microbial colonization of plant roots, *Annu. Rev. Phytopathol.*, 14, 121, 1976.
4. Bagyaraj, D. J. and Rangaswami, G., Studies on the rhizosphere microflora of *Eleusine coracana* under the influence of foliar chemical sprays. I. Quantitative incidence of microorganisms, *Madras Agric. J.*, 59, 517, 1972.
5. Bagyaraj, D. J. and Rangaswami, G., Studies on the rhizosphere microflora of *Eleusine coracana* under the influence of foliar chemical sprays. III. Actinomycete flora of the rhizosphere, *Madras Agric. J.*, 39, 529, 1972.

6. Rovira, A. D., Biology of soil-root interface, in *The Soil-Root Interface*, Harley, J. L. and Russell, S. R., Eds., Academic Press, London, 1979, 145.
7. Tinker, P. B., Role of rhizosphere microorganisms in phosphorus uptake by plants, in *The Role of Phosphorus in Agriculture*, Khasawneh, F. E., Sample, E. C., and Kamprath, E. J., Eds., American Society of Agronomy, Madison, Wisc., 1980, 617
8. Rovira, A. D. and McDougall, B. M., Microbiological and biochemical aspects of the rhizosphere, in *Soil Biochemistry*, Vol. 1., McLaren, A. D. and Peterson, G. H., Eds., Marcel Dekker, New York, 1967, 417.
9. Nye, P. H., The rate limiting step in plant nutrient absorption from soil, *Soil Sci.*, 123, 292, 1977.
10. Rangaswami, G., The factors influencing rhizosphere microflora of crop plants and their significance in plant disease control, *Indian Phytopathol. Soc. Bull.*, 4, 120, 1968.
11. Brown, M., Seed and root bacterization, *Annu. Rev. Phytopathol.*, 12, 181, 1974.
12. Rovira, A. D. and Davey, C. B., Biology of the rhizosphere, in *The Plant Root and its Environment*, Carson, E. W., Ed., University Press of Virginia, Charlottesville, 1971, 153.
13. Brockwell, J., Experiments with crop and pasture legumes - principles and practice, in *Methods for Evaluating Biological Nitrogen Fixation*, Bergersen, F. J., Ed., John Wiley & Sons, New York, 1980, 417.
14. Moore, L. W. and Cooksey, D. A., Biology of *Agrobacterium tumefaciens:* plant interactions, in *Biology of the Rhizobiaceae*, Giles, K. L. and Atherly, A. G., Eds., Academic Press, New York, 1981, 15.
15. Bowen, G. D., Misconceptions, concepts and approaches in rhizosphere biology, in *Contemporary Microbial Ecology*, Ellwood, D. C., Hedger, J. N., Latham, M. J., Lynch, J. M., and Slater, J. H., Eds., Academic Press, London, 1980, 283.
16. Mosse, B., Stribley, D. P., and Le Tacon, F., Ecology of mycorrhizae and mycorrhizal fungi, *Adv. Microb. Ecol.*, 5, 137, 1981.
17. Tinker, P. B., Mycorrhizas: the present position, in *Whither Soil Research, Trans. 12th Int. Congr. Soil Sci.*, New Delhi, 1982, 150.
18. Janse, J. M., Les endophytes radicaux des quelques plantes Javanaises, *An. Jard. Bot. Buitenz*, 14, 53, 1896.
19. Jones, F. R., A mycorrhizal fungus in the roots of legumes and some other plants, *J. Agric. Res.*, 29, 459, 1924.
20. Asai, T., Ubermykorrhizenbildung der leguminosen Pflanzen, *Jpn. J. Bot.*, 13, 463, 1944.
21. Ross, J. P. and Harper, J. A., Effect of *Endogone* mycorrhiza on soybean yields, *Phytopathology*, 60, 1552, 1970.
22. Schenck, N. C. and Hinson, K., Response of nodulating and non-nodulating soybeans to a species of *Endogone* mycorrhiza, *Agron. J.*, 65, 849, 1973.
23. Crush, J. R., Plant growth responces to vesicular-arbuscular mycorrhiza. VII. Growth and nodulation of some herbage legumes, *New Phytol.*, 73, 745, 1974.
24. Mosse, B., Powell, C. L., and Hayman, D. S., Plant growth responses to vesicular-arbuscular mycorrhiza. IX. Interactions between VA mycorrhiza, rock phosphate and symbiotic nitrogen fixation, *New Phytol.*, 76, 331, 1976.
25. Mosse, B., Plant growth responses to vesicular-arbuscular mycorrhiza. X. Responses of *Stylosanthes* and maize to inoculation in unsterile soils, *New Phytol.*, 78, 277, 1977.
26. Sanni, S. O., Vesicular-arbuscular mycorrhiza in some Nigerian soils and their effect on the growth of cowpea *(Vigna unguiculata)*, tomato *(Lycopersicon esculentum)* and maize *(Zea mays)*, *New Phytol.*, 77, 667, 1976.
27. Islam, R., Ayanaba, A., and Sanders, F. E., Response of cowpea *(Vigna unguiculata)* to inoculation with VA mycorrhizal fungi and to rock phosphate fertilization in some unsterilized Nigerian soils, *Plant Soil*, 54, 107, 1980.
28. Barea, J. M., Escudero, J. L., and Azcon-G. de Aguilar, C., Effects of introduced and indigenous VA mycorrhizal fungi on nodulation, growth and nutrition of *Medicago sativa* in phosphate fixing soils as affected by P fertilizers, *Plant Soil*, 54, 283, 1980.
29. Waidyanatha, U. P. de S., Yogaratnam, N., and Ariyaratne, W. A., Mycorrhizal infection on growth and nitrogen fixation of *Pueraria* and *Stylosanthes* and uptake of phosphorus from two rock phosphates, *New Phytol.*, 82, 147, 1979.
30. Munns, D. N. and Mosse, B., Mineral nutrition of legume crops, in *Advances in Legume Science*, Summerfield, R. J. and Bunting, A. H., Eds., University of Reading Press, England, 1980, 115.
31. Daft, M. J. and El-Giahmi, A. A., Effect of *Endogone* mycorrhiza on plant growth. VII. Influence of infection on the growth and nodulation in French bean *(Phaseolus vulgaris)*, *New Phytol.*, 73, 1139, 1974.
32. Daft, M. J. and El-Giahmi, A. A., Studies on nodulated and mycorrhizal peanuts, *Ann. Appl. Biol.*, 83, 273, 1976.

33. Carling, D. E., Riehle, W. G., Brown, M. F., and Johnson, D. R., Effects of a vesicular-arbuscular mycorrhizal fungus on nitrate reductase and nitrogenase activities in nodulating and non-nodulating soybeans, *Phytopathology*, 68, 1590, 1978.
34. Smith, S. E. and Daft, M. J., The effect of mycorrhizas on the phosphate content, nitrogen fixation and growth of *Medicago sativa*, in *Microbial Ecology*, Loutit, M. W. and Miles, J. A. R., Eds., Springer Verlag, Basel, 1978, 314.
35. Godse, D. B., Wani, S. P., Patil, R. B., and Bagyaraj, D. J., Response of cowpea [*Vigna unguiculata* (L.) Walp.] to *Rhizobium*-VA mycorrhiza dual inoculation, *Curr. Sci.*, 47, 784, 1978.
36. Manjunath, A. and Bagyaraj, D. J., Response of pigeonpea and cowpea to phosphate and dual inoculation with vesicular arbuscular mycorrhiza and Rhizobium, *Trop. Agric.*, 61, 48, 1984.
37. Redente, E. F. and Reeves, F. B., Interactions between vesicular-arbuscular mycorrhiza and *Rhizobium* and their effect on sweetvetch growth, *Soil Sci.*, 132, 410, 1981.
38. Abbott, L. K. and Robson, A. D., Growth stimulation of subterranean clover with vesicular-arbuscular mycorrhizas, *Aust. J. Agric. Res.*, 28, 639, 1977.
39. Smith, S. E. and Daft, M. J., Interactions between growth, phosphate content and N_2 fixation in mycorrhizal and non-mycorrhizal *Medicago sativa*, *Aust. J. Plant Physiol.*, 4, 403, 1977.
40. Asimi, S., Gianinazzi-Pearson, V., and Gianinazzi, S., Influence of increasing soil phosphorus levels on interactions between vesicular-arbuscular mycorrhizae and *Rhizobium* in soybeans, *Can. J. Bot.*, 58, 2200, 1980.
41. Smith, S. E. and Bowen, G. D., Soil temperature, mycorrhizal infection and nodulation of *Medicago truncalata* and *Trifolium subterraneum*, *Soil Biol. Biochem.*, 11, 469, 1979.
42. Chambers, C. A., Smith, S. E., Smith, F. A., Ramsey, M. D., and Nicholas, D. J. D., Symbiosis of *Trifolium subterraneum* with mycorrhizal fungi and *Rhizobium trifolii* as affected by ammonium sulphate and nitrification inhibitors, *Soil Biol. Biochem.*, 12, 93, 1980.
43. McLachlan, K. D. and Norman, B. W., Phosphorus and symbiotic nitrogen fixation in subterranean clover, *J. Aust. Inst. Agric. Sci.*, 27, 244, 1961.
44. Gibson, A. H., Limitations to nitrogen fixation in legumes, in *Proc. 1st Int. Symp. Nitrogen Fixation*, Vol. 2, Newton, W. E. and Nyman, C. J., Eds., Washington State University Press, Pullman, 1976, 400.
45. Bergersen, F. J., Biochemistry of symbiotic nitrogen fixation in legumes, *Annu. Rev. Plant Physiol.*, 22, 121, 1971.
46. Stewart, W. D. P. and Alexander, G., Phosphorus availability and nitrogenase activity in aquatic blue gree algae, *Freshwater Biol.*, 1, 389, 1971.
47. Hayman, D. S. and Mosse, B., Clover in grassland, *Rothamsted Exp. Stan. Annu. Rep.*, 1, 241, 1978.
48. Hayman, D. S. and Mosse, B., Improved growth of white clover in hill grasslands by mycorrhizal inoculum, *Ann. Appl. Biol.*, 93, 141, 1979.
49. Daft, M. J., Nitrogen fixation in nodulated and mycorrhizal crop plants, *Ann. Appl. Biol.*, 88, 461, 1978.
50. Mosse, B., The role of mycorrhiza in legume nutrition on marginal soils, in *Exploiting the Legume-Rhizobium Symbiosis in Tropical Agriculture*, Vincent, J. M., Whitney, A. S., and Bose, J., Eds., University of Hawaii College of Tropical Agriculture Misc. Publ. No. 145, Honolulu, 1977, 175.
51. Smith, S. E., Nicholas, D. J. D., and Smith, F. A., Effect of early mycorrhizal infection on nodulation and nitrogen fixation in *Trifolium subteraneum* L., *Aust. J. Plant Physiol.*, 6, 305, 1979.
52. Pate, J. S., Physiology of the reaction of nodulated legumes to environment, in *Symbiotic Nitrogen Fixation in Plants*, Nutman, P. S., Ed., Cambridge University Press, Cambridge, 1976, 335.
53. Nicholas, D. J. D., The functions of trace elements in plants, in *Trace Elements in Soil-Plant-Animal Systems*, Nicholas, D. J. D. and Egan, A. R., Eds., Academic Press, New York, 1975, 181.
54. Munns, D. N., Mineral nutrition and the legume symbiosis, in *A Treatise on Dinitrogen Fixation*, Vol. 4, Hardy, R. W. F. and Gibson, A. H., Eds., John Wiley & Sons, New York, 1977, 353.
55. Nutman, P. S., The relation between nodule bacteria and the legume host in the rhizosphere and in the infection process, in *Ecology of Soil Borne Plant Pathogen*, Baker, K. F. and Snyder, W. C., Eds., University of California Press, Berkeley, 1965, 231.
56. Bagyaraj, D. J., Legume-*Rhizobium* symbiosis, *Curr. Res.*, 4, 19, 1975.
57. Daft, M. J. and El-Giahmi, A. A., Effects of vesicular-arbuscular mycorrhiza on plant growth. VII. Effects of defoliation and light on selected hosts, *New Phytol.*, 80, 365, 1978.
58. Gilmore, A. E., The influence of endotrophic mycorrhizae on the growth of peach seedlings, *J. Am. Soc. Hortic. Sci.*, 96, 35, 1971.
59. Daft, M. J. and Hacskaylo, E., Growth of endomycorrhizal and non-mycorrhizal red maple seedlings in sand and anthracite spoil, *For. Sci.*, 23, 207, 1977.
60. Bagyaraj, D. J. and Manjunath, A., Response of crop plants to VA mycorrhizal inoculation in an unsterile Indian soil, *New Phytol.*, 85, 33, 1980.

61. Allen, M. F., Moore, T. S., Jr., and Christensen, M., Phytohormone changes in *Bouteloua gracilis* infected by vesicular-arbuscular mycorrhizae. I. Cytokinin increases in the host plant, *Can. J. Bot.*, 58, 371, 1980.
62. Allen, M. F., Moore, T. S., Jr., and Christensen, M., Phytohormone changes in *Bouteloua gracilis* infected by vesicular-arbuscular mycorrhiza. II. Altered levels of gibberllin-like substances and abscisic acid in the host plant, *Can. J. Bot.*, 60, 468, 1982.
63. Bagyaraj, D. J. and Menge, J. A., Interaction between a VA mycorrhiza and *Azotobacter* and their effects on the rhizosphere microflora and plant growth, *New Phytol.*, 80, 567, 1978.
64. Barea, J. M., Navarro, E., and Montoya, E., Production of plant growth regulators by rhizosphere phosphate solubilizing bacteria, *J. Appl. Bacteriol.*, 40, 129, 1976.
65. Azcon-G. de Aguilar, C. and Barea, J. M., Effects of interactions between different culture fractions of "phosphobacteria" and *Rhizobium* on mycorrhizal infection, growth, and nodulation of *Medicago sativa, Can. J. Microbiol.*, 24, 520, 1978.
66. Bagyaraj, D. J., Manjunath, A., and Patil, R. B., Interaction between a VA mycorrhiza and *Rhizobium* and their effects on soybean in the field, *New Phytol.*, 82, 141, 1979.
67. Islam, R. and Ayanaba, A., Effect of seed inoculation and pre-infecting cowpea *(Vigna unguiculata)* with *Glomus mosseae* on growth and seed yield of the plants under field conditions, *Plant Soil*, 61, 341, 1981.
68. Bagyaraj, D. J., Manjunath, A., Kulkarni, J. H., and Venugopal, N., Effect of grass-legume association and legume inoculation on native VA Mycorrhiza and grass yield, *Curr. Res.*, 9, 123, 1980.
69. Azcon-G. de Aguilar, C., Azcon, R., and Barea, J. M., Endomycorrhizal fungi and *Rhizobium* as biological fertilizers for *Medicago sativa* in normal cultivation, *Nature (London)*, 279, 325, 1979.
70. Bowen, G. D., Mycorrhizal roles in tropical plants and ecosystems, in *Tropical Mycorrhiza Research*, Mikola, P., Ed., Oxford University Press, Oxford, 1980, 165.
71. Hayman, D. S., Endomycorrhizas, in *Interactions Between Non-Pathogenic Soil Microorganisms and Plants*, Dommergues, Y. R. and Krupa, S. V., Eds., Elsevier, Amsterdam, 1978, 400.
72. Powell, C. L. and Bagyaraj, D. J., VA mycorrhizal inoculation of field crops, *Proc. N.Z. Agron. Soc.*, 12, 85, 1982.
73. Powell, C. L., Field inoculation with VA mycorrhizal fungi, in *VA Mycorrhiza*, Powell, C. L. and Bagyaraj, D. J., Eds., CRC Press, Boca Raton, Fla., 1984.
74. Hattingh, M. J. and Gerdemann, J. W., Inoculation of Brazilian sour orange seed with an endomycorrhizal fungus, *Phytopathology*, 65, 1013, 1975.
75. Hayman, D. S., Morris, E. J., and Page, R. J., Methods for inoculating field crops with mycorrhizal fungi, *Ann. Appl. Biol.*, 99, 247, 1981.
76. Johnson, C. R. and Michelini, S., Effect of mycorrhizae on container grown acacia, *Proc. Fla. State Hortic. Soc.*, 87, 520, 1974.
77. Daft, M. J., Hacskaylo, E., and Nicolson, T. H., Arbuscular mycorrhizas in plants colonising coal spoils in Scotland and Pennsylvania, in *Endomycorrhizas*, Sanders, F. E., Mosse, B., and Tinker, P. B., Eds., Academic Press, London, 1975, 561.
78. Duke, J. A., *Handbook of Legumes of World Economic Importance*, Plenum Press, New York, 1981, 120.
79. Dalton, H., The cultivation of diazotrophic microorganisms, in *Methods for Evaluating Biological Nitrogen Fixation*, Bergerson, F. J., Ed. John Wiley & Sons, New York, 1980, 13.
80. Dobereiner, J., Day, J. M., and Dart, P. J., Nitrogenase activity and oxygen sensitivity of *Paspalum notatum-Azotobacter paspali* association, *J. Gen. Microbiol.*, 71, 103, 1972.
81. Neyra, C. A. and Dobereiner, J., Nitrogen fixation in grasses, *Adv. Agron.*, 29, 1, 1977.
82. Barea, J. M., Brown, M. E., and Mosse, B., Association between VA mycorrhiza and *Azotobacter*, *Rothamsted Exp. Stn. Annu. Rep.*, 1, 82, 1973.
83. Ho, I. and Trappe, J. M., Interaction of a VA-mycorrhizal fungus and a free-living nitrogen fixing bacterium on growth of tall fescue, in *Abst. 4th N. Am. Conf. Mycorrhizae*, Fort Collins, Colorado, 1979.
84. Brown, M. E. and Carr, G. H., Effects on plant growth of mixed inocula of VA endophytes and root-microorganisms, *Rothamsted Exp. Stn. Annu. Rep.*, 1, 189, 1980.
85. Manjunath, A., Mohan, R., and Bagyaraj, D. J., Interactions between *Beijerinckia mobilis*, *Aspergillus niger* and *Glomus fasciculatus* and their effects on growth of onion, *New Phytol.*, 87, 723, 1981.
86. Gaskins, M. H. and Hubbell, D. H., Response of non-leguminous plants to root inoculation with free-living diazotrophic bacteria, in *The Soil-Root Interface*, Harley, J. L. and Russell, R. S., Eds., Academic Press, London, 1979, 175.
87. Okon, Y., Recent progress in research on biological nitrogen fixation with non-leguminous crops, *Phosphorus Agric.*, 82, 3, 1982.
88. Mishustin, E. N. and Naumova, A. N., Bacterial fertilizers, their effectiveness and mode of action, *Mikrobiologiya*, 31, 543, 1962.

89. Barea, J. M., Azcon, R., and Hayman, D. S., Possible synergistic interactions between *Endogone* and phosphate-solubilizing bacteria in low-phosphate soils, in *Endomycorrhizas,* Sanders, F. E., Mosse, B., and Tinker, P. B., Eds., Academic Press, London, 1975, 408.
90. Azcon, R., Barea, J. M., and Hayman, D. S., Utilization of rock phosphate in alkaline soils by plants inoculated with mycorrhizal fungi and phosphate solubilizing bacteria, *Soil Biol. Biochem.,* 8, 135, 1976.
91. Raj, J., Bagyaraj, D. J., and Manjunath, A., Influence of soil inoculation with vesicular-arbuscular mycorrhiza and a phosphate-dissolving bacterium on plant growth and ^{32}P-uptake, *Soil Biol. Biochem.,* 13, 105, 1981.
92. Gopalakrishna, N. N., Interaction Between VA Mycorrhiza and Phosphate Solubilizing Fungi and Their Effects on Rhizospere Microflora and Growth of Finger Millet, M.Sc. (Agric.) thesis, University of Agricultural Sciences, Bangalore, India, 1980.
93. Baya, A. M., Boethling, R. S., and Ramos-Cormenzana, A., Vitamin production in relation to phosphate solubilization by soil bacteria, *Soil Biol. Biochem.,* 13, 527, 1981.
94. Harley, J. L., *The Biology of Mycorrhiza,* Leonard Hill, London, 1969, 73 and 211.
95. Cook, F. D. and Lochhead, A. G., Growth factor relationships of soil microorganisms as affected by proximity to the plant root, *Can. J. Microbiol.,* 5, 323, 1959.
96. Krishna, K. R., Balakrishna, A. N., and Bagyaraj, D. J., Interaction between VA mycorrhiza and *Streptomyces cinnamomeus* and their effect on finger millet, *New Phytol.,* 92, 401, 1982.
97. Balakrishna, A. N., Establishment of Azotobacters in the Rhizosphere of Maize and Wheat: Effect of Root Exudates and Associative Microorganisms, M.Sc. (Agric.) thesis, University of Agricultural Sciences, Bangalore, India, 1979.
98. Becking, J. H., Plant-endophyte symbiosis in non-leguminous plants, *Plant Soil,* 32, 611, 1970.
99. Bond, G. and Wheeler, C. T., Non-legume nodule systems, in *Methods for Evaluating Biological Nitrogen Fixation,* Bergersen, F J., Ed., John Wiley & Sons, New York, 1980, 185.
100. Williams, W. E. and Aldon, E. F., Endomycorrhizal (vesicular-arbuscular) associations of some arid zone shrubs, *Southwest Nat.,* 20, 437, 1976.
101. Hall, R. B., McNabb, H. S., Jr., Maynard, C. A., and Green, T. L., Toward development of optimal *Alnus glutinosa* symbiosis, *Bot. Gaz. (Chicago),* 40(Suppl.), 121, 1979.
102. Rose, S. L. and Youngberg, C. T., Tripartite associations in snowbrush *(Ceanothus velutinus)*: effect of vesicular-arbuscular mycorrhizae on growth, nodulation, and nitrogen fixation, *Can. J. Bot.,* 59, 34, 1981.
103. Rose, S. L., Mycorrhizal associations of some actinomycete nodulated nitrogen-fixing plants, *Can. J. Bot.,* 58, 1449, 1980.
104. Safir, G., The Influence of Vesicular-Arbuscular Mycorrhiza on the Resistance of Onion to *Pyrenochaeta terrestris,* M. S. thesis, University of Illinois, Urbana, 1968.
105. Ross, J. P., Influence of *Endogone* mycorrhiza on *Phytophthora* rot of soybean, *Phytopathology,* 62, 896, 1972.
106. Davis, R. M., Menge, J. A., and Erwin, D. C., Influence of *Glomus fasciculatus* and soil phosphorus on *Verticillium* wilt of cotton, *Phytopathology,* 69, 453, 1979.
107. Ramirez, B. N., Influence of Endomycorrhizae on the Relationship of Inoculum Density of *Phytophthora palmivora* in Soil to Infection of Papaya Roots, M.S. thesis, University of Florida, Gainesville, 1974.
108. Menge, J. A., Nemec, S., Davis, R. M., and Minassian, V., Mycorrhizal fungi associated with citrus and their possible interactions with pathogens, in *Proc. Int. Soc. Citric.,* Vol. 3, William, G., Ed., University of Florida Press, Lake Alfred, 1979, 872.
109. Davis, R. M., Influence of *Glomus fasciculatus* on *Thielaviopsis basicala* root rot of citrus, *Plant Dis.,* 64, 839, 1980.
110. Winter, A. G., Studies on the distribution and importance of mycorrhiza in cultivated Gramineae and some other agricultural economic plants, *Phytopathol. Z.,* 17, 421, 1951.
111. Hall, J. B. and Finch, H. C., Mycorrhiza in roots of avocado: effect upon chemotaxis of *Phytophthora cinnamomi* zoospores, *Proc. Am. Phytopathol. Soc.,* 1, 86, 1974.
112. Schonbeck, F. and Dehne, H. W., Damage to mycorrhizal and non-mycorrhizal cotton seedlings by *Thielaviopsis basicola, Plant Dis. Rep.,* 61, 266, 1977.
113. Baltruschat, H. and Schonbeck, F., Influence of endotrophic mycorrhiza on chlamydospore production of *Thielaviopsis basicola* in tobacco roots, *Phytopathol. Z.,* 74, 358, 1972.
114. Baltruschat, H. and Schonbeck, F., The influence of endotrophic mycorrhiza on the infestation of tobacco by *Thielaviopsis basicola, Phytopathol Z.,* 84, 172, 1975.
115. Chou, L. G. and Schmitthenner, A. F., Effect of *Rhizobium japonicum* and *Endogone mosseae* on soybean root rot caused by *Pythium ultimum* and *Phytophthora megasperma* var. *sojae, Plant Dis. Rep.,* 58, 221, 1974.
116. Woodhead, S. H., Gerdemann, J. W., and Paxton, J. D., Mycorrhizal infection of soybean roots reduces *Phytophthora* root rot, in *Abst. 3rd N. Am. Conf. Mycorrhizae,* Athens, Ga., 1977, 10.

117. Schenck, N. C., Ridings, W. H., and Cornell, J. A., Interaction of two vesicular-arbuscular mycorrhizal fungi, and *Phytophthora parasitica* on two citrus root stocks, in *Abst. 3rd N. Am. Conf. Mycorrhizae*, Athens, Ga., 1977, 9.
118. Paget, D. K., The effect of *Cylindrocarpon* on plant growth responses to vesicular-arbuscular mycorrhiza, in *Endomycorrhizas*, Sanders, F. E., Mosse, B. and Tinker, P. B., Eds., Academic Press, London, 1975, 593.
119. Dehne, H. W. and Schonbeck, F., The influence of the endotrophic mycorrhiza on the fusarial wilt of tomato, *Z. Pflanzenkr. Pflanzenschutz*, 82, 630, 1975.
120. Stewart, E. L. and Pfleger, F. L., Development of *Poinsettia* as influenced by endomycorrhizae, fertilizer and root rot pathogens *Pythium ultimum* and *Rhizoctonia solani*, *Florist Rev.*, 159, 37, 1977.
121. Becker, W. N., Quantification of Onion Vesicular-Arbuscular Mycorrhizae and Their Resistance to *Pyrenochaeta terrestris*, Ph.D. thesis, University of Illinois, Urbana, 1976.
122. Schonbeck, F. and Dehne, H. W., Investigations on the influence of endotrophic mycorhiza on plant diseases. IV. Fungal parasites on shoots, *Olipidium brassicae*, TMV, *Z. Pflanzenkr. Pflanzenschutz*, 86, 103, 1979.
123. Iqbal, S. H., Qureshi, K. S., and Ahmad, J. S., Influence of vesicular-arbuscular mycorrhiza on damping off caused by *Rhizoctonia solani* in *Brassica napus*, *Biologia (Lahore)*, 23, 197, 1977.
124. Cooper, K. M., Mycorrhizal fungi act as deterrents to soil-borne root pathogens, in *Proc. N. Z. Microbiol. Soc. Annu. Conf.*, Auckland, 1981.
125. Davis, R. M. and Menge, J. A., *Phytophthora parasitica* inoculation and intensity of vesicular-arbuscular mycorrhiza in citrus, *New Phytol.*, 87, 705, 1981.
126. Graham, J. H. and Menge, J. A., Influence of vesicular-arbuscular mycorrhizae and soil phosphorus on take-all disease of wheat, *Phytopathology*, 72, 95, 1982.
127. Krishna, K. R. and Bagyaraj, D. J., Interaction between *Glomus fasciculatus* and *Sclerotium rolfsii* in peanut, *Can. J. Bot.*, 61, 2349, 1983.
128. Barnard, E. L., The Mycorrhizal Biology of *Liriodendron tulipifera* L. and Its Relationship to *Cylindrocladium* Root Rot, Ph.D. thesis, Duke University, Durham, N.C., 1977.
129. Weaver, D. J. and Wehunt, E. J., Effect of soil pH on susceptibility of peach to *Pseudomonas syringae*, *Phytopathology*, 65, 984, 1975.
130. Halos, P. M. and Zorilla, R. A., Vesicular-arbuscular mycorrhize increase growth and yield of tomatoes and reduce infection by *Pseudomonas solanacearum*, *Philipp. Agric.*, 62, 309, 1979.
131. Schonbeck, F. and Schinzer, U., Investigations on the influence of endotrophic mycorrhiza on TMV lesion formation in *Nicotiana tabacum* L. var. Xanthi-nc., *Phytopathol. Z.*, 73, 78, 1972.
132. Daft, M. J. and Okusanya, B. O., Effect of *Endogone* mycorrhiza on plant growth. V. Influence of infection on the multiplication of viruses in tomato, petunia and strawberry, *New Phytol.*, 72, 975, 1973.
133. Schonbeck, F. and Spengler, G., Detection of TMV in mycorrhizal cells of tomato by immunofluorescence, *Phytopathol. Z.*, 94, 84, 1979.
134. Fox, J. A. and Spasoff, L., Interaction of *Heterodera solanaceraum* and *Endogone gigantea* on tobacco, *J. Nematol.*, 4, 224, 1972.
135. Schenck, N. C. and Kellam, M. K., The influence of vesicular-arbuscular mycorrhizae on disease development, *Fla. Agric. Exp. Stn. Tech. Bull.*, 798, 1978.
136. Schonbeck, F., Endomycorrhiza in relation to plant diseases, in *Soil-Borne Plant Pathogens*, Schippers, B. and Gams, W., Eds., Academic Press, New York, 1979, 271.
137. Schenck, N. C., Can mycorrhizae control root disease, *Plant Dis.*, 65, 230, 1981.
138. Hussey, R. S. and Roncadori, R. W., Vesicular-arbuscular mycorrhizae may limit nematode activity and improve plant growth, *Plant Dis.*, 66, 9, 1982.
139. Atilano, R. A., Rich, J. R., Ferris, H., and Menge, J. A., Effect of *Meloidogyne arenaria* on endomycorrhizal grape *(Vitis vinifera)* rootings, *J. Nematol.*, 8, 278, 1976.
140. Schenck, N. C., Kinlock, R. A., and Dickson, D. W., Interaction of endomycorrhizal fungi and root-knot nematode on soybean, in *Endomycorrhizas*, Sanders, F. E., Mosse, B., and Tinker, P. B., Eds., Academic Press, London, 1975, 605.
141. Baltruschat, H., Sikora, R. A., and Schonbeck, F., Effect of VA mycorrhizae *(Endogone mosseae)* on the establishment of *Thielaviopsis basicola* and *Meloidogyne incognita* in tobacco, in *Proc. 2nd Int. Congr. Plant Pathol.*, Minneapolis, 1973.
142. Sikora, R. A. and Schonbeck, F.., Effect of vesicular-arbuscular mycorrhiza *(Endogone mosseae)* on the population dynamics of the root-knot nematodes (*Meloidogyne incognita* and *Meloidogyne hapla*), in *Proc. 8th Int. Congr. Plant Protection*, Moscow, 1975, 158.
143. Kellam, M. K., and Schenck, N. C., Interactions between a vesicular-arbuscular mycorrhizal fungus and root-knot nematode on soybean, *Phytopathology*, 70, 293, 1980.
144. Roncadori, R. W. and Hussey, R. S., Interaction of the endomycorrhizal fungus *Gigaspora margarita* and root knot nematode on cotton, *Phytopathology*, 67, 1507, 1977.

145. Bagyaraj, D. J., Manjunath, A., and Reddy, D. D. R., Interaction of vesicular-arbuscular mycorrhiza with root knot nematodes in tomato, *Plant Soil*, 51, 397, 1979.
146. O'Bannon, J. H., Inserra, R. N., Nemec, S., and Vovlas, N., The influence of *Glomus mosseae* on *Tylenchulus semipenetrans* infected and uninfected *Citrus lemon* seedlings, *J. Nematol.*, 11, 247, 1979.
147. Atilano, R. A., Menge, J. A., Grundy, S. D. V., Interaction between *Meloidogyne arenaria* and *Glomus fasiculatus* in grape, *J. Nematol.*, 13, 52, 1981.
148. Sikora, R. A., Predisposition to *Meloidogyne* infection by the endotrophic mycorrhizal fungus *Glomus mosseae*, in *Root-Knot Nematodes (Meloidogyne Species) Systematics, Biology and Control*, Lamberti, F. and Taylor, C. E., Eds., Academic Press, New York, 1979, 399.
149. Suresh, C. K., Interaction Between Vesicular-Arbuscular Mycorrhiza and Root-Knot Nematode in Tomato, M.Sc. (Agric.) thesis, University of Agricultural Sciences, Bangalore, India, 1980.
150. Hussey, R. S. and Roncadori, R. W., Interactions of *Pratylenchus brachyurus* and *Gigaspora margarita* on cotton, *J. Nematol.*, 10, 16, 1978.
151. O'Bannon, J. H. and Nemec, S., The response of *Citrus lemon* seedlings to a symbiont, *Glomus etunicatus*, and a pathogen, *Radopholus similis*, *J. Nematol.*, 11, 270, 1979.
152. Wilhelm, S., Principles of biological control of soil-borne plant diseases, *Soil Biol. Biochem.*, 4, 53, 1973.
153. Dehne, H. W. and Schonbeck, F., Investigations on the influence of endotrophic mycorrhiza on plant diseases. II. Phenol metabolism and lignification, *Phytopathol. Z.*, 95, 210, 1979.
154. Trudgill, D. L. and Parrott, D. M., The behaviour of the population of potato cyst nematode *Heterodera rostochinensis* towards three resistant potato hybrids, *Nematologica*, 15, 381, 1969.
155. Fassuliotis, G., Resistance in *Cucumis* spp. to the root knot nematode, *Meloidogyne incognita acrita*, *J. Nematol.*, 2, 174, 1970.
156. Ratnayake, M., Leonard, R. T., and Menge, J. A., Root exudation in relation to supply of phosphorus and its possible relevance to mycorrhizal formation, *New Phytol.*, 81, 543, 1978.
157. Fuchs, W. H. and Grossman, F., Ernahrung und Resistenz von Kulturpflanzen gegenuber Krankheitserregern und Schadlingen, in *Handbuch der Pflanzenernahrung und Dungung*, Bd. 1, Linser, H., Ed., Springer-Verlag, New York, 1972, 1007.
158. Sikora, R. A., Effect of the endotrophic mycorrhizal fungus, *Glomus mosseae*, on the hostparasite relationship of *Meloidogyne incognita* in tomato, *Z. Pflanzenkr. Pflanzenschutz*, 85, 197, 1978.
159. Setty, K. G. H., Studies on Biology and Host Parasite Relationships on Root Knot Nematode (*Meloidogyne* spp.) on tomato, Ph.D. thesis, University of London, 1969.
160. Parvatha Reddy, P., Studies on the Action of Amino Acids on the Root Knot Nematode, *Meloidogyne incognita*, Ph.D. thesis, University of Agricultural Sciences, Bangalore, India, 1974.
161. Krishna Prasad, K. S., Effects of amino acid and plant growth substance on tomato and its root knot nematode *Meloidogyne incognita* Chitwood, M.Sc. (Agric.) thesis, University of Agricultural Sciences, Bangalore, India, 1971.
162. Williams, P. G., The Prospects of Using Vesicular Arbuscular Mycorrhiza in Broadscale Agriculture to Improve the Efficiency of Phosphate Fertiliser Use, Rep. Aust. Meat Res. Counc. Workshop, Cambridge, 1981.
163. Gerdemann, J. W. and Trappe, J. M., The Endogonaceae in the Pacific Northwest, *Mycol. Mem.*, 5, 1, 1974.
164. Mosse, B., The establishment of vesicular-arbuscular mycorrhiza under aseptic conditions, *J. Gen. Microbiol.*, 27, 509, 1962.
165. Flentje, N. T., The physiology of penetraton and infection, in *Plant Pathology — Problems and Progress 1908-58*, Holten, C. S., Fischer, G. W., Fulton, R. W., Hart, H., and McCallan, S. E. A., Eds., University of Wisconsin Press, Madison, 1959, 76.
166. Mosse, B., The influence of soil type and *Endogone* strain on the growth of mycorrhizal plants in phosphate deficient soils, *Rev. Ecol. Biol. Soil*, 9, 529, 1972.
167. Wilson, J. F. and Hanton, W. K., Bacteria-like structure in fungi, in *Viruses and Plasmids in Fungi*, Ser. on Mycology, Vol. 1, Lemke, P. E., Ed., Marcel Dekker, New York, 1979, 525.
168. Mosse, B., Honey-coloured, sessile *Endogone* spores. II. Changes in fine structure during spore development, *Arch. Mikrobiol.*, 74, 129, 1970.
169. Macdonald, R. M. and Chandler, M. R., Bacterium-like organelles in the vesicular-arbuscular mycorrhizal fungus *Glomus caledonius*, *New Phytol.*, 89, 241, 1981.
170. Bonfante-Fasolo, P. and Scannerini, S., The ultrastructure of the zygospores in *Endogone flammicorana* Trappe and Gerdemann, *Mycopathologia*, 59, 117, 1976.
171. Bonfante-Fasolo, P. and Scannerini, S., Cytological observations on the mycorrhiza '*Endogone flammicorana*' -'*Pinus strobus*', *Allionia*, 22, 23, 1977.
172. Bonfante-Fasolo, P. and Scannerini, S., A cytological study of vesicular-arbuscular mycorrhiza in '*Ornithogalum umbellatum*', *Allionia*, 22, 5, 1977.

173. Bonfante-Fasolo, P., Some ultrastructural features of the vesicular arbuscular mycorrhiza in the grapevine, *Vitis*, 17, 386, 1978.
174. Protzenko, M. A., Microorganisms in the hyphae of mycorrhiza-forming fungus, *Mikrobiologiya*, 44, 1121, 1975.
175. Scannerini, S., Bonfante-Fasolo, P., and Fontana, A., An ultrastructural model for the host-symbiont interaction in the endotrophic mycorrhizae of *Ornithogalum umbellatum* L., in *Endomycorrhizas*, Sanders, F. E., Mosse, B., and Tinker, P. B., Eds., Academic Press, London, 1975, 314.
176. Smith, D. C., From extracellular to intracellular: the establishment of a symbiosis, *Proc. R. Soc. London Ser. B.*, 204, 115, 1979.
177. Thaxter, R., A revision of the Endogonaceae, *Proc. Am. Acad. Arts Sci.*, 57, 291, 1922.
178. Malencon, M. C., Etudes de parasitisme mycopathologique, *Rev. Mycol.*, 7, 27, 1947.
179. Mosse, B., Studies on the Endotrophic Mycorrhiza of Some Fruit Plants, Ph.D. thesis, University of London, 1954.
180. Godfrey, R. M., Studies of British species of *Endogone*, II. Fungal parasites, *Trans. Br. Mycol. Soc.*, 40, 136, 1957.
181. Gerdemann, J. W. and Nicolson, T. H., Spores of mycorrhizal *Endogone* species extracted from soil by wet sieving and decanting, *Trans. Br. Mycol. Soc.*, 46, 235, 1963.
182. Mosse, B. and Bowen, G. D., A key to the recognition of some *Endogone* spore types, *Trans. Br. Mycol. Soc.*, 51, 469, 1968.
183. Schenck, N. C. and Nicolson, T. H., A zoosporic fungus occurring on species of *Gigaspora margarita* and other vesicular-arbuscular mycorrhizal fungi, *Mycologia*, 69, 1049, 1977.
184. Sparrow, F. K., A *Rhizidiomycopsis* on azygospores of *Gigaspora margarita*, *Mycologia*, 69, 1053, 1977.
185. Ross, J. P., Effect of non-treated field soil on sporulation of vesicular-arbuscular mycorrhizal fungi associated with soybean, *Phytopathology*, 70, 1200, 1980.
186. Bhattacharjee, M., Mukerji, K. G., Tewari, J. P., and Skoroped, W. P., Structure and hyperparasitism of a new species of *Gigaspora*, *Trans. Br. Mycol. Soc.*, 78, 184, 1982.
187. Ross, J. P. and Ruttencutter, R., Population dynamics of two vesicular-arbuscular endomycorrhizal fungi and the role of hyperparasitic fungi, *Phytopathology*, 67, 490, 1977.
188. Daniels, B. A. and Menge, J. A., Hyperparasitization of vesicular-arbuscular mycorrhizal fungi, *Phytopathology*, 70, 585, 1980.
189. Menge, J. A., Lembright, H., and Johnson, E. L. V., Utilization of mycorrhizal fungi in citrus nurseries, in *Proc. Int. Soc. Citric.*, Vol. 1, William, G., Ed., University of Florida Press, Lake Alfred, 1977, 129.
190. Menge, J. A., Johnson, E. L. V., and Minassian, V., Effect of heat treatment and three pesticides upon the growth and reproduction of the mycorrhizal fungus *Glomus fasciculatus*, *New Phytol.*, 82, 473, 1979.
191. Warnock, A. J., Fitter, A. H., and Usher, M. B., The influence of a springtail *Folsomia Candida* (Insect, Collembola) on the mycorrhizal association of leek *Allium porrum* and the vesicular-arbuscular mycorrhizal endophyte *Glomus fasciculatus*, *New Phytol.*, 90, 285, 1982.
192. Schafer, S. R., Unpublished data, 1982.

Chapter 8

PHYSIOLOGY OF VA MYCORRHIZAL ASSOCIATIONS

Karen M. Cooper

TABLE OF CONTENTS

I.	Introduction		156
II.	Mineral Nutrition		157
	A.	Utilization of Soil Phosphorus	157
		1. Phosphorus Uptake from Soil Solution	157
		a. Mechanisms of Phosphorus Uptake	158
		2. Uptake from Insoluble Phosphorus Sources	159
		a. Possible Mechanisms for Improved Uptake from Insoluble Phosphorus Sources	160
		3. Storage and Translocation of Phosphorus	161
		4. Transfer to Host	162
	B.	Effect of Phosphorus on Mycorrhizal Infection	162
	C.	Nitrogen Nutrition and Metabolism	164
		1. Assimilation of Nitrogen and Ammonium	164
		2. Amino Acid Metabolism	164
		3. Dinitrogen Fixation	165
	D.	Uptake of Other Nutrients	165
III.	Carbon Physiology and Biochemistry		166
	A.	Factors Affecting Mycorrhizal Development	166
	B.	Carbon Transfer from Host to Fungus	167
		1. Transport and Storage of Carbon Compounds	167
		a. Storage as Lipid	167
		b. Storage as Glycogen	167
		c. Growth and Storage "Sinks"	168
		2. Carbohydrate Transfer to Fungus	168
		3. Metabolic Pathways	169
	C.	Carbon Transfer from Fungus to Host	169
	D.	Lipid Biochemistry	169
		1. Phospholipids	169
		2. Fatty Acids	170
		3. Sterols	170
	E.	Carbon Cost of the Symbiosis	170
		1. Carbon Demand of the Fungal Symbiont	170
		2. Growth Depressions and Compensation Effect	171
IV.	Hormonal Effects		173
V.	Water Relations		174
	A.	Water Transport Under Mesic Conditions	174
	B.	Water Transport Under Dry Conditions	176
VI.	Conclusions		177

Acknowledgments 178

References 178

I. INTRODUCTION

Investigations of vesicular-arbuscular mycorrhizal (VAM) associations have generally centered on responses of the host to fungal infection. Positive host growth responses are frequently found, particularly in soils of low nutrient status. This effect is usually attributed to enhanced nutrient uptake by mycorrhizal roots.[1] It is commonly accepted that the fungi involved in the symbiosis are obligate symbionts.[2]

In comparison with other classes of mycorrhizae, very little work has been done on the physiology of the VAM symbiosis. Much of the work already published on the physiological implications has been extrapolated from experiments designed for other purposes and is consequently of limited value. Several fundamental problems exist which make physiological experiments difficult to design and more often difficult to interpret.

First, because the fungi cannot be grown in axenic culture in the absence of host root tissue, information on the activity of the fungi is frequently inferred by difference from the different responses of mycorrhizal and uninfected plants. Some information on fungal structures can be gleaned by stripping mycelium from root surfaces and studying it, but it is difficult to separate it from fragments of host root tissue. In addition, physiological properties of external mycelium and spores are likely to differ from those of the internal mycelium, vesicles, and arbuscules. Extramatrical mycelium can now be produced in dual-membered solution culture,[3-5] and internal structures may be studied after enzymatic degradation of host tissue.[6] Both techniques may assist in elucidating the physiological properties of fungal tissue.

Second, mycorrhizal infection generally increases P uptake by the plant. Although this increase may only be temporary,[7] it is generally difficult to distinguish host responses from factors other than those linked with improved P nutrition.

Third, if growth conditions are such that an absence of mycorrhizal infection causes a P deficiency in nonmycorrhizal plants, then a direct comparison of mycorrhizal and nonmycorrhizal plants cannot be made. Mycorrhizal plants, because of their difference in size, development, and growth, may be of a different physiological age to nonmycorrhizal plants.[8] This difficulty can be partially overcome by adjusting the soil P levels to allow mycorrhizal and nonmycorrhizal plants to develop at similar rates and to allow nonmycorrhizal plants to achieve tissue P contents similar to those in infected plants. However, mycorrhizal infection can also change the distribution of dry matter between root and shoot, resulting in lower root:shoot ratios[8] and sometimes a lower percentage of dry matter in leaves of mycorrhizal plants,[7] even in plants of approximately equivalent size.

Fourth, sequential harvesting of plants is desirable in order to detect physiological changes which may occur in the plant as a result of the development of mycorrhizal infection. The infection of a root and the spread of the fungus in the root follows a sigmoid progression curve.[9] An initial lag phase, which reflects the slow establishment of infection, is followed by a period of rapidly increasing infection, to a final plateau indicating a balanced relationship between root growth and fungal spread. Effective mycorrhizal infection (>20 to 30%) may not occur for 3 to 4 weeks and may not reach a stable plateau until 5 to 6 weeks.[7,10] The maximum infection attained also differs with host and the various fungal endophytes.[10,11] Because of the resulting uncertainty as to the appropriate sampling times, many experiments really need to be both lengthy and large to accommodate the development phases in the establishment of the symbiosis.

Fifth, the ability of the mycorrhiza to absorb nutrients is closely linked with the development of external mycelium.[12] However, the effects of the fungus on host nutrition and carbon metabolism are also likely to be influenced by the proportion of the fungus which is metabolically active. Evidence on this is scanty. Enzymatic techniques[13] and staining of live roots with fluorescein diacetate accompanied by UV microscopy[14] may prove useful in distinguishing metabolically active fungal structures. Still lacking are estimations of the amount of fungal mycelium which needs to be active in order to maintain adequate transfer processes between host and fungus.

The purpose of this chapter is to discuss some of the more recent physiological findings with respect to the VAM symbiosis and to elucidate some of the implications for the fungal and plant partners of the association.

II. MINERAL NUTRITION

Ion uptake by plant roots from soil is governed by two major factors: transfer of ions through the soil, and the absorbing power of the root.[15] Whether the transfer of an ion to the roots occurs primarily by mass flow or by diffusion depends on the mobility of that particular ion in the soil. Under adequate nutrition regimes, ions such as NO_3^-, $SO_4^=$ and Ca^{++} may move rapidly through the soil to the root by mass flow, and uptake would be limited by the absorbing capacity of the root or mycorrhiza. In contrast, where ions (such as $H_2PO_4^-$, NH_4^+, Zn^{++} and Cu^{++}) are poorly mobile and move to the root by diffusion, the rate limiting step is likely to be movement to the root surface. Under these conditions, it is to be expected that uptake would be increased if ions could be moved rapidly through the fungal hyphae rather than diffusing slowly through the soil.[16,17]

The role of VAM in mineral nutrition has been discussed several times recently.[8,18-23] It is now well established that mycorrhizae can improve the P nutrition of a host, particularly in low fertility soils, and it is likely that they can also assist in the uptake of other ions. However, no attempt will be made in this chapter to discuss the numerous references which substantiate these views. Rather, it is the intention to summarize recent physiological findings.

A. Utilization of Soil Phosphorus

Phosphate is present in the soil in three forms: soluble inorganic P in the soil solution, insoluble inorganic P found in crystal lattices, and organic P compounds such as phytate.[19] Phosphate is relatively immobile in soil and diffuses only slowly to the plant root.[15] As a result, in soils of low P availability, depletion zones soon develop around the root.[17]

1. Phosphorus Uptake from Soil Solution

It is now well established that mycorrhizal infection can enhance the uptake of P by plant roots.[20,24] Although particularly important in low P soils, an increased rate of P uptake can occur over a range of soil P levels even when mycorrhizal growth responses no longer occur.[25] This enhanced P uptake by mycorrhizal plants is most likely to be due to the external fungal hyphae acting as an extension of the root system, thereby providing a more efficient (more extensive and better distributed) absorbing surface for uptake of nutrients from the soil and for translocation to the host root.[17,18,21] In effect, the hyphae act as a bypass for phosphate through the depleted zone around the root, much in the same way as root hairs.[15] From measurements of P inflow rates of VAM and uninfected roots in soil, calculations have shown that the hyphal inflow of

P (uptake per unit length of hypha per unit time) is around 18×10^{-14} mol cm^{-1} s^{-1} or about 6 times the rate to a single uninfected root.[26]

Studies on P uptake from isotopically labeled media have shown that the external fungal hyphae are able to absorb P directly from the soluble P pools in the soil and translocate it to the host root.[21,27,28] Two techniques have been particularly useful in confirming this hyphal translocation of P under sterile conditions. Using split-plate cultures[27-29] or soil chambers,[30,31] it is possible to geographically isolate the fungal hyphae from the host root. When radioactive labeled P is fed to the external mycelium, it is rapidly translocated to the host root. Hyphae have been shown to translocate labeled P over distances of up to 8 cm, and it is likely that they are able to absorb available P from undepleted sources in the soil at considerable distances from the root.[30]

Mycorrhizal infection may create its own depletion zones. An autoradiographic study has shown that the effect of mycorrhizal infection was to increase the diameter of the P depletion zone around the roots from 1 to 2 mm.[32] Thus, the effect of infection was similar to that of a cylinder of root hairs 1-mm long. However, analysis of the P hyphal inflow values of Sanders and Tinker[26] by Tinker[19] and Beever and Burns[22] has shown that although the hyphal inflow is sufficient to account for increased P uptake rates by mycorrhizal plants it may not be sufficient to create a depletion zone in soil. The size and intensity of these depletion zones will undoubtedly be influenced by the rooting density and the extent of mycelial development in the soil.

The ability of the fungus to absorb nutrients may be closely linked with the development of external mycelium. P uptake rates and host growth enhancement have sometimes been directly correlated with the development of external mycelium.[10,12] However, Cooper and Tinker[27] have shown a poor correlation between P transport and the length of external hyphae, and suggested that uptake may be at least partially controlled by host plant demand. It is likely that the spatial distribution of hyphae may also have an effect on nutrient uptake beyond the root hair depletion zone. Hyphal inflows of P can differ with fungal species,[10] and this may in part explain differences in the "efficiency" of various fungi in promoting plant growth.

P uptake may also be influenced by the steepness of the phosphate gradient from root to leaves.[16] In mycorrhizal plants, the increased conversion of phosphate from inorganic to organic forms in the leaves[33] could increase the inorganic phosphate gradient from root to leaves, possibly increasing the phosphate sink and thus enhancing P uptake.

a. Mechanisms of Phosphorus Uptake

Most research on the physiology of P movement in mycorrhizal plants involves studies on translocation and transport, and very little is known about possible uptake mechanisms of P from soil solutions.

Calculations have shown that the P concentration in the external hypha is around 0.3% dry weight compared with soil concentrations around 1 to 3 μM orthophosphate (Pi).[18] Whether or not Pi moves into the hypha against an electrochemical gradient depends on the form of the P in the cell, since a conversion of Pi to polyphosphate in effect reduces the concentration of Pi in the cell and could create a sink. If mycorrhizal hyphae are like other fungi, then about 15% of the P may be present as Pi, and the hyphal Pi concentration will be around 2 to 4 mM,[22] so that the soil to cell concentration gradient will be at least 1:1000, requiring an active uptake mechanism at the plasmalemma (Figure 1).[34] It is not likely that P uptake is merely by diffusion into a zero concentration sink, as assumed in the calculations of Tinker.[19]

Uptake of Pi can be described in terms of the Michaelis-Menten parameters, Vmax and Km.[22] Such kinetic analysis has shown that two discrete uptake systems may be involved in Pi uptake by mycorrhizal plants.[35] Biphasic Pi uptake patterns have been

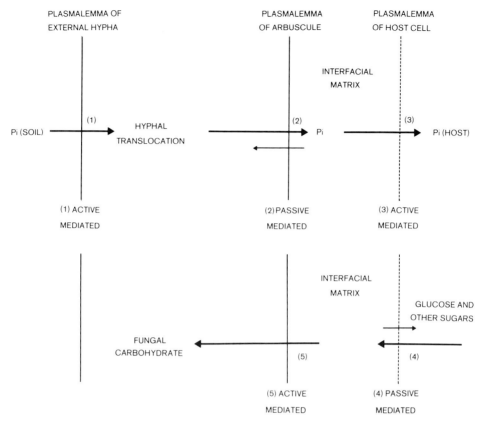

FIGURE 1. Possible pathways of phosphate (Pi) and carbohydrate transport in a VAM system.[34] (1) An active-mediated phosphate transport system at the plasmalemma of the external hyphae; (2) a passive-mediated phosphate transport system at the plasmalemma of the arbuscule; (3) an active-mediated phosphate system at the plasmalemma of the host cell; (4) a passive-mediated sugar transport system at the plasmalemma of the host cell; (5) an energy-linked (active) sugar transport system at the plasmalemma of the arbuscule. (With permission from *Endomycorrhizas*, Sanders, F. E., Mosse, B., and Tinker, P. B., Eds., 1975. Copyright: Academic Press Inc. (London) Ltd.)

recorded for both fungi and higher plants separately, so the pattern could arise from the behavior of either component of the mycorrhizal association. At low solution concentrations (1 to 20 μM Pi) the higher uptake rate in mycorrhizal plants was associated with a lower Km value. This indicates an increased absorption site affinity for Pi, which could be important in many P-deficient soils. However, at solution concentrations above 30 μM, increased absorption was related to a greater Vmax. This suggests that when P is readily available, increased uptake by mycorrhizal plants is due to the increased number of absorbing sites provided by the greater surface area of the external mycelium. Unfortunately, in these studies, no distinction was made between uptake into the host root and into the fungus. The kinetics of P uptake by the fungal component require further investigation.

2. Uptake from Insoluble Phosphorus Sources

It is widely believed that mycorrhizal plants absorb P only from the soluble P pools in the soil and that they are unable to utilize sources of P which are unavailable to uninfected roots.[18,19] However, in soils the amount of P which is readily available to plants is small and probably accounts for around only 1 to 5% of the total P content. This has led to the suggestion that VAM may be able to solubilize sources of P which would previously be unavailable to nonmycorrhizal plants.

Several techniques have been developed to test this possibility, but results are conflicting. One such technique involves labeling soil with ^{32}P and allowing it to equilibrate with the soil solution P (labile pool). The size of this pool is determined by the isotopic dilution of ^{32}P which occurs.[19] Calculations of the specific activities (ratios of ^{32}P to ^{31}P) in mycorrhizal and nonmycorrhizal plants grown in labeled soil and analyses of isotopic dilution kinetics of such soils have shown that the soluble pool size in soil is generally unaffected by the presence of mycorrhizal fungi, and that both mycorrhizal and nonmycorrhizal plants absorb P from the same labile pool (i.e., the soil solution or absorbed P in equilibrium with it).[8,21,36,37] Similar conclusions were reached by Barrow et al.[38] who compared the growth of mycorrhizal and nonmycorrhizal plants in soils where the amount of nonlabile P had been artificially increased by heating to high temperatures. However, evidence from other soil isotopic labeling experiments is contradictory, and indicates that mycorrhizal fungi may indeed be able to mobilize less available forms of soil P, thereby increasing the desorption or solubilization of P, and thus increasing the size of the labile pool.[36,39] Mycorrhizal infection has also improved plant growth and P uptake from highly phosphate-fixing soils and from latosols which contain insoluble iron and aluminum phosphates.[40,41]

Plants infected with VAM often respond to the addition of insoluble forms of phosphate, such as tricalcium phosphate and rock phosphate, which are not readily available to uninfected plant roots.[8,21,42,43] In some instances, growth and P content of plants given rock phosphate is enhanced by mycorrhizal infection, although the growth response is usually affected by the amount of fertilizer added, and growth increases may be obtained only at extremely high rates of application of rock phosphate.[42-44] Other experiments have demonstrated that mycorrhizal infection did not improve the P availability from insoluble rock phosphate or calcium phosphate sources even though plant growth responses occurred.[37,45-47] Furthermore, Powell[48] has shown that the increased uptake of P from various rock phosphate sources following mycorrhizal infection decreased over successive harvests as the available P levels in the soil were depleted. These results tend to support the conclusion of Tinker[19] that mycorrhizae do not mobilize insoluble soil phosphates, but rather they greatly increase the utilization of that which is available. The growth responses of mycorrhizal plants in the presence of relatively insoluble P sources could merely be due to a more efficient uptake of the chemically dissociated ions which would be drawn into solution as the P in the soil solution is depleted.

Nevertheless, there is some evidence to suggest that mycorrhizae may be able to utilize some organic P sources in soil. The major organic P source in the soil is phytate.[22] Although both mycorrhizal and nonmycorrhizal plants can utilize calcium phytate as a P source, there is more active hydrolysis of phytate by mycorrhizal plants in both soil[37] and in axenic culture.[33] Herrera et al.[49] showed a direct transfer of ^{32}P to mycorrhizal plants from leaf litter. However, in this study it is unclear if the mycorrhizal fungi are absorbing organic P from the litter or if they are absorbing previously degraded P from a soluble pool. Nevertheless, the study is interesting in that it shows possible transfer of P from a discrete insoluble P source: in all other studies, the insoluble P sources used are finely divided and well mixed into the soil.

a. Possible Mechanisms for Improved Uptake from Insoluble Phosphorus Sources

If indeed, VAM are capable of utilizing insoluble P sources, then several mechanisms are possible.

First, organic phosphates may be made available to mycorrhizal plants through the action of phosphatases.[22] Phosphatase activity is generally distributed through the soil by virtue of the presence of soil microorganisms. However, the acid phosphatase present in actively growing VAM hyphae[13] and the increased phosphatase activity of root surfaces as a result of mycorrhizal infection[33] would result in Pi being liberated from organic phosphates in a localized way (immediately adjacent to the cell surface) to be captured by the uptake mechanisms. This is a possible reason why mycorrhizal plants utilized more phytate than nonmycorrhizal plants.[33]

Second, inorganic P sources could be solubilized by the action of organic acids. As part of their normal carbohydrate metabolism, many fungi produce oxalic acid which is able to chelate calcium ions or to remove calcium from soil solution by precipitation as calcium oxalate.[22,50] Oxalic acid could also facilitate release of phosphate ions adsorbed on iron and aluminium hydroxides although there is as yet no direct evidence for this for VAM.

Third, some phosphate-solubilizing bacteria and soil fungi can release P into the soluble P pool of the soil, and mycorrhizae might utilize this extra source. However, unless the mycorrhizal hyphae were to specifically stimulate this activity by some synergistic interaction (Chapter 7), the effect would come down to better exploration of existing pools, since the nonmycorrhizal root would also use any such P released adjacent to it.

3. Storage and Translocation of Phosphorus

Polyphosphate (poly P) is a major P reserve in many fungi[22] and in ectomycorrhizae,[51] and it occurs in VAM onion roots although not in uninfected roots.[52] Root cortical cells containing arbuscules often have higher concentrations of P than do uninfected cells,[53-55] and intracellular granules containing poly P are commonly found in young and proliferating arbuscules although they are absent from collapsed arbuscules.[56-60] Indeed, poly P can form up to 40% of the total P present in the fungal component.[52] Transfer of mycorrhizal roots from low to high P media results in a rapid accumulation of poly P.[6,52] Enzymes of poly P synthesis[22] have been found in mycorrhizal tissue.[6] Polyphosphate kinase, which catalyzes the transfer of the terminal phosphate from ATP to poly P, was detected in both external hyphae and mycorrhizal roots but not in uninfected roots, indicating that poly P can be synthesized only by the fungal component of the mycorrhiza.

As has been found with ectomycorrhizae[51] the poly P granules in VAM could also have a role in the storage of P in the fungal vacuoles prior to transfer of P to the host plant. In the complete mycorrhizal system, the VA fungal biomass forms only a small proportion — probably less than 10% — of the root dry weight,[20] and so the total amount of mycorrhizal P stored in this form will be small, despite the importance of poly P to the fungal component. Experimental evidence suggests that VAM store very little of the absorbed P;[24,52] instead, after a short time lag,[27] they transfer it rapidly to the host plant. Some support for this rapid transfer of P comes from the high calculated P flux rates in fungal hyphae. Experimental values range between 0.1 to 2×10^{-9} mol cm^{-2} s^{-1}.[27-29] The flux value of 3.8×10^{-8} mol cm^{-2} s^{-1} calculated for hyphal entry points into the root[26] is probably atypically high owing to a funnelling effect at the points of entry into the root.

Of the various mechanisms usually proposed for fungal translocation (such as diffusion, bulk flow, and protoplasmic streaming), calculation of the energy requirements indicate that cytoplasmic streaming is the most likely mechanism to account for the observed P flux rates.[18] Cox et al.[56] and Callow et al.[52] proposed that P may be translocated from the mycelium in the soil to the arbuscule and hence into the root, by the transport of poly P granules in streaming cytoplasm. Subsequently Cox et al.[57] calculated cytoplasmic streaming rates of 12.6 cm hr^{-1}, which is adequate to account for the calculated flux values. Cooper and Tinker[28] have further shown that hyphal translocation can occur in the absence of host transpiration, and is slowed by low temperatures and stopped by a cytoplasmic streaming inhibitor (cytochalasin B), thereby supporting the hypothesis that translocation of P occurs in the hyphae by cytoplasmic streaming.

4. Transfer to Host

Histochemical and electron microscope studies have shown the arbuscule to be a specialized haustorial organ capable of tremendous metabolic activity and particularly well adapted as a site of nutrient exchange between fungus and host.[61-64] It is likely that the exchange occurs in the arbuscule branches.[63,65-67] The average arbuscule has a life of around 4 days[68] and calculated P flux values are high enough to explain P inflow rates, supporting the theory that transfer of P from fungus to host occurs across living membranes, rather than by arbuscule decay as previously suggested.[69]

Although it now seems likely that P is translocated along the hyphae as poly P granules, little is yet known of the biochemical mechanisms involved. In the scheme proposed by Woolhouse[34] (Figure 1), a passive phosphate carrier mediates the Pi secretion at the plasmalemma of the arbuscule. The various transport and unloading steps within the arbuscule may be linked to poly P metabolism. Activity of enzymes of polyphosphate breakdown (exopolyphosphatase and endopolyphosphatase) is greater (134%) in mycorrhizal roots than in uninfected roots.[6] Both enzymes have been detected in extracts of internal hyphae, but not external hyphae, thus supporting the suggestion of Callow et al.[52] that on reaching the arbuscule, poly P may be broken down by a polyphosphatase enzyme, liberating Pi. Alternatively, the reaction catalyzed by polyphosphate kinase is readily reversible, so there is also the possibility that poly P could be broken down in this way, liberating ATP.[22] The Pi (or ATP) so released would then be transferred across the arbuscule into the host. The nature of this step remains uncertain. However, the situation is very similar to that which applies in the uninfected root, when phosphate must be transferred from the cortical cells (equivalent to the arbuscule) to the xylem lumen (equivalent to the interfacial matrix), thence to the absorbing cells (host cell) in the leaf or fruit (Figure 1). Evidence to date does not rule out some form of active transport across the arbuscule plasmalemma.

Support for Woolhouse's[34] hypothesis of a carrier-mediated active transport mechanism for P uptake at the host plasmalemma comes from the presence of intense ATPase activity along the host plasmalemma in living arbuscule branches.[67] This concentration of plasmalemma-bound ATPase activity is not observed in cells containing either very young or degenerated arbuscules, implying that it is a specialization of the host membrane which occurs when the structure of the arbuscule is favorable for nutrient exchange.

Other enzymes have also been implicated in P metabolism: mycorrhizal-specific alkaline phosphatases are located in the vacuoles of mature arbuscular and intercellular hyphae.[67,70-73] Maximum activity occurs while infection is young (100% arbuscular) coinciding with the start of the mycorrhizal growth response, but disappears with degeneration and collapse of the arbuscule. These enzymes appear to be of fungal origin. Alkaline phosphatases may be involved in active P transport and transfer processes in VAM fungi as has been proposed for certain basidiomycetes,[74] though they could alternatively be involved in the penetration of the hypha into the host cells. However, further experiments are needed to discover their exact role.

B. Effect of Phosphorus on Mycorrhizal Infection

Many environmental and edaphic factors can influence the development of mycorrhizal infection in a root system. However, one of the single most important factors involved in controlling infection is soil and plant P. High soil P levels frequently inhibit infection.[1,9] Many publications have failed to determine whether P affects fungal development in the soil or the infection and development process inside the root.[26,75,76] Some authors suggest that mycorrhizal infection is regulated by soil P and that there may be a threshold level of soil P below which infection can be maintained at a high level.[76,77] However, there is strong evidence that infection is more likely to be regulated by the P content of the plant tissue rather than by the P in the soil.[78-81]

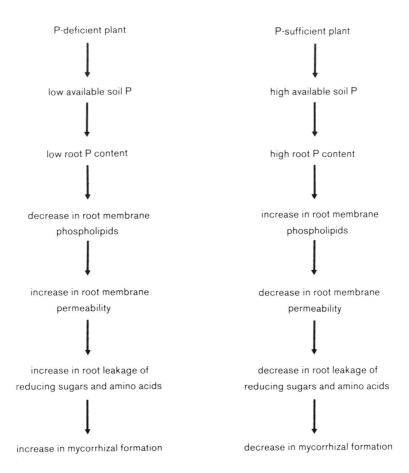

FIGURE 2. A proposed scheme for membrane-mediated phosphorus inhibition of mycorrhizal infection.[83-85]

The mechanism whereby the internal P content of the host regulates mycorrhizal infection is not clear. Mycorrhiza-specific phosphatases, which are associated with the most active phase of mycorrhizal infection, may be involved (Section II. A.). Activity of these phosphatases declines at high P levels indicating a possible mechanism whereby P acts to inhibit mycorrhizal infection in the root tissue.

Plant P status can also affect the soluble carbohydrate content of the root and root exudation, thereby regulating the carbon source and supply which would be available to the fungus.[81,82] There is increasing evidence to suggest that the mechanism of P control of infection is more likely to be associated with the permeability of the root cell membranes rather than with any changes in the root carbohydrate content.[83-85] These authors propose that under low P nutrition, increased root cell membrane permeability leads to increased loss of metabolites, sufficient to sustain the growth of the mycorrhizal fungus during pre- and postinfection (Figure 2). Subsequently, mycorrhizal infection leads to improved root P nutrition and a reduced loss of metabolites. However, care is needed not to take too simplistic a view. For example, the efflux of Pi from plant cells is decreased by P deficiency,[86] arguing against a general effect and pointing to a selective increase in carbohydrate permeability. The increased carbohydrate supply in the host, which is a known response to P deficiency, would reinforce the effect. Thus, exudation of reducing sugars from mycorrhizae is higher with a longer photoperiod.[85] Conversely, decreased light intensity would tend to reduce root exudation and hence to reduce mycorrhizal formation. It may be possible to grow a P-deficient plant in low light intensities and cause it to have the mycorrhizal status of a P-sufficient plant.

Irrespective of the mechanism by which plant P status affects mycorrhizal infection, it is clear that different VAM fungi differ in their sensitivity and tolerance to applied P.[9] Mycorrhizal infection is often found in plants growing in well-fertilized soils, and it seems that some VAM fungi are quite tolerant of high P fertilizer regimes.[87] Mycorrhizal fungi derived from low P soils are much more sensitive to P supply than those from high P soils, as an increase in P supply causes a decrease in plant infection.[81,87,88]

C. Nitrogen Nutrition and Metabolism

Plants infected with VAM fungi generally have a lower tissue nitrogen concentration (as % dry weight) than those without mycorrhizae;[89-91] this is probably due to the increased growth of infected plants.[8] Mycorrhizal plants sometimes do have a higher N content but it is usually not clear whether this is due to increased nitrogen uptake or is merely a side effect of relief from another deficiency, such as P.[8,23] However, a recent experiment has suggested that VAM may affect the nitrogen nutrition of plants more directly. Mycorrhizal celery plants were able to obtain ^{15}N-labeled nitrogen from inorganic and organic nitrogen sources placed well out of reach of the root system, with more ^{15}N appearing in shoots and roots of infected plants than in the controls.[92] It seems likely that this increased uptake of nitrogen occurred by hyphal translocation.

1. Assimilation of Nitrogen and Ammonium

High concentrations of inorganic nitrogen compounds reduce mycorrhizal infection, and this reduction is generally greater with NH_4^+ than with NO_3^-.[93,94] However, the plant growth response to mycorrhizal infection may be greater in the presence of NH_4^+ than NO_3^-,[93,95] possibly reflecting preferential uptake and metabolism of NH_4^+ by the symbiont. Although most fungi can utilize both NO_3^- and NH_4^+ they often have a preference for ammonium-N,[96] and ammonium-N also appears to be a major N source for ectomycorrhizal fungi.[8,23] Because NO_3^- is highly mobile and moves readily in the soil solution whereas NH_4^+ is comparatively immobile, it might be predicted that mycorrhizae would have more influence on N uptake and translocation if ammonium rather than nitrate were the nitrogen source.[8,21]

Ammonium-N and nitrate-N have different pathways for metabolism, cation-carboxylate storage, and pH regulation, and hence they have rather different biochemical and physiological implications for the host.[8,23,93] Nitrate is reduced, first to nitrite by nitrate reductase, then to ammonium by nitrite reductase. Information on nitrate assimilation by VAM is scanty, although spores of *Glomus mosseae* have been shown to reduce nitrate to nitrite.[97] However, there have been no studies of nitrate assimilation by the mycelium. Ammonium-N, once inside the cell, becomes directly incorporated into the various pathways for amino acid synthesis.[23,96] Assimilation may be by glutamate dehydrogenase (GDH), or else via glutamine synthetase (GS) and glutamate synthase (GOGAT) with the formation of glutamate. The enzymes of both these pathways occur in bacteria and higher plants but the GS/GOGAT system may not be present in most fungi.[23] The presence of NADP-linked GDH in *Glomus mosseae*[13] suggests that amino acid synthesis could occur by this pathway, although the GOGAT system has not yet been investigated.

2. Amino Acid Metabolism

The formation of VAM may result in changes in plant N metabolism,[91,98] probably as a response to relief from deficiency of other elements. Amino acid content is generally lower in leaves of mycorrhizal or P-supplemented citrus,[91] suggesting increased utilization as a response to improved P supply. In contrast, some authors found a much higher amino acid content (3 to 10 times),[99] and a much greater predominance of arginine and citrulline[100,101] in mycorrhizal roots as compared with uninfected ones. Arginine also accumulated in the leaves, possibly reflecting transport from the roots, or even the hyphae (arginine is a major constituent in the mycelium of *Glomus etunicatum*[91]). At present, direct evidence on the nitrogen metabolism of the fungal component is lacking.

It is not too surprising that there are conflicting results on relative N levels. In a nonmycorrhizal plant that is restricted in its growth by limited P supply, N can easily be in oversupply, leading to a characteristic formation of amides and arginine. If the P supply is increased, then growth can increase to the point where N is no longer in oversupply or is limiting, and the amides will disappear. If however the N supply is much more restricted, the nonmycorrhizal plant may not accumulate N. In this case, mycorrhizal infection, by stimulating P supply and thus root growth, could improve the N supply beyond the additional needs of growth. What happens is a consequence of the interplay between two major plant nutrients.

3. Dinitrogen Fixation

Despite some early suggestions that mycorrhizae can fix atmospheric nitrogen, there is no evidence that they can indeed do so. However, by interaction with other microorganisms, VA endophytes are able to enhance the nitrogen nutrition of the host (see Chapter 7). They can do this in two ways: (1) by stimulating the number and activity of free-living nitrogen fixers (such as *Beijerinckia* or *Azotobacter*) in the soil, or (2) by increasing nodulation and the nitrogen-fixing ability of symbiotic nitrogen fixers of legume or actinorhizal hosts. The main effects appear to be indirect, and to result from the improved P uptake and plant growth that results from mycorrhizal infection. Increased uptake of minor elements, such as copper, zinc, and molybdenum by mycorrhizal fungi, may also influence nodulation and nitrogen fixation.[102]

D. Uptake of Other Nutrients

Uptake of nutrients such as K, Ca, Mg, Fe, Cu, Mn, Zn, Na, or B has often been implicated in mycorrhiza-assisted nutrition, but results are generally variable.[1,103] Mycorrhizal plants may contain higher total quantities of some elements than uninfected plants because of their greater biomass, but more often nutrient concentrations of mycorrhizal and nonmycorrhizal plants are similar on a dry weight basis.[104-107] In general, nutrient uptake is likely to be affected by P deficiency or whatever is limiting plant growth. Unfortunately, there are very few critical experiments which have been designed to investigate specifically the uptake and translocation of other minerals in VAM. Most work is complicated by P-deficiency, and any increased uptake of other nutrients is merely a side effect of improved P nutrition.

Nutrient uptake can be influenced by both soil P level and mycorrhizal infection. Element concentrations in mycorrhizal plants can decrease with increasing levels of applied P, so that concentrations, particularly of Zn, Cu, K, and S, converge in mycorrhizal and nonmycorrhizal plants as soil P is increased.[44,107-110] This convergence has been attributed to the decrease in mycorrhizal infection which occurs at high P levels,[108] but it is of course possible that P was not limiting.

Mycorrhizal growth responses are sometimes recorded when trace element deficiencies limit plant growth. For example, mycorrhizal inoculation can relieve zinc deficiency in peach and apple.[111-113] However, micronutrient deficiencies, particularly of Cu and Zn, can also be induced at high P levels.[21,114] It is likely that at least some of these P-induced micronutrient deficiencies occur as a result of a reduction in mycorrhizal infection at high soil P levels.[107,110,115]

Direct uptake of Zn has been observed in VAM plants,[116] although experiments on radioactive labeling of the Zn pools in soil suggest that VAM have access only to the more soluble Zn fractions.[117] Furthermore, using split-plate cultures or soil chambers similar to those used for P uptake studies, Cooper and Tinker[27] and Rhodes et al.[118] have demonstrated hyphal translocation of Zn from soil to the host plant. However, this translocation of Zn is influenced by P nutrition; at high P levels, plants were unable to obtain Zn, possibly because of the suppression of mycorrhizal infection and the elimination of hyphal translocation,[118] although interactions between P and Zn are complex.[114]

Enhanced sulfur (S) uptake by VAM plants and hyphal translocation of S have also been demonstrated.[27,31,119] However, in soil chambers severing of the fungal hyphae made little difference to ^{35}S uptake by mycorrhizal plants.[31] Mobility of SO_4 ions in soils is much greater than for H_2PO_4 ions and hence hyphal translocation is probably not as important for S as for P. S present at the root surface may be taken up more rapidly by mycorrhizal roots because of their greater absorbing power. Hyphal translocation is most likely to occur in soils which are deficient in S, although at least some of the increased absorption of S by roots of mycorrhizal plants can be attributed to improved P nutrition.[120]

Although the hyphae of VAM are able to translocate other nutrients besides P, the relative amounts of other nutrients may be low. For example, the molar amounts of P, S, and Zn transported by hyphae of *Glomus mosseae* in 10 days were in the ratio of 35:6:1.[27] This suggests a high relative efficiency in the uptake and translocation systems for P, at least for that particular endophyte. However, endophytes do differ in their tolerance to heavy metals[121] and in their ability to assist in Cu and Zn nutrition in plants,[106] and it is likely that they may differ in their capacity to translocate other nutrients, rather as they do for P.

Hyphal translocation can account for the movement of some Ca to mycorrhizal roots.[122] The amount moved is small, and though the relevance of hyphal transport to the Ca supply of plants is unclear, the effect is likely to be a minor one.[44,108] The significance of Ca is likely to lie in its role as an essential nutrient of the fungus. X-ray microanalysis of elements in cells of mycorrhizal roots has shown calcium to be a strong secondary constituent of poly P granules,[53,59,60] presumably balancing the strong anionic charge of poly P. Calcium may also be involved in P transfer to the host because of its ability to stimulate alkaline phosphatase activity (Section II. A.).[71] Furthermore, Ca is required for the maintenance of membrane integrity.[22]

III. CARBON PHYSIOLOGY AND BIOCHEMISTRY

The carbon nutrition of mycorrhizal fungi and their symbiotic associations has been discussed recently.[2,8,123-125] Information on carbon nutrition of VAM is still scanty, in contrast to the situation with ecto- and orchidaceous mycorrhizae. This is unfortunate, as carbohydrate metabolism of VA endophytes may differ from that of other mycorrhizal classes.[126] It is commonly accepted that the VAM fungi are obligate symbionts and that carbohydrates are transferred from autotroph to heterotroph.[2]

A. Factors Affecting Mycorrhizal Development

In 1942, Bjorkman[127] proposed that the extent of ectomycorrhizal infection was positively correlated with the amount of soluble carbohydrates in the root. Since then, this hypothesis has caused a good deal of investigation and controversy (see Lewis[2]). For VAM, there is a variable correlation between mycorrhizal infection and total or reducing sugar content of root exudates and extracts,[84,128,129] although an increase in the soluble carbohydrate content of P-deficient clover roots was directly associated with an increase in infection.[81] Alteration of root carbohydrates as a result of changes in P nutrition can have a major effect on mycorrhizal infection (Section II. B.).

Infection is likely to be affected by any factors which influence host photosynthesis such as light intensity, photoperiod, or defoliation. In general, mycorrhizal infection and development is reduced under low light intensity, short daylengths, and with defoliation,[85,130-132] probably because of a reduction in supply of photosynthate to the fungus. Vesicle frequency also decreases under conditions of reduced carbohydrate supply to the host[130,132] although this is not surprising if indeed vesicles are carbohydrate storage organs for the fungus (Section III. B.).

B. Carbon Transfer from Host to Fungus
1. Transport and Storage of Carbon Compounds

It is likely that the fungus obtains the bulk of its carbon from host sugars as short-term $^{14}CO_2$ labeling experiments have shown transport of photosynthate from the host to the fungus.[133-135] Most (70 to 90%) of the ^{14}C label present in both the roots and mycelium was in the form of soluble carbohydrates. Though sucrose and glucose are the major components of the labeled (and unlabeled) carbohydrate pools in the uninfected roots, mycorrhizal roots, and fungal mycelium, we do not know if sucrose is the form transported into the hypha from the root. Polyols may be involved. In ectomycorrhizae, which similarly form a sink for carbohydrate from the host,[136] the carbohydrate transfer involves the conversion by the fungus of host-derived glucose into polyols such as mannitol, fungal sugars such as trehalose, and ultimately the storage polysaccharide glycogen, which are unavailable to the host plant.[124,125,133] Although trehalose and mannitol have not been located in VAM roots,[130,133,137] Cooper and MacRae[135] have readily detected both polyols (inositol, mannitol/sorbitol) and trehalose in VA fungal mycelium stripped from root surfaces. These compounds were detected only in very small quantities in mycorrhizal root tissue, presumably because the high ratio of root to fungal tissue masks their presence.

a. Storage as Lipid

The accumulation of lipid in fungal structures points to synthesis of lipid by VAM fungi as providing a possible alternative sink for photosynthate.[56,133,138] Oil droplets and lipid globules have frequently been observed in hyphae, spores, and vesicles.[138-141] Electron micrographs show abundant lipid droplets in arbuscules, vesicles, and hyphae, particularly in older infections and in mature arbuscules, but none in younger parts of the internal hyphae and arbuscules at early stages of development.[56,61,69,142,143]

^{14}C derived from host photosynthate can accumulate in lipid-filled spores and in lipid-rich hyphae, vesicles, and arbuscules.[56,144] The lipid component of mycorrhizal roots became labeled when plants photosynthesized in the presence of $^{14}CO_2$ or were supplied with exogenous sources of labeled sucrose, acetate, and glycerol.[134] However, the amount of ^{14}C from photosynthate incorporated into lipids was unexpectedly low (only 5 to 10% of the total fixed) in well-infected onions[134,135] although more (20 to 30%) label accumulated in lipid fractions in mycorrhizal roots of onion under conditions of P stress at early stages of mycorrhizal infection.[135]

b. Storage as Glycogen

Some storage of carbohydrate may occur in the hyphae as glycogen.[64,133,142] Glycogen accumulation is greatest in hyphae containing large concentrations of lipid and in the arbuscular hyphae.[66,69,142,143]

Generally starch is absent or scarce in infected host cells, particularly in those occupied by arbuscules[69,140,142,143] and this is usually associated with an accumulation of glycogen.[69] The absence of starch in infected cortical cells may be evidence that translocated sugar is synthesized into insoluble glycogen in the fungus. In orchid, ecto- and eucalypt mycorrhizae, a similar disappearance of starch occurs in the root tissue, and this has been correlated with movement of sugars to the fungus.[145-147]

Accumulation of glycogen in arbuscular hyphae, and of glycogen and lipid in the intracellular hyphae which subtend arbuscules, strongly suggests that transfer of sugars to the fungus occurs during arbuscular development and precedes senescence of the arbuscule.[69,140] However, Strullu[148] and Bonfante-Fasolo[143] have observed amyloplasts in infected cells of *Taxus* and grapevine, suggesting that only a portion of the host sugar is used by the fungus and some stored in the host cell as starch. It is possible that some of the starch reserves are used by the host cell to meet the energy requirements for the intense metabolic activity occurring in the cytoplasm of infected cells.[66,142]

c. Growth and Storage "Sinks"

It is relevant to distinguish between possible "growth sinks" and "storage sinks" for host photosynthate. For example, storage of lipid may indicate a high carbohydrate supply but no growth, whereas high levels of soluble carbohydrates may be more typical of active growth. Some support for this comes from the data of Cooper and MacRae.[135] ^{14}C-labeled lipid accumulated in mycorrhizal roots only when mycorrhizal infection was low or plants were P-stressed (i.e., both low growth conditions). However, when infection was well established and plants were growing vigorously, ^{14}C label accumulated in carbohydrates. The proportion of ^{14}C-labeled photosynthate incorporated into the soluble carbohydrate components of roots of rapidly growing mycorrhizal and nonmycorrhizal plants was similar,[133,135] implying that under adequate growth conditions, the fungus has little effect on the overall carbon metabolism of the plant. It is likely that any storage sinks would be small, and possibly transitory.

Fungal mycelium, both in the soil and root, might be expected to constitute significant growth and storage sinks, particularly during the exponential and lag phases of infection (see also Section III. E.). Furthermore, the increased synthesis of cytoplasm and membrane material and the increased root respiration which occurs in the host as a response to infection may also constitute a photosynthetic sink (see also Section III. E.).[68,142] Alternatively, a carbohydrate pooling system may operate as is observed in some biotrophic associations.[149] In these cases, the soluble carbohydrate transport pools are separate from the more slowly turning over storage pools in fungal tissue.

2. Carbohydrate Transfer to Fungus

Little is yet known about the process whereby carbon compounds are transferred from the host to the fungus. Bevege et al.[133] have proposed that a concentration gradient is maintained by the fungus converting the soluble sucrose from the host into insoluble glycogen and subsequently to lipid. These metabolic steps would require the consumption of ATP for production of phosphorylated intermediates. Fungi can have both active and passive sugar transport systems.[150] Woolhouse[34] has proposed that in VAM, sugar is passively released into the interfacial matrix by the host plasmalemma and is then actively taken up by an energy-linked sugar transport system located in the fungal plasmalemma (Figure 1). The process envisaged is essentially the same as that which is believed to occur in the leaves of higher plants, where sugar is released from the photosynthetic cells and actively accumulated by the transfer cells of the vein endings.

The arbuscule is probably the site of P transfer between fungus and host (Section II. A.) and is likely that carbon transfer between host and fungus also occurs here. Polyphosphates may serve in phosphorylation for the active transport of carbon skeletons into the arbuscule from the host either through the ATP produced by degradative polyphosphate kinase action or through a direct phosphorylation of sugars by enzymes of the polyphosphate glucokinase type.[34,52] The recent discovery of poly P glucokinase activity (presumed to be of fungal origin) in mycorrhizal roots[6] and ATPase activity located along the host plasmalemma in living arbuscule branches[67] (Section II. A.) supports the concept of an active sugar transfer system between host and fungus.

3. Metabolic Pathways

The possible metabolic pathways of carbon utilization in VAM are largely uninvestigated. Dehydrogenases indicative of glycolysis are found in hyphae, vesicles, arbuscules, and spore germ tubes.[13,140] From this, MacDonald and Lewis[13] have inferred that VAM fungi possess the Embden-Meyerhof-Parnas glycolytic scheme, the hexose monophosphate shunt (or Pentose phosphate cycle), and the tricarboxylic acid cycle.

A cyanide-insensitive respiratory pathway has been noted in VAM roots.[151] Such a pathway of electron transport to oxygen has been established in the sheath tissue of ectomycorrhizal roots.[152,153] This pathway is not coupled to oxidative phosphorylation and may operate when oxidative phosphorylation is reduced by ADP limitations. It is likely that the operation of such a pathway would increase the overall utilization of carbohydrate in mycorrhizal tissues.[154]

C. Carbon Transfer from Fungus to Host

Although the net flux of carbohydrate is from host to fungus this does not preclude some movement of carbon in the opposite direction, from fungus to host. Such reverse transfer of ^{14}C-labeled compounds has been demonstrated in ectomycorrhizae,[155] in ericoid mycorrhizae,[156] and orchidaceous mycorrhizae.[157] VA endophytes are capable of uptake and translocation of ^{14}C (from ^{14}C-glucose) in the soil, and carbon compounds can be transferred from one plant to another via fungal hyphae.[158] However, it is uncertain whether the carbon was released to the recipient plant or retained in fungal tissues in the root. Other suggestions that carbon compounds are absorbed by the external mycelium have never been fully substantiated.[159]

Lewis[2] and Harley[126] have pointed out that carbon which moves from host to fungus as carbohydrate could return as other compounds. For example, the products of carbohydrate catabolism would provide the necessary skeletal materials for transfer of amino acids between fungus and host. In ectomycorrhizae, glutamine is the principal mobile nitrogen compound between fungus and host.[2] In VAM, although glutamine, glutamic acid, and γ-amino-butyric acid were the first amino acids labeled with ^{14}C,[135] there is as yet no direct evidence that these were transferred from fungus to host.

D. Lipid Biochemistry

Because of their possible role as storage compounds (Section III. B.), some attention has been given to the nature of lipids occurring in fungal tissue and in mycorrhizal roots. Infected roots contained more total lipid than uninfected roots.[138] Spores generally contain around 45% lipid on a dry weight basis but this can increase to 70% during spore germination.[160,161] This is high compared to most fungal spores,[162] although accumulation of lipid in fungal structures is encountered in a range of host-parasite systems.[163] Lipid composition varies during spore germination and germ tube growth, but triglycerides are generally the most abundant neutral lipids in fungal mycelium and mycorrhizal roots[138,164] and in fungal spores.[160,161,165]

1. Phospholipids

Phospholipid content of spores is generally low (3 to 12%),[161] although mycorrhizal roots have a substantially higher phospholipid content than uninfected roots.[138,164] Some of the additional phospholipid may be located in the host membrane although proliferation of this, particularly around arbuscules as a result of infection,[61] may mask the smaller amount of fungal phospholipid present. Phospholipids in fungal tissue (spores and mycelium) differed from those of the host plant, in particular by the higher proportions of phosphatidyl ethanolamine and phosphatidyl serine and the absence of phosphatidyl inositol and glycolipids.[138,161] However, the arbuscule wall appears to be primarily glycolipid in composition in contrast to the walls of vesicles and hyphae which are chitinous.[140]

2. Fatty Acids

Fatty acids of VAM spores appear similar to those of other fungi and plants in that the predominant fatty acids are in the 16:0, 16:1, 18:1, and 18:2 classes.[160,161] However, mycorrhizal spores also contain a range of polyunsaturated fatty acids which are rare in other fungi of the Mucorales. Striking differences in the fatty acid composition between mycorrhizal and nonmycorrhizal citrus roots have been noted. Four fatty acids, unusual to plant material (in the 16:1, 18:3, 20:3, and 20:4 classes), occurred in mycorrhizal citrus roots but not in control roots and their accumulation seemed to be at the expense of C_{18} fatty acids.[164,165] At present it is not known whether these unusual fatty acids are confined to the hyphae in the mycorrhizal root tissue or also occur in the host cells. Peroxidase and catalase activities have been noted in fungal tissue[166] and in senescing arbuscules.[140] These enzymes may be associated with the α and β oxidation of fatty acids in the arbuscules. Both enzymes are implicated in the degradation of fatty acids in plants,[167] but little is known about their activities in fungi.[162]

3. Sterols

Mycorrhizal roots generally have a higher sterol content than uninfected roots, and the major components of infected roots and mycorrhizal spores are 24-methylcholesterol, cholesterol, and camposterol.[161,164,165,168,169] The increased metabolism of sterols in mycorrhizal roots could influence plant growth. As components of cell membranes[170] sterols may affect cell permeability, thereby enhancing nutrient or water absorption and translocation. Because sterols also exhibit hormonal activity[171] their increased presence in mycorrhizal roots may help to explain some of the observed hormonal activities of mycorrhizae (Section IV).

E. Carbon Cost of the Symbiosis

As the fungal component derives carbohydrate for its growth from the host, the question of carbon cost of the fungus is important to an overall understanding of the VAM symbiosis. Indeed, Harley[123] has suggested that mycorrhizal growth responses are in fact determined by two opposing processes: a stimulating effect due to enhanced P uptake, and a detrimental effect caused by fungal drain of host photosynthate.

1. Carbon Demand of the Fungal Symbiont

This raises the question of how much carbon the fungus requires for its growth. ^{14}C-labeling experiments have shown that extramatrical hyphae removed only 1 to 4% of carbon from the root system,[133,135] but these experiments did not take respiration into account. Evidence concerning comparative respiratory metabolism of mycorrhizal and nonmycorrhizal roots is conflicting. On one hand, mycorrhizal inoculation increased transport of fixed carbon to the roots of leek and faba bean by about 7 to 10%,[7,172] much being accounted for in below-ground respiration. Indeed, Pang and Paul[172] showed a 74% increase in CO_2 loss from roots as a result of mycorrhizal infection (Figure 3). Although the actual amount of host-derived carbon incorporated into fungal material is small, probably only around 1 to 5% of that fixed by the plant,[7,173,174] it is likely that most, if not all, of the differences found in below-ground respiration between mycorrhizal and uninfected plants can be attributed to fungal respiration. In contrast, Kucey and Paul[174] and Silsbury et al.[175] detected little difference in root respiration rates between mycorrhizal and nonmycorrhizal plants. At least some of this discrepancy may be explained in terms of fungal development. Cooper and MacRae[135] have shown that the amount of ^{14}C incorporated into mycorrhizal roots and fungal mycelium varies with the stage of development of mycorrhizal infection. Roots of young plants with less than 10% infection but well-developed mycelium incorporated 2.5 times more label than uninfected plants. However, at later stages of infection (more than 20%) the two groups of plants incorporated similar amounts of radioactivity.

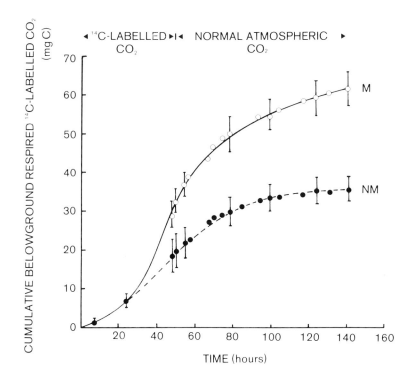

FIGURE 3. Below ground ^{14}C-labeled CO_2 respired by mycorrhizal (M) and nonmycorrhizal (NM) faba beans. $^{14}CO_2$ generated from Na_2 $^{14}CO_3$ (specific activity = 50 µCi/g) with lactic acid. Plants labeled for 48 hr after which $^{14}CO_2$ was replaced with nonlabeled CO_2.[172] Reproduced with kind permission of the *Canadian Journal of Soil Science.*

2. Growth Depressions and Compensation Effect

Inoculation of plants with VAM can cause growth depressions which may be either transitory or persistent.[8,176] Growth depressions have variously been attributed to a pathogenic phase in the establishment of the endophyte, competition between mycorrhizal fungi and plant roots for soil P, P toxicity due to internal plant P levels reaching toxic levels, or to a drain of host photosynthate into the fungus. Although the first three may be appropriate explanations in certain circumstances, none of them has application to all the observed instances. Transient plant growth depressions may perhaps occur as a result of the increased synthesis of host cytoplasm which occurs as a response to infection.[68,69,142] However, the simplest explanation of long-lasting growth depressions, and the one which is gaining in credibility, is that of some form of carbon drain from the host to the fungus. Although the amount of fungal material in and external to the root may be small and actual storage of host-derived photosynthate may be low (Sections II. A., III. B.), it is nevertheless likely that the endophyte can act as a respiratory and growth sink thus causing considerable drain on the host's carbon resources (see elsewhere this section). Metabolic processes of either of the partners in the symbiosis may also affect the host carbon balance. The reduced supply of carbon for the host as a result of competition between the fungus and host may impair host protein synthesis.[177] Carbohydrate supply could also be affected by any change in the method of pH regulation in host cells[8] or by changes in the mobilization of carbohydrate reserves owing to altered hormone levels in the host as a result of mycorrhizal infection.[132]

When plants of equal size are compared, mycorrhizal plants often contain higher internal P concentrations than uninfected plants (Figure 4).[176,178] One explanation is

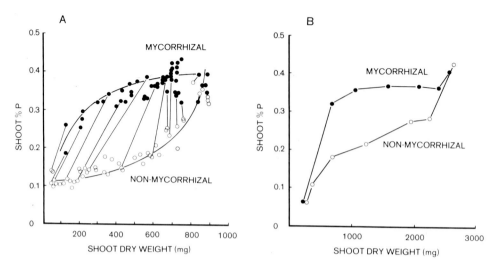

FIGURE 4. Relationship between shoot yield per pot and phosphorus concentration in shoots (% P) of mycorrhizal (•) and nonmycorrhizal (o) plants, grown at different levels of added P. (A) For leeks. Thin lines link data from a selection of results for mycorrhizal and nonmycorrhizal plants grown on the same soil at the same P level.[176,178] (B) For subterranean clover.[44] Both graphs are reproduced with kind permission of the *New Phytologist*.

that the increased carbohydrate demand by the mycorrhizal hyphae increases the P/C ratio. On this basis, the hypothetical dry weight "loss" of shoots from mycorrhizal plants shown in Figure 4 could have been 40 to 60%. Yokom[179,180] has applied similar cost/benefit theories to an experimental situation. He considered that when nonmycorrhizal plants were supplied with levels of P equivalent to those available to mycorrhizal plants, the potential growth improvement was equivalent to the cost accrued by the plants for the symbiotic association. Using this method he found that the estimated cost to the mycorrhizal plants could also be around 40%. However, evidence from ^{14}C-labeling experiments with leek and legumes has shown that the loss of ^{14}C from shoots of mycorrhizal plants relative to that retained in nonmycorrhizal shoots is likely to be much lower (around 10 to 12%), i.e., the ^{14}C retained by mycorrhizal shoots is only 88 to 90% of that in shoots of uninfected plants.[7,172]

If the presence of mycorrhizal hyphae increased the efflux of carbohydrate from the root into the soil, either directly or through the hyphae, loss from the root would exceed utilization by the fungus. Despite these uncertainties, it is clear that differences in shoot dry weight between mycorrhizal and nonmycorrhizal plants of equivalent P content can represent a cost to the plant of mycorrhizal infection.

Under normal growing conditions, mycorrhizal plants may, however, compensate for this carbohydrate drain. Although roots of mycorrhizal faba beans had higher respiration rates than nonmycorrhizal roots there was relatively no change in plant dry matter.[172] Photosynthetic rates and CO_2 fixation rates increase CO_2 fixation in mycorrhizal plants[181-183] by as much as 8 to 17%.[174] Higher chlorophyll levels in mycorrhizal plants with no accompanying change in chlorophyll a/b ratio may indicate a greater number of photosynthetic units.[33,182] Indirect effects of mycorrhizal infection on leaf thickness,[184] percentage dry matter,[7] and stomatal opening[182,185] (see also Section V) may also improve the carbon balance of the host plant. Such effects could explain the conclusions of Silsbury et al.[175] that mycorrhizal and nonmycorrhizal subterranean clover plants have essentially similar carbon economies when grown as swards — both used about 30% of the net CO_2 gain during the light period for purposes of biosynthesis. Increased photosynthesis and subsequent transport of photosynthate may merely reflect improved P nutrition as a result of mycorrhizal infection.[85,134] However, the percentage dry matter in leaves may be inversely affected by shoot P content[7] which is consistent with the C-loss hypothesis of Stribley et al.[176]

The greater loss of carbon from the shoots of mycorrhizal plants is of considerable interest, and may go some way toward explaining the host growth depressions which are frequently observed. Under adequate growth conditions, this loss of yield would probably not occur either because the plant would be gaining benefit from the additional P being transported by the fungus, or because the plant was able to compensate in some way for the loss of carbon. If no compensation occurred then a carbon loss of 10 to 12% could represent a serious loss in crop productivity.

IV. HORMONAL EFFECTS

Auxins, gibberellins, and cytokinins are involved in root development and many basic processes of plant growth, and thus can mobilize nutrients and control their translocation.[186] Mycorrhizal fungi may be able to influence host growth by the production of hormonal compounds. Various auxins, cytokinins, gibberellins, and vitamins are produced by ectomycorrhizal fungi in pure culture,[187,188] although there is no direct evidence for transfer of these fungal hormones to the host plant in the symbiotic association. In contrast, the possible production of growth-promoting compounds by VAM fungi has been little investigated and studies are limited by the inability to grow the fungi in culture.

Host responses which could be attributed to hormonal effects are rare and usually cannot be separated from overriding nutritional components. Infection by VA endophytes has little effect on gross root morphology. Reproductive and anatomical development is sometimes stimulated or altered in mycorrhizal plants;[105,184,189,190] although this has been attributed to hormonal effects, it may merely be due to differences in growth and development which result in infected and uninfected plants being of different physiological ages.[8]

Hormone production by mycorrhizal fungi has also been implicated in the improved rooting of cuttings. Ectomycorrhizal fungi are able to enhance rooting of some woody cuttings,[191] and to stimulate shoot and root growth in the absence of any mycorrhizal infection.[187] VA endophytes have also stimulated rooting of poinsettia.[192] Although in these experiments, treatment with a rooting compound was essential to promote the development of root initials, mycorrhizal inoculation can improve initiation and development of roots of tamarillo and meyer lemon in the absence of rooting hormone, and prior to any mycorrhizal infection.[193] However, care should be taken in the interpretation of such experiments, as substances responsible for improved rooting may have been added with the mycorrhizal inoculum, perhaps from an associated microbial population.[194,195]

There is now evidence to suggest that hormones are directly involved in the interaction between host and fungus. Arbuscule development in roots of cowpea was decreased by disbudding but increased by applications of exogenous IAA, suggesting that auxin influenced their formation.[132] However, as disbudding of plants increases cytokinin production and its export from the roots,[196] arbuscule development could also be influenced by host cytokinin levels. Mycorrhizal infection can substantially increase cytokinin activity in leaves and roots of mycorrhizal plants.[197,198] These increases have been cited as a probable cause of the altered growth habits of mycorrhizal plants in a grazing environment.[183] Higher levels of cytokinins in mycorrhizal roots may be attributed to increased P uptake, or may merely reflect increased hormone production by a larger root mass. Nevertheless, in mycorrhizal plants of *Bouteloua* grown in sterile medium, Allen et al.[197] detected increases in cytokinin activity of 57% in leaves and 111% in roots on a fresh weight basis. The information is particularly valuable as any

Table 1
EFFECT OF MYCORRHIZAL INFECTION ON RESISTANCE TO WATER TRANSPORT

Water regime	Host	Resistance to water transport in mycorrhizal plants	Site or cause of resistance to water transport	P nutritional regime	Ref.
Well-watered plants	Bouteloua gracilis	Decreased (50%)	Whole plant	Low	201
	Red clover	Decreased	Root (fungal hyphae) / Leaf (stomata)	Low—moderate	202
	Onion	Decreased (75%) / No difference	Whole plant / —	Low / High	203
	Soybean	Decreased (40%) / No difference	Root / —	Low / High	200
Water-stressed plants	Bouteloua gracilis	Decreased (53—89%)	Whole plant	Low	182
	Citrus	Decreased	Leaf (stomata)	High	181
	Red clover	Increased	Leaf (stomata)	Low—moderate	202
	Onion	More drought resistant	Improved P nutrition	High	214
	Wheat	More drought resistant	Regulation of stomatal closure	High	219

hormonal effects which could be attributed to soils, root exudates, or other rhizosphere organisms were eliminated by the sterile culture conditions.

Increased hormone levels in mycorrhizal plants would certainly have a significant impact on plant growth and development, and could substantially affect the host growth response to mycorrhizal infection. However, there is as yet insufficient evidence to determine whether VAM fungi can indeed produce hormones, or if they do, if there is a direct transfer of fungal hormones to the host, or whether the symbiotic association stimulates host hormone production.

V. WATER RELATIONS

VAM fungi can tolerate a wide range of soil water regimes and can be found in habitats as diverse as arid deserts and aquatic environments.[9] In mesic to wet habitats where P is limiting, increased plant growth is probably due to improved P uptake alone (Section II). In arid regions, the low level of soil moisture will reduce the P diffusion rate considerably,[17] so that mycorrhizal infection could improve the P nutrition of plants even where the soil P content is quite high. There is the further possibility that mycorrhizal infection improves the water relations of the host.

A. Water Transport Under Mesic Conditions

VAM infection has been found to improve the water relations of many plants (Table 1). However, the mechanisms whereby mycorrhizal infection might increase drought resistance or improve water flow through the plant are still uncertain. One problem is to distinguish between nutritional advantages and those conferred by improved water uptake alone.

Mycorrhizal plants can have a lower resistance to water flow (i.e., higher hydraulic conductivity) than nonmycorrhizal plants.[185] In soybeans in unfertilized soils, mycorrhizal infection reduced resistance to water flow by about 40%, apparently as a result of reduced root resistance.[199,200] In mycorrhizal plants under low P conditions, higher hydraulic conductivity may be associated with changes in other parameters such as higher leaf water potentials, higher transpiration rates, and lower stomatal resistances,

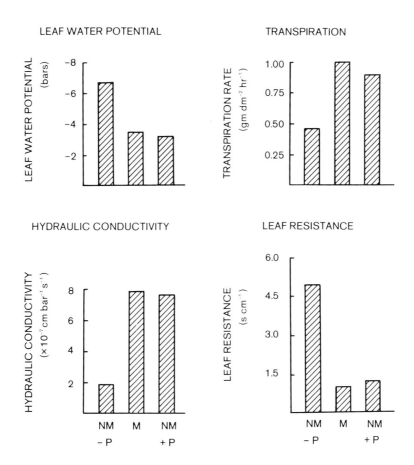

FIGURE 5. Effect of mycorrhizal infection or added P on water relation parameters in well-watered onions (M = mycorrhizal plants; NM = uninfected plants.).[203] Reproduced with kind permission of the *Journal of the American Society of Horticultural Science*.

than those measured in nonmycorrhizal plants.[181,182,201-203] However, at high soil P levels, these parameters were similar in infected and uninfected plants (Figure 5). Furthermore, *p*-chloronitrobenzene (PCNB), a fungal toxicant which inhibits mycorrhizal development[204] and can reduce nutrient uptake by mycorrhizae,[205] did not affect hydraulic conductivity.[200] These results suggest that the enhanced nutrition of infected plants may be more effective in altering hydraulic conductivity than the fungus itself, providing a low resistance pathway for water within the root. However, improved nutrition may not be the only operating factor in some cases. Hardie and Leyton[202] found hydraulic conductivities and water flow rates in red clover roots were higher in mycorrhizal than nonmycorrhizal plants despite additions of moderate amounts of P fertilizer.

If there is the possibility of mycorrhizal fungi lowering resistance to water flow in roots under low P conditions, then the site of such an effect needs to be considered.

Root hairs can indirectly influence water uptake by preventing the development of significant gaps between root and soil, thereby maintaining liquid continuity across the soil-root interface.[206] Hyphae ramifying into the soil are likely to increase the absorbing area for water uptake even further and may also be able to bypass the dry zones that often surround slow growing roots during periods of drought.[15] The low diffusivity of water in soil at low water contents makes this plausible. Hyphae may also bypass any cortical resistance in infected roots.[207] Improved P uptake with increasing host transpiration[28] implies that some water movement does occur in hyphae, and calculations by Allen[201] give an estimated fungus to root water transport rate of 2.8×10^{-5} mg s^{-1} or 100 nℓ h^{-1} per hyphal entry point. This is comparable with evapotranspiration rates measured in other coenocytic fungi.[208] However, calculations by Sanders and Tinker[26] indicate that if water transport through fungal hyphae were to account for the observed differences in P inflow between mycorrhizal and nonmycorrhizal roots, then an unrealistically high flow rate of around 5 cm min^{-1} would be required in the hyphae. Experiments with tritiated water showed no differences in water uptake which could be attributed entirely to hyphal translocation.[185] Some bulk liquid flow obviously does occur in hyphae but it would appear that under adequate water supply at least, the amount is insufficient to affect the water economy of a plant.

Besides having hyphae which penetrate beyond water depletion zones around the root surface, the mycorrhizal roots have greater surface absorbing areas because of their greater root length and diameter[202] and increased branching.[33] Such alterations to root morphology are known to change hydraulic conductivity and water flow rates.[15,209] However, Hardie and Leyton[202] concluded that the small increase of 8 to 18% in the root surface area that they found could not account for the observed increases in hydraulic conductivities in mycorrhizal plants without some alteration in root anatomy which would decrease resistance to water flow. Significant alterations to the stele or cortex of roots have not been found although there is sometimes increased vascular tissue in stems and leaves following infection.[184,190,210] Since water flow through growing roots involves crossing of membrane barriers, water absorption can be affected by membrane permeability.[211] It is likely that mycorrhizal infection can alter membrane permeability through improved P nutrition (Section II) or through increased production of phytosterols (Section III), thereby influencing water transport.

B. Water Transport Under Dry Conditions

Of particular interest are reports that VAM can improve drought tolerance of plants.[212-214] Mycorrhizal plants frequently appear to be less susceptible to wilting and transplant shock than uninfected plants.[181,192,193,202,215-218] Indeed, Schultz et al.[105] have concluded that cessation of diurnal wilting is an indication of mycorrhizal formation on sweetgum. When soil moisture is low, the increased drought resistance resulting from mycorrhizal infection cannot necessarily be duplicated by added P.[214]

Under conditions of water stress, resistance to water transport was decreased by up to 90% as a result of mycorrhizal infection.[182] As moisture stress increased, leaf water potentials of mycorrhizal plants dropped more rapidly than those of nonmycorrhizal controls, and throughout, stomatal resistance remained lower (by 50 to 70%) in inoculated plants. This reduced resistance in mycorrhizal plants could have resulted from improved water uptake, increased photosynthesis, or elevated cytokinin levels which stimulate stomatal opening. However, mycorrhizal plants may also have a greater tolerance for continued drought. When nonmycorrhizal wheat plants and mycorrhizal ones infected with *Glomus fasciculatum* were watered to soil saturation then allowed to continue transpiring while the soil dried, stomata in nonmycorrhizal plants began to close at leaf water potentials of -1.8 MPa and were closed after 4 days, but in mycorrhizal plants they did not begin to close until water potentials of -2.2 to -2.7 MPa were reached and were still transpiring after 6 to 7 days.[219] In another study, nonmycorrhizal plants wilted more easily at a higher soil water potential (-0.8 to -1.2 MPa) than mycorrhizal ones (-1.8 to -2.4 MPa), suggesting that the mycorrhizal plants were better able to extract soil moisture.[202] However, in yet another instance the rate of water depletion did not differ between mycorrhizal and nonmycorrhizal plants,[182] implying that the mycorrhizal plants may merely be more efficient at obtaining available moisture.

Rapid accumulation of proline can be a metabolic consequence of water deficit and salinity stress in some plants.[220] Mycorrhizal plants have lower concentrations of proline in both shoots[91] and roots,[135] implying a lowered physiological water stress. Under conditions of water stress, Levy and Krikun[181] noted a slight drop in proline levels in mycorrhizal plants. In general, not much is known about the capacity of VAM fungi to tolerate water stress although Cooper and Tinker[28] showed that transport processes in *Glomus mosseae* were insensitive to small changes in osmotic pressures of 1 to 2 bars. Some ectomycorrhizal fungi grow vigorously in laboratory media with low water potentials,[221-223] which implies that they can readily extract water against large potential gradients. If VAM fungi can withstand low soil water potentials, then fungal hyphae may have an advantage over root hairs when soil moisture is reduced.

It seems likely that VA endophytes may have a role in improving the water relations of plants. However, it is not clear whether this effect is a direct result of fungal invasion and attributable to improved water flow through hyphae, or a secondary response due to improved nutrition or physiological alterations of the host.

VI. CONCLUSIONS

The most consistent host responses to VAM infection occur as a result of improved P nutrition. Although at first sight infection appears to increase uptake of other nutrients, to improve host water relations and to alter plant hormone levels and host metabolic responses, much of the evidence is largely circumstantial. In many cases, it is still unclear whether the observed effects are due directly to the fungus itself or indirectly to some alteration in host physiology as a result of improved P nutrition. If the benefits are directly attributable to mycorrhizae, then the site and mechanism of action needs to be determined. The demand of the host plant may partly determine the mycorrhiza-induced response, but this aspect is largely uninvestigated. Nevertheless, if the mycorrhizal condition can indeed improve the water relations of the host and increase uptake and utilization of nitrogen and other nutrients, albeit through improved P nutrition, then the capacity of mycorrhizal plants to be exploited in commercial situations is increased.

One of the current aims of mycorrhizal research is to manipulate the symbiosis to improve plant growth under a variety of conditions and in a variety of habitats. In contrast to most pathogens, biotrophs, and indeed some types of mycorrhizal fungi, VA endophytes show little specificity for infection of their hosts.[9] However, the growth responses of a particular host as a result of infection by different fungi can vary considerably. This may reflect the differing efficiencies of the endophytes in their ability to infect roots, to compete with other organisms in the rhizosphere or to transfer nutrients to the host. It may also reflect possible interactions between the host, fungus, and soil type. Endophytes also differ in their tolerance of and in their ability to adapt to various nutritional, environmental, and soil conditions.[9] However, the physiological basis for the differing behavior of various endophytes is still unknown. Such information is needed for selecting superior fungal strains or species appropriate for commercial inoculation procedures.

In view of the commercial possibilities, the probability of VA endophytes depressing host growth is important and requires attention. Not all endophytes depress host growth and some may indeed depress the growth of one host but stimulate the growth of another under the same set of conditions.[193] The growth reduction appears due to some form of increased carbohydrate drain from the host as a result of infection (Section III). A loss of even 10 to 12% carbon from the shoot would be sufficient to cause serious losses in crop productivity in the absence of compensating advantages. The possibility that host can compensate for this drain through increases photosynthesis and carbon assimilation requires further study.

An interesting feature of all mycorrhiza associations is that, in contrast with pathogenic fungi, the symbionts fail to activate the host's defense mechanisms on infection. Even the depressions of growth as a result of VAM infection are more likely to be attributed to some form of carbon drain than to any pathogenic phase in the establishment of the endophyte. Nevertheless, as a result of the host's metabolic response to infection, there are alterations in chitinase activity and phenol metabolism, and increased production of phenolic compounds, oxidative enzymes, phytosterols, cytokinins, and amino acids (Chapter 7), all of which are implicated in disease resistance and which could reasonably be expected to reduce the host's susceptibility to mycorrhizal infection. However, none of these host defense mechanisms appears to be active against the endophyte; furthermore, the endophyte increases the host's tolerance to invading fungal and nematode root pathogens.

Electron microscopy and histochemical studies have pinpointed the arbuscule as the most probable site for the exchange of nutrients and carbon compounds between host and fungus. However, our understanding of the way in which transport is mediated at the host-arbuscule interface is limited largely because of the difficulties associated with the size, isolation, and access of this intracellular region. The arbuscule is similar in many morphological and functional aspects to the haustorial organs of some plant pathogenic fungi. Significant advances have been made possible in the study of haustoria with the development of a technique for isolating haustorial complexes from leaves infected with powdery mildew (see Manners and Gay[149]). Similar techniques may be useful in studying the transport of nutrients and photosynthates across the host-arbuscule interface in VAM and in elucidating the biochemistry of the transport processes.

ACKNOWLEDGMENTS

I am grateful to my colleagues, Drs. R. E. Beever, R. L. Bieleski, A. R. Ferguson, and I. B. Ferguson for their comments and constructive criticism of the manuscript. Thanks are also due to D. H. Yokom for permission to use unpublished data.

REFERENCES

1. Mosse, B., Advances in the study of the vesicular-arbuscular mycorrhiza, *Annu. Rev. Phytopathol.*, 11, 171, 1973.
2. Lewis, D. H., Comparative aspects of the carbon nutrition of mycorrhizas, in *Endomycorrhizas*, Sanders, F. E., Mosse, B., and Tinker, P. B., Eds., Academic Press, London, 1975, 119.
3. Crush, J. R. and Hay, M. J. M., A technique for growing mycorrhizal clover in solution culture, *N. Z. J. Agric. Res.*, 24, 371, 1981.
4. Howeler, R. H., Asher, C. J., and Edwards, D. G., Establishment of an effective endomycorrhizal association on *Cassava* in flowing solution culture and its effects on phosphorus nutrition, *New Phytol.*, 90, 229, 1982.
5. MacDonald, R. M., Routine production of axenic vesicular-arbuscular mycorrhizas, *New Phytol.*, 89, 87, 1981.
6. Capaccio, L. C. M. and Callow, J. A., The enzymes of polyphosphate metabolism in vesicular-arbuscular mycorrhizas, *New Phytol.*, 91, 81, 1982.
7. Snellgrove, R. C., Splittstoesser, W. E., Stribley, D. P., and Tinker, P. B., The distribution of carbon and the demand of the fungal symbiont in leek plants with vesicular-arbuscular mycorrhizas, *New Phytol.*, 92, 75, 1982.
8. Smith, S. E., Mycorrhizas of autotrophic higher plants, *Biol. Rev.*, 55, 475, 1980.
9. Mosse, B., Stribley, D. P., and LeTacon, F., Ecology of mycorrhizae and mycorrhizal fungi, *Adv. Microb. Ecol.*, 5, 137, 1981.

10. Sanders, F. E., Tinker, P. B., Black, R. L. B., and Palmerley, S. M., The development of endomycorrhizal root systems. I. Spread of infection and growth promoting effects with four species of vesicular-arbuscular endophyte, *New Phytol.*, 78, 257, 1977.
11. Warner, A., Spread of Vesicular-Arbuscular Mycorrhizal Fungi in Soil, Ph.D. thesis, University of London, 1980.
12. Graham, J. H., Linderman, R. G., and Menge, J. A., Development of external hyphae by different isolates of mycorrhizal *Glomus* spp. in relation to root colonisation and growth of Troyer citrange, *New Phytol.*, 91, 183, 1982.
13. MacDonald, R. M. and Lewis, M., The occurrence of some acid phosphatases and dehydrogenases in the vesicular-arbuscular mycorrhizal fungus, *Glomus mosseae*, *New Phytol.*, 80, 135, 1978.
14. Ames, R. N., Ingham, E. R., and Reid, C. P. P., Ultraviolet induced autofluorescence of arbuscular mycorrhizal root infections: an alternative to clearing and staining methods for assessing infections, *Can. J. Microbiol.*, 28, 351, 1982.
15. Nye, P. H. and Tinker, P. B., *Solute Movement in the Soil-Water System*, Blackwell Scientific, Oxford, 1977.
16. Bieleski, R. L., Phosphate pools, phosphate transport and phosphate availability, *Annu. Rev. Plant Physiol.*, 24, 225, 1973.
17. Bieleski, R. L., Passage of phosphate from soil to plant, in *Reviews in Rural Science. III. Prospects for Improving Efficiency of Phosphorus Utilisation*, Blair, G. J., Ed., University of New England, Armidale, Australia, 1976, 124.
18. Tinker, P. B. H., Effects of vesicular-arbuscular mycorrhizas on higher plants, in *Symbiosis*, Jennings, D. H. and Lee, D. L., Eds., Cambridge University Press, Cambridge, 1975, 325.
19. Tinker, P. B. H., Soil chemistry of phosphorus and mycorrhizal effects on plant growth, in *Endomycorrhizas*, Sanders, F. E., Mosse, B., and Tinker, P. B., Eds., Academic Press, London, 1975, 353.
20. Tinker, P. B. H., Effects of vesicular-arbuscular mycorrhizas on plant nutrition and plant growth, *Physiol. Veg.*, 16, 743, 1978.
21. Rhodes, L. H. and Gerdemann, J. W., Nutrient translocation in vesicular-arbuscular mycorrhizae, in *Cellular Interactions in Symbiosis and Parasitism*, Cook, C. B., Pappas, P. W., and Rudolph, E. D., Eds., The Ohio State University Press, Columbus, 1980, 173.
22. Beever, R. E. and Burns, D. J. W., Phosphorus uptake, storage and utilisation by fungi, *Adv. Bot. Res.*, 8, 127, 1980.
23. Bowen, G. D. and Smith, S. E., The effects of mycorrhizas on nitrogen uptake by plants, *Ecol. Bull. (Stockholm)*, 33, 237, 1981.
24. Bowen, G. D., Bevege, D. I., and Mosse, B., Phosphate physiology of vesicular-arbuscular mycorrhizas, in *Endomycorrhizas*, Sanders, F. E., Mosse, B., and Tinker, P. B., Eds., Academic Press, London, 1975, 241.
25. Smith, S. E., Smith, A. F., and Nicholas, D. J., Effect of endomycorrhizal infection on phosphate and cation uptake by *Trifolium subterraneum*, *Plant Soil*, 63, 57, 1981.
26. Sanders, F. E. and Tinker, P. B., Phosphate flow into mycorrhizal roots, *Pestic. Sci.*, 4, 385, 1973.
27. Cooper, K. M. and Tinker, P. B., Translocation and transfer of nutrients in vesicular-arbuscular mycorrhizas. II. Uptake and translocation of phosphorus, zinc and sulphur, *New Phytol.*, 81, 43, 1978.
28. Cooper, K. M. and Tinker, P. B., Translocation and transfer of nutrients in vesicular-arbuscular mycorrhizas. IV. Effect of environmental variables on movement of phosphorus, *New Phytol.*, 88, 327, 1981.
29. Pearson, V. and Tinker, P. B., Measurement of phosphorus fluxes in the external hyphae of endomycorrhizas, in *Endomycorrhizas*, Sanders, F. E., Mosse, B., and Tinker, P. B., Eds., Academic Press, London, 1975, 277.
30. Rhodes, L. H. and Gerdemann, J. W., Phosphate uptake zones of mycorrhizal and nonmycorrhizal onions, *New Phytol.*, 75, 555, 1975.
31. Rhodes, L. H. and Gerdemann, J. W., Hyphal translocation and uptake of sulphur by vesicular-arbuscular mycorrhizae of onion, *Soil Biol. Biochem.*, 70, 355, 1978.
32. Owusu-Bennoah, E. and Wild, A., Autoradiography of the depletion zone of phosphate around onion roots in the presence of vesicular-arbuscular mycorrhiza, *New Phytol.*, 82, 133, 1979.
33. Allen, M. F., Sexton, J. C., Moore, T. S., and Christensen, M., Influence of phosphate source on vesicular-arbuscular mycorrhizae of *Bouteloua gracilis*, *New Phytol.*, 87, 687, 1981.
34. Woolhouse, H. W., Membrane structure and transport problems considered in relation to phosphorus and carbohydrate movement and the regulation of endotrophic mycorrhizal associations, in *Endomycorrhizas*, Sanders, F. E., Mosse, B., and Tinker, P. B., Eds., Academic Press, London, 1975, 209.
35. Cress, W. G., Throneberry, G. O., and Lindsey, D. L., Kinetics of phosphorus absorption by mycorrhizal and nonmycorrhizal tomato roots, *Plant Physiol.*, 64, 484, 1979.

36. Owusu-Bennoah, E. and Wild, A., Effects of vesicular-arbuscular mycorrhiza on the size of the labile pool of soil phosphate, *Plant Soil,* 54, 233, 1980.
37. Gianinazzi-Pearson, V., Fardeau, J. C., Asimi, S., and Gianinazzi, S., Source of additional phosphorus absorbed from soil by vesicular-arbuscular mycorrhizal soybeans, *Physiol. Veg.,* 19, 33, 1981.
38. Barrow, N. J., Malajczuk, N., and Shaw, T. C., A direct test of the ability of vesicular-arbuscular mycorrhiza to help plants take up fixed soil phosphate, *New Phytol.,* 78, 269, 1977.
39. Swaminathan, K. and Verma, B. C., Symbiotic effect of vesicular-arbuscular mycorrhizal fungi on the phosphate nutrition of potatoes, *Proc. Indian Acad. Sci.,* 85B, 310, 1977.
40. Swaminathan, K., Nature of the inorganic fraction of soil phosphate fed on by vesicular-arbuscular mycorrhizae of potatoes, *Proc. Indian Acad. Sci. Sect. B.,* 88, 423, 1979.
41. Powell, C. L., Phosphate response curves of mycorrhizal and nonmycorrhizal plants. I. Responses to superphosphate, *N.Z. J. Agric. Res.,* 23, 225, 1980.
42. Islam, R., Ayanaba, A., and Sanders, F. E., Response of cowpea *(Vigna unguiculata)* to inoculation with VA mycorrhizal fungi and to rock phosphate fertilisation in some unsterilised Nigerian soils, *Plant Soil,* 54, 107, 1980.
43. Powell, C. L., Metcalfe, D. M., Buwalda, J. G., and Waller, J. E., Phosphate response curves of mycorhizal and nonmycorrhizal plants. II. Responses to rock phosphate, *N. Z. J. Agric. Res.,* 23, 477, 1980.
44. Pairunan, A. K., Robson, A. D., and Abbott, L. K., The effectiveness of vesicular-arbuscular mycorrhizas in increasing growth and phosphorus uptake of subterranean clover from phosphorus sources of different solubilities, *New Phytol.,* 84, 327, 1980.
45. Mosse, B., Powell, C. L., and Hayman, D. S., Plant growth responses to vesicular-arbuscular mycorrhiza. IX. Interactions between VA mycorrhiza, rock phosphate and symbiotic nitrogen fixation, *New Phytol.,* 76, 331, 1976.
46. Mosse, B., Plant growth responses to vesicular-arbuscular mycorrhiza. X. Responses of *Stylosanthes* and maize to inoculation in unsterile soils, *New Phytol.,* 78, 277, 1977.
47. Barea, J. M., Escudero, J. L., and Azcon-G. de Aguilar, C., Effects of introduced and indigeous VA mycorrhizal fungi on nodulation, growth and nutrition of *Medicago sativa* in phosphate-fixing soils as affected by P fertiliser, *Plant Soil,* 54, 283, 1980.
48. Powell, C. L., Effect of mycorrhizal fungi on recovery of phosphate fertiliser from soil by ryegrass plants, *New Phytol.,* 83, 681, 1979.
49. Herrera, R., Merida, T., Stark, N., and Jordon, C. F., Direct phosphorus transfer from leaf litter to roots, *Naturwissenchaften,* 65, 208, 1978.
50. Sollins, P., Cromack, L., Fogel, R., and Li, C. Y., Role of low molecular-weight organic acids in the inorganic nutrition of fungi and higher plants, in *The Fungal Community — Its Organization and Role in the Ecosystems,* Mycol. Ser., Vol. 2, Wicklow, D. T. and Carroll, G. C., Eds., Marcel Dekker, New York, 1981, 607.
51. Chilvers, G. A. and Harley, J. L., Visualisation of phosphate accumulation in beech mycorrhizas, *New Phytol.,* 84, 319, 1980.
52. Callow, J. A., Capaccio, L. C. M., Parish, G., and Tinker, P. B., Detection and estimation of polyphosphate in vesicular-arbuscular mycorrhizas, *New Phytol.,* 80, 125, 1978.
53. Schoknecht, J. D. and Hattingh, M. J., X-ray microanalysis of elements in cells of VA mycorrhizal and nonmycorrhizal onions, *Mycologia,* 68, 296, 1976.
54. Walker, G. D. and Powell, C. L., Vesicular-arbuscular mycorrhizas in white clover: a scanning electron microscope and x-ray microanalytical study, *N. Z. J. Bot.,* 17, 55, 1979.
55. Bartschi, H. and Garrec, J. P., Microlocalisation of some mineral elements in cortical cells of mycorrhizal and nonmycorrhizal *Vitis vinifera* L. roots, *C. R. Acad. Sci. Ser. D.,* 290, 919, 1980.
56. Cox, G., Sanders, F. E., Tinker, P. B., and Wild, J. A., Ultrastructural evidence relating to host endophyte transfer in a vesicular-arbuscular mycorrhiza, in *Endomycorrhizas,* Sanders, F. E., Mosse, B., and Tinker, P. B., Eds., Academic Press, London, 1975, 297.
57. Cox, G., Moran, K. J., Sanders, F. E., Nockolds, C., and Tinker, P. B., Translocation and transfer of nutrients in vesicular-arbuscular mycorrhizas. III. Polyphosphate granules and phosphorus translocation, *New Phytol.,* 84, 649, 1980.
58. Ling-Lee, M., Chilvers, G. A., and Ashford, A. E., Polyphosphate granules in three different kinds of tree mycorrhiza, *New Phytol.,* 75, 551, 1975.
59. White, J. A. and Brown, M. F., Ultrastructure and x-ray analysis of phosphorus granules in a vesicular-arbuscular mycorrhizal fungus, *Can. J. Bot.,* 57, 2812, 1979.
60. Strullu, D. G., Gourret, J. P., Garrec, J. P., and Fourcy, A., Ultrastructure and electron-probe microanalysis of the metachromatic vacuolar granules occurring in *Taxus* mycorrhizas, *New Phytol.,* 87, 537, 1981.
61. Cox, G. and Sanders, F. E., Ultrastructure of the host-fungus interface in a vesicular-arbuscular mycorrhiza, *New Phytol.,* 73, 901, 1974.
62. Bonfante-Fasolo, P. and Scannerini, S., A cytological study of the vesicular-arbuscular mycorrhiza in *Ornithogalum umbellatum* L., *Allionia,* 22, 5, 1977.

63. Dexheimer, J., Gianinazzi, S., and Gianinazzi-Pearson, V., Ultrastructural cytochemistry of the host-fungus interface in the endomycorrhizal association Glomus mosseae/Allium cepa, Z. Pflanzenphysiol., 92, 191, 1979.
64. Holley, J. D. and Peterson, R. L., Development of a vesicular-arbuscular mycorrhiza in bean roots, Can. J. Bot., 57, 1960, 1979.
65. Bonfante-Fasolo, P., Dexheimer, J., Gianinazzi, S., Gianinazzi-Pearson, V., and Scannernini, S., Cytochemical modifications in the host-fungus interface during intracellular interactions in vesicular-arbuscular mycorrhizae, Plant Sci. Lett., 22, 13, 1981.
66. Gianinazzi-Pearson, V., Morandi, D., Dexheimer, J., and Gianinazzi, S., Ultrastructural and ultracytochemical features of a Glomus tenuis mycorrhiza, New Phytol., 88, 633, 1981.
67. Marx, C., Dexheimer, J., Gianinazzi-Pearson, V., and Gianinazzi, S., Enzymatic studies on the metabolism of vesicular-arbuscular mycorrhiza. IV. Ultracytoenzymological evidence (ATPase) for active transfer processes in the host-arbuscule interface, New Phytol., 90, 37, 1982.
68. Cox, G. and Tinker, P. B., Translocation and transfer of nutrients in vesicular-arbuscular mycorrhizas. I. The arbuscule and phosphorus transfer: a quantitative ultrastructural study, New Phytol., 77, 371, 1976.
69. Kinden, D. A. and Brown, M. F., Electron microscopy of vesicular-arbuscular mycorrhizae of yellow poplar. III. Host-endophyte interactions during arbuscular development, Can. J. Microbiol., 21, 1930, 1975.
70. Gianinazzi-Pearson, V. and Gianinazzi, S., Enzymatic studies on the metabolism of vesicular-arbuscular mycorrhiza. I. Effects of mycorrhiza formation and phosphorus nutrition on soluble phosphatase activities in onion roots, Physiol. Veg., 14, 833, 1976.
71. Gianiniazzi-Pearson, V. and Gianinazzi, S., Enzymatic studies on the metabolism of vesicular-arbuscular mycorrhiza. II. Soluble alkaline phosphatase specific to mycorrhizal infection in onion roots, Physiol. Plant Pathol., 12, 45, 1978.
72. Gianinazzi-Pearson, V., Gianinazzi, S., Dexheimer, J., Bertheau, Y., and Asimi, S., Les phosphatases alcalines solubles dans l'association endomycorrhizienne à vésicles et arbuscules, Physiol., Veg., 16, 671, 1978.
73. Gianinazzi, S., Gianinazzi-Pearson, V., and Dexheimer, J., Enzymatic studies on the metabolism of vesicular-arbuscular mycorrhiza. III. Ultrastructural localisation of acid and alkaline phosphatases in onion roots infected by Glomus mosseae (Nicol. and Gerd.), New Phytol., 82, 127, 1979.
74. Gianinazzi-Pearson, V. and Gianinazzi, S., Role of endomycorrhizal fungi in phosphorus cycling in the ecosystem, in The Fungal Community — Its Organization and Role in the Ecosystem, Mycol. Ser., Vol. 2, Wicklow, D. T. and Carroll, G. C., Eds., Marcel Dekker, New York, 1981, 637.
75. Mosse, B., Plant growth responses to vesicular-arbuscular mycorrhiza. IV. In soils given additional phosphate, New Phytol., 72, 127, 1973.
76. Cooper, K. M., Growth responses to the formation of endotrophic mycorrhizas in Solanum, Leptospermum and New Zealand ferns, in Endomycorrhizas, Sanders, F. E., Mosse, B., and Tinker, P. B., Eds., Academic Press, London, 1975, 392.
77. Nelson, C. E., Bolgiano, N. C., Furutani, S. C., Safir, G. R., and Zandstra, B. H., The effect of soil phosphorus levels on mycorrhizal infection of field grown onion plants and on mycorrhizal reproduction, J. Am. Soc. Hort. Sci., 106, 786, 1981.
78. Sanders, F. E., The effect of foliar-applied phosphate on mycorrhizal infection of onion roots, in Endomycorrhizas, Sanders, F. E., Mosse, B., and Tinker, P. B., Eds., Academic Press, London, 1975, 261.
79. Menge, J. A., Steirle, D., Bagyaraj, D. J., Johnson, E. L. V., and Leonard, R. T., Phosphorus concentrations in plants responsible for inhibition of mycorrhizal infection, New Phytol., 80, 575, 1978.
80. Azcon, R., Marin, A. D., and Barea, J. M., Comparative role of phosphate in soil or inside the host on the formation and effects of endomycorrhiza, Plant Soil, 49, 561, 1978.
81. Jasper, D. A., Robson, A. D., and Abbott, L. K., Phosphorus and the formation of vesicular-arbuscular mycorrhizas, Soil Biol. Biochem., 11, 501, 1979.
82. Bowen, G. D., Nutrient status effects on loss of amides and amino acids from pine roots, Plant Soil, 30, 139, 1969.
83. Ratnayake, M., Leonard, R. T., and Menge, J. A., Root exudation in relation to supply of phosphorus and its possible relevance to mycorrhizal formation, New Phytol., 81, 543, 1978.
84. Graham, J. H., Leonard, R. T., and Menge, J. A., Membrane-mediated decrease in root exudation responsible for phosphorus inhibition of vesicular-arbuscular mycorrhiza formation, Plant Physiol., 68, 548, 1981.
85. Johnson, C. R., Menge, J. A., Schwab, S., and Ting, I. P., Interaction of photoperiod and vesicular-arbuscular mycorrhizae on growth and metabolism of sweet orange, New Phytol., 90, 665, 1982.
86. McPharlin, I. R., Phosphorus Transport and Phosphorus Nutrition in Lemna and Spirodela, Ph.D. thesis, University of Auckland, Auckland, New Zealand, 1981.

87. Cooper, K. M., Adaptation of VA mycorrhizal fungi to phosphate fertilisers, in *Plant Nutrition 1978*, Ferguson, A. R., Bieleski, R. L., and Ferguson, I. B., Eds., New Zealand Department of Scientific and Industrial Research Inf. Ser. No. 134, Government Printer, Wellington, 1978, 107(Abst.).
88. Abbott, L. K. and Robson, A. D., Growth of subterranean clover in relation to the formation of endomycorrhizas by introduced and indigenous fungi in a field soil, *New Phytol.*, 81, 575, 1978.
89. Lanowska, J., Influence of different sources of nitrogen on the development of mycorrhiza in *Pisum sativum, Pamiet. Pulawski*, 21, 365, 1966.
90. Daft, M. J., Hacskaylo, E., and Nicolson, T. H., Arbuscular mycorrhizas in plants colonising coal spoils in Scotland and Pennsylvania, in *Endomycorrhizas*, Sanders, F. E., Mosse, B., and Tinker, P. B., Eds., Academic Press, London, 1975, 561.
91. Nemec, S. and Meredith, F. I., Amino acid content of leaves in mycorrhizal and nonmycorrhizal citrus rootstocks, *Ann. Bot. (London)*, 47, 351, 1981.
92. Ames, R., Reid, C., Porter, L., and Cambardella, C., Hyphal uptake and transport of nitrogen from two ^{15}N-labelled sources by *Glomus mosseae*, a vesicular-arbuscular mycorrhizal fungus, *New Phytol.*, 95, 381, 1983.
93. Chambers, C. A., Smith, S. E., and Smith, F. A., Effects of ammonium and nitrate ions on mycorrhizal infection, nodulation and growth of *Trifolium subterraneum*, *New Phytol.*, 85, 47, 1980.
94. Chambers, C. A., Smith, S. E., Smith, F. A., Ramsay, M. D., and Nicholas, D. J. D., Symbiosis of *Trifolium subterraneum* with mycorrhizal fungi and *Phizobium trifolii* as affected by ammonium sulphate and nitrification inhibitors, *Soil Biol. Biochem.*, 12, 93, 1980.
95. Brown, R. W., Schultz, R. C., and Kormanik, P. P., Response of vesicular-arbuscular endomycorrhizal sweetgum seedlings to three nitrogen fertilizers, *For. Sci.*, 27, 413, 1981.
96. Pateman, J. A. and Kinghorn, J. R., Nitrogen metabolism, in *The Filamentous Fungi, Vol. 2, Biosynthesis and Metabolism*, Smith, J. E. and Berry, D. R., Eds., Edward Arnold, London, 1976, chap. 7.
97. Ho, I. and Trappe, J. M., Nitrate reduction by vesicular-arbuscular mycorrhizal fungi, *Mycologia*, 67, 886, 1975.
98. Krishna, K. R. and Bagyaraj, D. J., Changes in the free amino nitrogen and protein fractions of groundnut caused by inoculation with VA mycorrhiza, *Ann. Bot. (London)*, 51, 399, 1983.
99. Young, J. L., Ho, I., and Trappe, J. M., Endomycorrhizal invasion and effect of free amino acids content of corn roots, *Agron. Abstr.*, p. 102, 1972.
100. Baltruschat, H. and Schonbeck, F., Untersuchungen uber den Einflus der endotrophen mykorrhiza auf den befall von Tabak mit *Thielaviopsis basicola, Phytopathol. Z.*, 84, 172, 1975.
101. Dehne, H. W., Schonbeck, F., and Baltruschat, H., The influence of endotrophic mycorrhiza on plant disease. III. Chitinase activity and ornithene cycle, *Z. Pflanzenkr. Pflanzenschutz*, 85, 666, 1978.
102. Smith, S. E., Nicholas, D. J. D., and Smith F. A., Effects of early mycorrhizal infection on nodulation and nitrogen fixation in *Trifolium subterraneum* L., *Aust. J. Plant Physiol.*, 6, 305, 1979.
103. Gerdemann, J. W., Vesicular-arbuscular mycorrhizae, in *The Development and Function of Roots*, Torrey, J. G. and Clarkson, D. I., Eds., Academic Press, London, 1975, 575.
104. Ross, J. P., Effect of phosphate fertilisation on yield of mycorrhizal and nonmycorrhizal soybeans, *Phytopathology*, 61, 1400, 1971.
105. Schultz, R. C., Kormanik, P. P., Bryan, W. C., and Brister, G. H., Vesicular-arbuscular mycorrhiza influence growth but not mineral concentrations in seedlings of eight sweetgum families, *Can. J. For. Res.*, 9, 218, 1979.
106. Jensen, A., Influence of four vesicular-arbuscular mycorrhizal fungi on nutrient uptake and growth in barley *(Hordeum vulgare)*, *New Phytol.*, 90, 45, 1982.
107. Lambert, D. H., Baker, D. E., and Cole, H., The role of mycorrhizae in the interactions of phosphorus with zinc, copper and other elements, *J. Soil Sci. Soc. Am.*, 43, 976, 1979.
108. Vander Zaag, P., Fox, R. L., De La Pena, R. S., and Yost, R. S., P nutrition of Cassava including mycorrhizal effects on P, K, S, Zn and Ca uptake, *Field Crops Res.*, 2, 253, 1979.
109. Krikun, J. and Levy, Y., Effect of vesicular-arbuscular mycorrhiza on citrus growth and mineral composition, *Phytoparasitica*, 8, 195, 1980.
110. Timmer, L. W. and Leyden, R. F., The relationship of mycorrhizal infection to phosphorus-induced copper deficiency in sour orange seedlings, *New Phytol.*, 85, 15, 1980.
111. Gilmore, A. E., The influence of endotrophic mycorrhizae on the growth of peach seedlings, *J. Am. Soc. Hortic. Sci.*, 96, 35, 1971.
112. La Rue, J. H., McClellan, W. D., and Peacock, W. L., Mycorrhizal fungi and peach nursery nutrition, *Calif. Agric.*, 29, 6, 1975.
113. Benson, N. R. and Covey, R. P., Response of apple seedlings to zinc fertilization and mycorrhizal inoculation, *Hortic. Sci.*, 11, 252, 1976.
114. Loneragan, J. F., Anomalies in the relationship of nutrient concentrations to plant yield, in *Plant Nutrition*, Ferguson, A. R., Bieleski, R. L., and Ferguson, I. B., Eds., N. Z. D. S. I. R. Inf. Ser. No 134, Government Printer, Wellington, 1978, 283.

115. Timmer, L. W. and Leyden, R. F., Stunting of citrus seedlings in fumigated soils in Texas and its correction by phosphorus fertilization and inoculation with mycorrhizal fungi, *J. Am. Soc. Hortic. Sci.,* 103, 533, 1978.
116. Bowen, G. D., Skinner, M. F., and Bevege, D. I., Zinc uptake by mycorrhizal and uninfected roots of *Pinus radiata* and *Araucaria cunninghamii, Soil Biol. Biochem.,* 6, 141, 1974.
117. Swaminathan, K. and Verma, B. C., Responses of three crop species to vesicular-arbuscular mycorrhizal infection on zinc deficient Indian soils, *New Phytol.,* 82, 481, 1979.
118. Rhodes, L. H., Hirrel, M. C., and Gerdemann, J. W., Influence of soil phosphorus on translocation of ^{65}Zn and ^{32}P by external hyphae of vesicular-arbuscular mycorrhizae of onion, *Phytopathol. News,* 12(Abstr.), 197, 1978.
119. Gray, L. E. and Gerdemann, J. W., Uptake of sulphur-35 by vesicular-arbuscular mycorrhizae, *Plant Soil,* 39, 687, 1973.
120. Rhodes, L. H. and Gerdemann, J. W., Influence of phosphorus nutrition on sulphur uptake by vesicular-arbuscular mycorrhizae of onion, *Soil Biol. Biochem.,* 10, 361, 1978.
121. Gildon, A. and Tinker, P. B., A heavy metal tolerant strain of a mycorrhizal fungus, *Trans. Br. Mycol. Soc.,* 77, 648, 1981.
122. Rhodes, L. H. and Gerdemann, J. W., Translocation of calcium and phosphate by external hyphae of vesicular-arbuscular mycorrhizae, *Soil Sci.,* 126, 125, 1978.
123. Harley, J. L., *The Biology of Mycorrhiza,* 2nd ed., Leonard Hill, London, 1969.
124. Smith, D., Muscatine, L., and Lewis, D., Carbohydrate movement from autotrophs to heterotrophs in parasitic and mutualistic symbioses, *Biol. Rev.,* 44, 17, 1969.
125. Hacskaylo, E., Carbohydrate physiology and ectomycorrhizae, in *Ectomycorrhizae,* Marks, G. C. and Kozlowski, T. T., Eds., Academic Press, New York, 1973, 207.
126. Harley, J. L., Problems of mycotrophy, in *Endomycorrhizas,* Sanders, F. E., Mosse, B., and Tinker, P. B., Eds., Academic Press, London, 1975, 1.
127. Bjorkman, E., Uber die Bedingungen der Mykorrhizabildung bei Kiefer und Fichte, *Symb. Bot. Ups.,* 6, 1, 1942.
128. Nemec, S. and Guy, G., Carbohydrate status of mycorrhizal and nonmycorrhizal citrus rootstocks, *J. Am. Soc. Hortic. Sci.,* 107, 177, 1982.
129. Azcon, R. and Ocampo, J. A., Factors affecting the vesicular-arbuscular infection and mycorrhizal dependency of thirteen wheat cultivars, *New Phytol.,* 87, 677, 1981.
130. Hayman, D. S., Plant growth responses to vesicular-arbuscular mycorrhiza. VI. Effect of light and temperature, *New Phytol.,* 73, 71, 1974.
131. Daft, M. J. and El-Giahmi A. A., Effect of arbuscular mycorrhiza on plant growth. VIII. Effects of defoliation and light on selected hosts, *New Phytol.,* 80, 365, 1978.
132. Gunze, C. M. B. and Hennessy, C. M. R., Effect of host-applied auxin on development of endomycorrhiza in cowpeas, *Trans. Br. Mycol. Soc.,* 74, 247, 1980.
133. Bevege, D. I., Bowen, G. D., and Skinner, M. F., Comparative carbohydrate physiology of ecto and endo-mycorrhizas, in *Endomycorrhizas,* Sanders, F. E., Mosse, B., and Tinker, P. B., Eds., Academic Press, London, 149, 1975.
134. Losel, D. M. and Cooper, K. M., Incorporation of ^{14}C-labelled substrates by uninfected and VA mycorrhizal roots of onions, *New Phytol.,* 83, 415, 1979.
135. Cooper, K. M. and MacRae, E. A., Assimilation of host photosynthate by vesicular-arbuscular mycorrhiza, unpublished data, 1982.
136. Harley, J. L. and Lewis, D. H., The physiology of ectotrophic mycorrhizas, *Adv. Microb. Physiol.,* 3, 53, 1969.
137. Hepper, C. M. and Mosse, B., Trehalose and mannitol in vesicular-arbuscular mycorrhiza, *Rothamsted Exp. Stn. Annu. Rep.,* 1, 81, 1973.
138. Cooper, K. M. and Losel, D. M., Lipid physiology of vesicular-arbuscular mycorrhiza. I. Composition of lipids in roots of onion, clover and ryegrass infected with *Glomus mosseae, New Phytol.,* 80, 143, 1978.
139. Scannerini, S. and Bonfante-Fasolo, P., Dati preliminari sull'ultrastructura di vesicole intracellulari nell'endomicorriza di *Ornithogalum umbellatum* L., *Atti Accad. Sci. Torino Cl. Sci. Fis. Mat. Nat.,* 109, 619, 1975.
140. Nemec, S., Histochemical characteristics of *Glomus etunicatus* infection of *Citrus limon* fibrous roots, *Can. J. Bot.,* 59, 609, 1981.
141. Sward, R. J., The structure of the spores of *Gigaspora margarita*. I. The dormant spore, *New Phytol.,* 87, 761, 1981.
142. Kinden, D. A. and Brown, M. F., Electron microscopy of vesicular-arbuscular mycorrhizae of yellow poplar. II. Intracellular hyphae and vesicles, *Can. J. Microbiol.,* 21, 1768, 1975.
143. Bonfante-Fasolo, P., Some ultrastructural features of the vesicular-arbuscular mycorrhiza in the grapevine, *Vitis,* 17, 386, 1978.

144. Ho, I. and Trappe, J. M., Translocation of ^{14}C from *Festuca* plants to their endomycorrhizal fungi, *Nature (London)*, 244, 30, 1973.
145. Strullu, D. C. and Gourret, J. P., Ultrastructure et evolution du Champignon symbiotique des racines de *Dactylorchis maculata* (L) verm., *J. Microsc. (Paris)*, 20, 285, 1974.
146. Foster, R. C. and Marks, G. C., The fine structure of the mycorrhizas of *Pinus radiata*, *Aust. J. Biol. Sci.*, 19, 1027, 1966.
147. Ling-Lee, M., Ashford, A. E., and Chilvers, G. A., A histochemical study of polysaccharide distribution in eucalypt mycorrhizas, *New Phytol.*, 78, 329, 1979.
148. Strullu, D. G., Histologie et cytologie des endomycorhizes, *Physiol. Veg.*, 16, 657, 1978.
149. Manners, J. M. and Gay, J. L., Transport, translocation and metabolism of ^{14}C photosynthate at the host-parasite interface of *Pisum sativum* and *Erysiphe pisi*, *New Phytol.*, 91, 221, 1982.
150. Blumenthal, H. J., Reserve carbohydrates in fungi in *The Filamentous Fungi, Vol. 2, Biosynthesis and Metabolism*, Smith, J. E. and Berry, D. R., Eds., Edward Arnold, London, 1976, 292.
151. Antibus, R. K., Trappe, J. M., and Linkins, A. E., Cyanide resistant respiration in *Salix nigra* endomycorrhizae *Can. J. Bot.*, 58, 14, 1980.
152. Coleman, J. O. D. and Harley, J. L., Mitochondria of mycorrhizal roots of *Fagus sylvatica*, *New Phytol.*, 76, 317, 1976.
153. Harley, J. L., McCready, C. C., and Wedding, R. T., Control of respiration of beech mycorrhizas during ageing, *New Phytol.*, 78, 147, 1977.
154. Harley, J. L., Salt uptake and respiration of excised beech mycorrhizas, *New Phytol.*, 87, 325, 1981.
155. Reid, C. P. P. and Woods, F. W., Translocation of ^{14}C-labelled compounds in mycorrhizae and its implications in interplant nutrient cycling, *Ecology*, 50, 179, 1969.
156. Stribley, D. and Read, D. J., The biology of mycorrhiza in the ericaceae. III. Movement of carbon-14 from host to fungus, *New Phytol.*, 73, 731, 1974.
157. Purves, S. and Hadley, G., Movement of carbon compounds between the partners in orchid mycorrhiza, in *Endomycorrhizas*, Sanders, F. E., Mosse, B., and Tinker, P. B., Eds., Academic Press, London, 1975, 175.
158. Hirrel, M. C. and Gerdemann, J. W., Enhanced carbon transport between onions infected with a vesicular-arbuscular mycorrhizal fungus, *New Phytol.*, 83, 731, 1979.
159. Warner, A. and Mosse, B., Independent spread of vesicular-arbuscular mycorrhizal fungi in soil, *Trans. Br. Mycol. Soc.*, 74, 407, 1980.
160. Beilby, J. P., Fatty acid and sterol composition of ungerminated spores of the vesicular-arbuscular mycorrhizal fungus, *Acaulospora laevis*, *Lipids*, 15, 949, 1980.
161. Beilby, J. P. and Kidby, D. K., Biochemistry of ungerminated and germinated spores of the vesicular-arbuscular mycorrhizal fungus, *Glomus caledonius:* changes in neutral and polar lipids, *J. Lipid Res.*, 21, 739, 1980.
162. Weete, J. D., *Fungal Lipid Biochemistry*, Plenum Press, New York, 1974.
163. Brennan, P. J. and Losel, D. M., Physiology of fungal lipids, *Adv. Microb. Physiol.*, 17, 47, 1978.
164. Nagy, S., Nordby, H. E., and Nemec, S., Composition of lipids in roots of six citrus cultivars infected with the vesicular-arbuscular mycorrhizal fungus, *Glomus mosseae*, *New Phytol.*, 85, 377, 1980.
165. Nordby, H. E., Nemec, S., and Nagy, S., Fatty acids and sterols associated with citrus root mycorrhizae, *J. Agric. Food Chem.*, 29, 396, 1981.
166. Bonfante-Fasolo, P. and Scannerini, S., Ultrastructural localisation of peroxidase activity in endomycorrhizal roots, *Caryologia*, 33, 126, 1980.
167. Hitchcock, C. and Nichols, B. W., *Plant Lipid Biochemistry*, Academic Press, London, 1971.
168. Ho, I., Phytosterols in root systems of mycorrhizal and nonmycorrhizal *Zea mays* L., *Lloydia*, 40, 476, 1977.
169. Beilby, J. P. and Kidby, D. K., Sterol composition of ungerminated and germinated spores of the vesicular-arbuscular mycorrhizal fungus *Glomus caledonius*, *Lipids*, 15, 375, 1980.
170. Grunwald, C., Effects of free sterols, sterol esters and sterol glucosides on membrane permeability, *Plant Physiol.*, 48, 653, 1971.
171. Heftman, E., Function of sterols in plants, *Lipids*, 6, 128, 1970.
172. Pang, P. C. and Paul, E. A., Effects of vesicular-arbuscular mycorrhiza on ^{14}C and ^{15}N distribution in nodulated faba beans, *Can. J. Soil Sci.*, 60, 241, 1980.
173. Paul, E. A. and Kucey, R. M. N., Carbon flow in plant microbial associations, *Science*, 213, 473, 1981.
174. Kucey, R. M. N. and Paul, E. A., Carbon flow, photosynthesis and N_2 fixation in mycorrhizal and nodulated faba beans (*Vicia faba* L.), *Soil Biol. Biochem.*, 14, 407, 1982.
175. Silsbury, J. H., Smith, S. E., and Oliver, A. J., *New Phytol.*, 93, 555, 1983.
176. Stribley, D. P., Tinker, P. B., and Rayner, J. H., Relation of internal phosphorus concentration and plant weight in plants infected by vesicular-arbuscular mycorrhizas, *New Phytol.*, 86, 261, 1980.

177. Buwalda, J. G. and Goh, K. M., Host-fungus competition for carbon as a cause of growth depressions in vesicular-arbuscular mycorrhizal ryegrass, *Soil Biol. Biochem.*, 14, 103, 1982.
178. Stribley, D. P., Tinker, P. B., and Snellgrove, R. C., Effect of vesicular-arbuscular mycorrhizal fungi on the relations of plant growth, internal phosphorus concentration and soil phosphate analyses, *J. Soil Sci.*, 31, 655, 1980.
179. Yokom, D. H., Quantification of costs and benefits accrued by onion plants from vesicular-arbuscular associations, in Abstr. 5th N. Am. Conf. Mycorrhizae, Quebec, 1981, 21.
180. Yokom, D. H., Personal communication, 1982.
181. Levy, J. and Krikun, J., Effect of vesicular-arbuscular mycorrhiza on *Citrus jambhiri* water relations, *New Phytol.*, 85, 25, 1980.
182. Allen, M. F., Smith, W. K., Moore, T. S., Jr., and Christensen, M., Comparative water relations and photosynthesis of mycorrhizal and nonmycorrhizal *Bouteloua gracilis* H.B.K. Lag ex Steud, *New Phytol.*, 88, 683, 1981.
183. Wallace, L. L., Growth morphology and gas exchange of mycorrhizal and nonmycorrhizal *Panicum coloratum* L., a C_4 grass species, under different clipping and fertilisation regimes, *Oecologia*, 49, 272, 1981.
184. Krishna, K. R., Suresh, H. M., Syamsunder, J., and Bagyaraj, D. J., Changes in the leaves of finger millet due to VA mycorrhizal infection, *New Phytol.*, 87, 717, 1981.
185. Safir, G. R. and Nelson, C. E., Water and nutrient uptake by vesicular-arbuscular mycorrhizal plants, in *Mycorrhizal Associations and Crop Production*, Myers, R., Ed., Rutgers University Press, New Brunswick, N.J., 1981.
186. Letham, D. S., Goodwin, P. B., and Higgins, T. J. W., Eds., *Phytohormones and Related Compounds — A Comprehensive Treatise*, Vol. 1 and 2, Elsevier, Amsterdam, 1978.
187. Slankis, V., Hormonal relationships in mycorrhizal development, in *Ectomycorrhizae*, Marks, G. C. and Kozlowski, T. T., Eds., Academic Press, New York, 1973, 232.
188. Ng, P. P., Cole, A. L. J., Jameson, P., and McWha, J. A., Cytokinin production by ectomycorrhizal fungi, *New Phytol.*, 91, 57, 1982.
189. Daft, M. J. and Nicolson, T. H., Effect of *Endogone* mycorrhiza on plant growth. III. Influence of inoculum concentration on growth and infection in tomato, *New Phytol.*, 68, 953, 1969.
190. Daft, M. J. and Okusanya, B. O., Effect of *Endogone* on plant growth. VI. Influence of infection on the anatomy and reproductive development in four hosts, *New Phytol.*, 72, 1333, 1973.
191. Linderman, R. G. and Call, C. A., Enhanced rooting of woody plant cuttings by mycorrhizal fungi, *J. Am. Soc. Hortic. Sci.*, 10, 629, 1977.
192. Barrows, J. B. and Roncadori, R. W., Endomycorrhizal synthesis by *Gigaspora margarita* in poinsettia, *Mycologia*, 69, 1173, 1977.
193. Cooper, K. M., Mycorrhizal fungi can improve growth of horticultural crops, *Orchardist N.Z.*, 56, 410, 1983.
194. Brown, M. E., Rhizosphere microorganisms — opportunists, bandits or benefactors, in *Soil Microbiology*, Walker, N., Ed., Butterworths, London, 1972, 21.
195. Azcon, R., Azcon-G. de Aguilar, C., and Barea, J. M., Effects of plant hormones present in bacterial cultures on the formation and responses to VA endomycorrhiza, *New Phytol.*, 80, 359, 1978.
196. Colbert, K. A. and Beever, J. E., Effect of disbudding on root cytokinin export and leaf senescence in tomato and tobacco, *Exp. Bot.*, 32, 121, 1981.
197. Allen, M. F., Moore, T. S., and Christensen, M., Phytohormone changes in *Bouteloua gracilis* infected by vesicular-arbuscular mycorrhizae. I. Cytokinin increases in the host plant, *Can. J. Bot.*, 58, 371, 1980.
198. Brown, R. W., Brown, C. L., and Kormanik, P. P., Effects of VA mycorrhizae and soil phosphorus on cytokinins and nutrient levels in *Plantanus occidentalis*, in Abstr. 5th N. Am. Conf. Mycorrhizae, Quebec, 1981, 12.
199. Safir, G. R., Boyer, J. S., and Gerdemann, J. W., Mycorrhizal enhancement of water transport in soybean, *Science*, 172, 581, 1971.
200. Safir, G. R., Boyer, J. S., and Gerdemann, J. W., Nutrient status and mycorrhizal enhancement of water transport in soybean, *Plant Physiol.*, 49, 700, 1972.
201. Allen, M. F., Influence of vesicular-arbuscular mycorrhizae on water movement through *Bouteloua gracilis* (H.B.K.) Lag ex Steud, *New Phytol.*, 91, 191, 1982.
202. Hardie, K. and Leyton, L., The influence of vesicular-arbuscular mycorrhiza on growth and water relations of red clover. I. In phosphate deficient soil, *New Phytol.*, 89, 599, 1981.
203. Nelson, C. E. and Safir, G. R., The water relations of well watered mycorrhizal and nonmycorrhizal onion plants, *J. Am. Soc. Hortic. Sci.*, 107, 271, 1982.
204. Nesheim, O. N. and Linn, M. B., Deleterious effect of certain fungitoxicants on the formation of mycorrhizae on corn by *Endogone fasciculatus* and on corn root development, *Phytopathology*, 59, 297, 1969.
205. Gray, L. E. and Gerdemann, J. W., Uptake of phosphorus-32 by vesicular-arbuscular mycorrhizae, *Plant Soil*, 30, 415, 1969.

206. Tinker, P. B., Roots and water. Transport of water to plant roots in soil, *Philos. Trans. R. Soc. London Ser. B.*, 273, 445, 1976.
207. Reid, C. P. P., Mycorrhizae and water stress, in *Root Physiology and Symbiosis*, Riedacker, A. and Gagnaire-Michard, J., Eds., International Union of Forestry Research Organizations Proceedings, Nancy, France, 1979, 392.
208. Cowan, M. C., Lewis, B. G., and Thain, J. F., Uptake of potassium by the developing sporangiosphore of *Phycomyces blakesleanus, Trans. Br. Mycol. Soc.*, 58, 113, 1972.
209. Fiscus, E. L. and Markhardt, A. H., Relationships between root system, water transport properties and plant size in *Phaseolus, Plant Physiol.*, 64, 770, 1979.
210. Daft, M. J. and Hacskaylo, E., Growth of endomycorrhizal and nonmycorrhizal red maple seedlings in sand and anthracite soil, *For. Sci.*, 23, 207, 1977.
211. Fiscus, E. L., The interaction between osmotic and pressure-induced water flow in plant roots, *Plant Physiol.*, 55, 917, 1975.
212. Aldon, E. F., Endomycorrhizae enhance survival and growth of four-wing saltbrush on coal mine spoils, *U.S. For. Serv. Res. Note RM*, 294, 1975.
213. Lindsey, D. L., Cress, W. A., and Aldon, E. F., The effects of endomycorrhizae on growth of rabbit brush, four-wing saltbrush and corn in coal mine spoil material, *U.S. For. Serv. Res. Note RM*, 343, 1977.
214. Nelson, C. E. and Safir, G. R., Increased drought tolerance of mycorrhizal onion plants caused by improved phosphorus nutrition, *Planta*, 154, 407, 1982.
215. Menge, J. A., Davis, R. M., Johnson, E. L. V., and Zentmyer, G. A., Mycorrhizal fungi increase growth and reduce transplant injury in avocado, *Calif. Agric.*, April, 6, 1978
216. Sieverding, E., Einfluss der Bodenfeuchte auf die Effektivitat der VA-Mykorrhiza, *Angew. Bot.*, 53, 91, 1979.
217. Sieverding, E., Influence of soil water regimes on VA mycorrhiza. I. Effect on plant growth, water utilisation and development of mycorrhiza, *Z. Acker Pflanzenbau*, 150, 400, 1981.
218. Janos, D. P., Vesicular-arbuscular mycorrhizae affect lowland tropical rain forest plant growth, *Ecology*, 61, 151, 1980.
219. Allen, M. F. and Boosalis, M. G., Effects of two species of vesicular-arbuscular mycorrhizal fungi on drought tolerance of winter wheat, *New Phytol.*, 93, 67, 1983.
220. Tal, M., Katz, A., Heikin, H., and Dehan, K., Salt tolerance in the wild relatives of the cultivated tomato: proline accumulation in *Lycopersicon esculentum* Mill., *L. peruvianum* Mill. and *Solanum pennelli* Cor. treated with NaCl and polyethylene glycol, *New Phytol.*, 82, 349, 1979.
221. Uhlig, S. K., Untersuchungen zur trockenresistenz mykorrhizabildender pilze (Investigations on the drought resistance of mycorrhiza-forming fungi), *Zentralbl. Bakteriol. Parasitenkd. Infektionskr. Hyg. Abt.*, 127, 124, 1972.
222. Mexal, J. and Reid, C. P. P., The growth of selected mycorrhizal fungi in response to induced water stress, *Can. J. Bot.*, 51, 1579, 1973.
223. Theodorou, C., Soil moisture and the mycorrhizal association of *Pinus radiata* D. Don., *Soil Biol. Biochem.*, 10, 33, 1978.

Chapter 9

INOCULUM PRODUCTION

John A. Menge

TABLE OF CONTENTS

I.	Introduction	188
II.	Methods for Production of VAM Inoculum	188
	A. Soil Culture	188
	1. Host Plant	190
	2. Growth Medium	190
	3. Fertilization	190
	a. Phosphorus	190
	b. Nitrogen	191
	c. Micronutrients	191
	4. Water-Aeration	192
	5. pH	192
	6. Light and Photoperiod	192
	7. Temperature	192
	8. Pruning	193
	9. Pot Size	193
	10. Chemical Applications	193
	B. Nutrient Film or Circulating Hydroponic Culture	193
III.	Controlling Microorganisms Which Contaminate Mycorrhizal Inoculum	194
IV.	Parasites of VAM Fungi	196
V.	Storage of Inoculum	196
VI.	Principles of Symbiosis Are Part of Inoculum Production	196
	A. Growth of Plant and the Nutrition of VAM Fungi	196
	B. VAM Growth Dynamics	197
	C. Inoculum Potential	198
VII.	Conclusions	199
References		199

I. INTRODUCTION

Because of the current inability to grow vesicular-arbuscular mycorrhizae (VAM) fungi in pure culture, many scientists believe that the difficulties involved in producing clean, high-quality VAM inoculum on a commercial scale may prevent any large-scale agricultural use of these fungi.[1] The obligate symbiotic nature of VAM presently dictates that all VAM inoculum must be grown on roots of an appropriate host plant. The production of VAM inoculum has evolved from the original use of infested field soils or sievings from these soils, to the use of soil and roots of greenhouse plants (pot-culture inoculum) inoculated with field inoculum, to the current practice of using pot culture inoculum derived from surface-disinfested spores of a single VAM fungus on a host plant grown in a sterilized growth medium. Because it is relatively easy to produce infective inoculum by these crude and often uncontrolled methods, production methods are as varied as the researchers themselves. Manipulations of VAM fungi, rooting media, hosts, fertilizers, and environment to produce VAM inoculum often resemble an art form rather than true science. Furthermore, each VAM fungal species requires slightly different cultural practices to achieve maximum inoculum production, with some species requiring highly specific procedures for culture. This review will not attempt to describe the culture methods for a variety of host-VAM fungi combinations. Instead, the purpose of this chapter is (1) describe some of the practical methods which have been suggested for the production and storage of commercial VAM inoculum; (2) indicate some of the common mechanisms which have been shown to inhibit or improve the production and storage of VAM inoculum; and (3) outline the theory and principles for producing and storing consistently high-quality, disease-free VAM inoculum.

II. METHODS FOR PRODUCTION OF VAM INOCULUM

A. Soil Culture

The most common and most dependable method for producing inoculum is the soil culture method described by several researchers (Figure 1).[2-6] Mycorrhizal fungi must first be isolated from the field. Roots or soil from the field can be used to inoculate roots of "trap plants" growing in sterilized soil, in order to obtain crude "field pot-culture" isolates. Suitable methods for soil sterilization include autoclaving [103 kPa (15 lb/in²) pressure, 121°C for two separate 1-hr periods], gamma irradiation (0.8 to 1 Mrad), fumigation with biocides such as methyl bromide (0.45 to 1.00 kg/m³), or uniform steaming (83 to 100°C for two separate 1-hr periods; time extended for larger quantities of soil). "Field pot-cultures" normally contain many undesirable organisms as well as mixed populations of VAM fungi. It is always advisable to begin inoculum production with a single species culture which is free of all other organisms. VAM cultures derived from single spores are preferable since variability in the inoculum will be reduced. To obtain starter cultures of this type, spores are removed from "field pot-cultures" using methods described in Chapter 5: wet-sieving,[7] elutriation,[8] or sucrose-density centrifugation.[9] Spores should then be carefully examined under the microscope for spores containing parasites (see below). Parasites inside spores will not be killed by surface disinfestation. Only healthy or undamaged spores should be selected. These spores should be surface disinfested using solutions such as 2% w/v chloramine T and 200 ppm streptomycin for 15 min or 2% sodium hypochlorite for 10 min and rinsed three times in sterile distilled water.[10] Ten (10) to 20 spores are usually sufficient to inoculate roots of young seedlings grown aseptically in the greenhouse, provided the spores are placed in contact with the roots. Methods for increasing the chances of colonization by low numbers of VAM spores under greenhouse conditions are de-

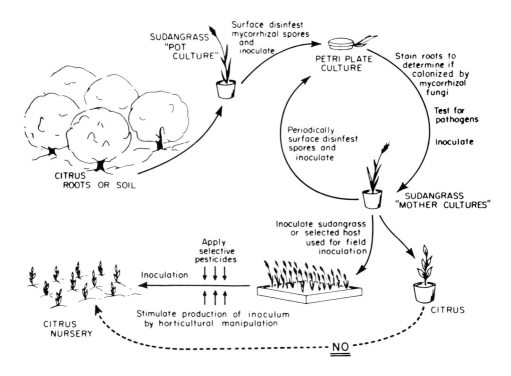

FIGURE 1. Proposed scheme for commercial production of VAM inoculum. With proper precautions, inoculum which is clean and free of plant pathogens can be produced under greenhouse conditions.

scribed by Menge and Timmer.[11] The establishment of axenically grown plants[12] (Chapter 5) infected by single spores appears to be the way for establishing pure cultures of VAM fungi. Pot cultures can be established using infected roots or even mycelium of those VAM fungi which do not sporulate, but one can never be sure of the purity of such isolates. Once a pure culture has been established, these greenhouse "mother cultures" (Figure 1) can be maintained as an inoculum source for all other cultures. The purity of these cultures should be checked regularly, and the cleaning process should be repeated every few years.

The growth medium in which mycorrhizal inoculum is produced should be chosen carefully. Fertilization is often critically important in producing VAM inoculum (see below). The medium must contain sufficient phosphorus (P) to allow the host plant to grow readily yet not inhibit colonization by VAM. Soils with a large amount of available P usually result in poor VAM inoculum production. Ideally, soils should be fertile enough so that only nitrogen need be applied during the 3 to 4 month period necessary to produce inoculum (see below). If a complete nutrition solution is necessary, there are several which are compatible with mycorrhizae under hydroponic conditions.[13,14] In soil, a $^1/_2$-strength Hoagland's solution[15] minus P can be applied once a week, usually with favorable results for VAM inoculum production.

To harvest VAM inoculum, plant tops are removed and roots and soil from the containers are chopped or ground into a uniform inoculum before storage. For small amounts of inoculum, hand chopping is usually adequate. Mechanical wood-chippers or shredders have been used on a commercial scale, but the blades are rapidly damaged by the soil mix. It is obvious that a more efficient mechanism for chopping and mixing VAM inoculum needs to be designed. Daniels and colleagues[16,17] have drawn attention to the commercial potential of *Glomus epigaeum*, which produces clean, uniform, and easily harvested chlamydospores in sporocarps on the soil surface. VAM inoculum produced in this manner may lack the hyphal components of inoculum potential, but ease in handling and the reduction in contamination problems makes such a fungus commercially desirable. Cost estimates for VAM inoculum production on host plants have been itemized by Johnson and Menge.[18]

There are many ecological preferences by VAM fungi. Therefore, some deviation from the simplified production scheme outlined above is expected for most species. Furthermore, there are many methods of horticultural manipulation which can increase or decrease the quality and quantity of VAM inoculum. Some of these methods are outlined below.

1. Host Plant

Choosing the host plant on which to produce VAM inoculum is often difficult. The plant should (1) be well adapted to the conditions under which it must be grown, (2) be an acceptable host for the VAM species which will be produced,[5,19,20] (3) grow rapidly and produce an abundance of roots, and (4) have no pathogens in common with the host for which the VAM inoculum is ultimately intended.

Since most plants are mycorrhizal with VAM fungi, a wide variety of plants are available for use in inoculum production. Some hosts which have been used include *Nardus stricta, Coprosoma robusta, Citrus, Sorghum, Stylosanthes, Coleus,* onion, pepper, strawberry, barley, corn, alfalfa, clover, peanut, cotton, and asparagus. Bagyaraj and Manjunath[19] screened eight grasses for suitability for mass production of VAM inoculum and Guinea grass (*Panicum maximum* Jacq.) was found to be superior to other species tested, including *Sorghum*. Ferguson[21] examined a variety of host plants for their suitability in producing VAM inoculum. This work indicated that both peanut and alfalfa were superior to *Sorghum* in producing VAM spores. However, in general Ferguson found that the health and vigor of the host plant may be more important in VAM inoculum production than the plant species.

2. Growth Medium

The most important aspect of VAM inoculum production may be the selection of an appropriate growth medium. It is usually the cause of most production difficulties. Soil is usually a critical component in the growth medium since it acts as the source of P and micronutrients and also functions as a natural buffer to mediate soil nutrient availability. A coarse-textured, low-nutrient, sandy soil with a high cation exchange capacity to reduce the availability of P is most frequently recommended. Clay soils or soils with poor structure or drainage usually create water and aeration problems. The soil used in commercially produced VAM inoculum can also create transport problems. The weight of soil is important in determining shipping costs and many countries restrict soil importation despite assurances of its sterility. It is, therefore, frequently advantageous to produce VAM inoculum in a partially, if not completely, artificial growth medium. VAM inoculum can be produced in vermiculite, peat, sawdust, bark, perlite, pumice, or in mixtures of these media. However, under these conditions a nutrient solution must be added and the availability of P and minor elements must be regulated more carefully (see section on hydroponic growth). Furthermore, redwood shavings, some barks, and certain types of peat moss can severely inhibit infection by VAM fungi (unpublished data).

3. Fertilization
a. Phosphorus

Phosphorus is usually considered to be the major problem when VAM infection is poor. Phosphorus additions are known to reduce VAM colonization of roots and spore production in most soils.[22-28] However, conclusive recommendations for specific soil P levels for mycorrhizal production cannot be made. There are several reasons for this.

First, it is recognized that it is not soil P per se that regulates mycorrhizal infection, but rather the amount of P absorbed by the host plant.[25,29] Thus, it is the amount of available P, but not the form or source of P fertilization which influences mycorrhizal infection.[27] Second, methods for evaluating available soil P often differ greatly, and plant tissue analysis is a far more reliable method for determining available soil P than most methods which analyze soil. Finally since host plants vary in their ability to absorb P and mycorrhizal fungi vary in their response to P, each plant-soil-VAM symbiont system must be evaluated separately.[23,26,28] For example, in one sandy soil, maximum spore production by *Glomus fasciculatum* on sour orange occurred at 50 ppm added P, but this amount of fertilizer applied to a different citrus variety, Troyer citrange, resulted in no spore production.[25] Jasper et al.[23] found that mycorrhizal colonization of ryegrass in a virgin soil and in an adjacent cultivated soil was similar (60% root colonization). However, when approximately 100 ppm P was added, the infection in the virgin soil was reduced by two thirds while infection in the cultivated soil was reduced very little.

Even tissue P is not always a good estimate for mycorrhizal colonization, since the mycorrhizae themselves influence that factor. It is thought that P influences VAM colonization by affecting concentrations of root carbohydrates[23] or the amount of root exudates,[31] and for that reason the effects of P concentrations may be partially overcome by other factors such as high light intensity.[30] The best indicator for identifying a soil that will provide good VAM colonization appears to be the percentage of P in plants at the time VAM colonization takes place.[23] Therefore, growth media which provide the following tissue P concentrations in nonmycorrhizal plants will probably result in good VAM inoculum: sour orange, 0.05%;[25] *Griselinia*, 0.04 to 0.06%;[24] *Coprosma*, 0.06 to 0.09%;[24] and sudangrass (root tissue), 0.06 to 0.08%.[31]

For practical production of VAM inoculum, the growth medium must contain sufficient P so that mycorrhizae can absorb it and allow the host plant to grow readily, but excess or luxury amounts of P will reduce VAM colonization and spore production.[23,25] If inoculum is produced in a soil medium, additional P may not be required. However, in some soils, judicious P additions improve inoculum production. Adding P as rock phosphate or bone meal is often desirable, since the P in these sources is largely unavailable and is released slowly rather than in an initial flush as with superphosphate.

b. Nitrogen

Nitrogen often is the only fertilizer required during the 3 to 4 month period necessary to produce VAM inoculum. Yet the effects of N on VAM inoculum production are not yet fully understood. It appears that high concentrations of N fertilizers can reduce mycorrhizal colonization in roots.[32-34] Daily fertilization of citrus and a variety of woody ornamentals with over 100 ppm N (as a mixture of NO_3 and NH_4) retarded mycorrhizal development (unpublished data). The ammonia form of N is more toxic to VAM than nitrate,[33] so nitrate should be used in mycorrhizal inoculum production whenever possible. Nitrogen can be mixed into the potting medium at rates of 100 to 200 mg/kg soil, but since it is likely to leach out, additions of N in the irrigation water are usually necessary. Applications of 25 to 75 ppm N as $CaNO_3$ or KNO_3 in water are recommended as needed.

c. Micronutrients

Micronutrients such as manganese and zinc have been shown to inhibit germination of spores of mycorrhizal fungi.[35] Zinc or copper have been shown to inhibit mycorrhizal infection in clover, onion, maize, soybean, and pinto bean.[36-38] Although these compounds are relatively insoluble in most growth media, they should be present only in required quantities in fertilizers which are applied to VAM inoculum and care should be taken to prevent their accumulation.

4. Water-Aeration

The quality of VAM inoculum appears to be severely reduced under conditions which are either too wet or too dry. Reid and Bowen[39] found that maximum VAM colonization occurred at −0.2 bar water potential. VAM colonization dropped with each further decrease in water potential. VAM colonization was reduced to 50% of the maximum when the soil was saturated. Redhead[40] reported similar results in that VAM spore production was excellent when *Khaya* plants were watered daily. Weekly watering reduced spore germination by 90% and watering twice per day (waterlogging) reduced spore production by 75%. Withholding water in an attempt to induce premature senescence does not improve VAM infection or spore production.[41]

Under water saturation conditions, O_2 diffusion is limited and anaerobiosis may develop which can result in the release of toxic compounds such as Mn, H_2S, and various organic acids. Reduced O_2 concentrations can severely inhibit VAM spore germination and root colonization.[38,42] For these reasons it is often better to under-water plants infected with VAM fungi than to over-water them. Continuous checks for drainage impedances and the use of soil tensiometers in the growth medium will help to improve the quantity and quality of VAM inoculum.

5. pH

Because pH affects so many other soil factors it is often difficult to determine what effects changes in pH will have on VAM inoculum. Many VAM fungi will produce excellent inoculum over a wide pH range. *Glomus fasciculatum*, for instance, will infect a variety of plants growing at pHs from 5.5 to 9.5. Evidence is accumulating, however, that growth of many VAM isolates are limited by pH. Several isolates of *Glomus tenuis* did not infect at low pH.[43] *Glomus mosseae* and many other *Glomus* species appear to prefer pHs above 5.0 while many *Acaulospora* species appear to require pHs below 5.0.[44-46] Care must be taken to adjust the pH of the plant growth medium to a pH which is conducive to mycorrhizal formation by the fungus being grown.

6. Light and Photoperiod

High light intensities and long daylengths improve mycorrhizal colonization or spore production in maize, alfalfa, sudangrass, citrus, onions, and many other plants.[30,47-51] The increased VAM colonization apparently results from increased photosynthesis which leads to increased carbohydrate concentrations in roots,[49] and/or increased exudation of these compounds.[30,50-51] Therefore, to maximize VAM inoculum production, it would seem reasonable to attempt to saturate host plant photosynthesis with light. Most greenhouses or growth chambers provide only a fraction of full sunlight. Photoperiod extension seems to increase VAM colonization to a greater degree than does increasing light intensity.[49-51] Reducing light intensity by 30% did not reduce mycorrhizal infection in either *Coprosoma* or sudangrass.[50,52] However, by increasing photosynthetically active radiation to sudangrass by threefold with high intensity metal halide or mercury vapor lamps and extended photoperiods, the spore production by *Glomus fasciculatum* was increased fivefold.[50] Furthermore, with this method, commercial VAM inoculum could be produced in 2 months instead of the normal 3 or 4 months. Commercial VAM inoculum is currently being produced using constant supplemental, low intensity fluorescent or incandescent light. This procedure appears to be essential during winter months.[51]

7. Temperature

Temperature has a severe effect upon both host plant and VAM fungus growth.[21,28,30,53-57] Optimal VAM infection, therefore, varies considerably with both host and fungus. However, it appears that in most cases VAM colonization and spore production increase with increasing temperature until growth of the host plant is severely inhibited.[21,30,53,55,57] Ferguson[21] found that spore production by *Glomus fasciculatum* on sudangrass increased as temperature increased up to 30°C which is the optimal temperature for growth of sudangrass. Temperatures below 15°C are usually inhibitory to VAM colonization. It is hypothesized that high temperature increases both VAM fungus growth[30,57] and root exudation which can lead to increased VAM colonization.[30]

It must be remembered that soil temperatures are far more important in the production of VAM inoculum than air temperatures. Soil heating cables can frequently enhance VAM colonization in cold soils. Cold shocks to induce senescence and improve mycorrhizal spore production have not given consistent results.[21] It is recommended that VAM inoculum be produced at temperatures which are at or slightly above optimal for the host plant.

8. Pruning

Pruning host plants reduces VAM infection or sporulation and is normally not conducive to the production of VAM inoculum.[21,41,47] This is probably due to the consequent reduction in flow of photosynthate to roots.

9. Pot Size

Host plant size and root mass can be influenced by the amount of available soil and the pot volume. Ferguson[21] found that VAM spore production both on a per gram soil and a per gram root basis increased with pot size. Plants in 15,000 cm³ pots produced 90 times as many spores per pot as 750 cm³ pots. It is therefore desirable to produce VAM inoculum in large containers. However, growth media greater than 1-m deep usually contains relatively few roots and is usually not highly infective when used as VAM inoculum.

10. Chemical Applications

Many pesticides are detrimental to VAM infection and sporulation, but some are quite compatible with VAM inoculum production.[58] Pesticides can be beneficial for VAM inoculum, since they reduce contamination by unwanted microorganisms and eliminate plant pathogens or parasites which compete with VAM fungi for space or nutrients. Some pesticides such as DBCP, 1-3D, ethazole, sodium azide, captan, and metalaxyl have increased VAM infection or spore production even when competing microorganisms are not obvious.[58] It is hypothesized that these chemicals may increase root exudation and thereby increase VAM colonization.

Most herbicides damage host plant growth and so are detrimental to mycorrhizal inoculum. However, herbicides such as paraquat, simazine, atrazine, and maleic hydrazide, have been found to increase VAM infection, probably because they increase root exudation of amino acids or sugars.[59,60] However, further work is required to verify that increased inoculum potential accompanies such artificially induced VAM infections.

B. Nutrient Film or Circulating Hydroponic Culture

Nutrient film or circulating hydroponic culture techniques can be used to produce high quality inoculum under rigorously defined conditions.[13,14,51] The nutrient film method for producing VAM inoculum is a patented process developed at Rothamsted Experimental Station in England.[61] Basically it involves placing plants which have been inoculated with specific VAM fungi on an inclined plane with roots between plastic sheets or capillary matting. A liquid solution culture is pumped onto the inclined plane where it flows down over the roots as a film and into a container where it is recycled. Roots can become heavily mycorrhizal with this technique and, after harvesting and air-drying, can function as effective soil culture inoculum.[61] This inoculum relies heavily upon mycelium both inside and outside the roots for its inoculum potential. Since spores are not abundant, the survival ability of inoculum produced by nutrient film culture must be examined carefully.

The circulating hydroponic culture systems,[13,14] in which plant roots are bathed in a circulating nutrient solution, are quite similar. Howeler et al.[13] placed their plants directly in the nutrient solution with the roots continuously submerged, while Ojala and Jarrell[14] grew their plants in pots of silica sand which were automatically filled with nutrient solution four times per day and allowed to drain. In both systems the nutrient solution was stored in large containers and monitored periodically so that fluctuations in nutrient concentrations and pH were minimized.

Critical factors in hydroponic production of inoculum appears to be aeration, phosphorus, and nitrogen supply. Aeration was maintained in the nutrient film technique by air exchange into the thin film of moving water. In the Howeler et al.[13] system, compressed air was systematically injected into nutrient solutions. In the Ojala and Jarrell[14] method, saturation of silica sand was limited to a few hours daily. Phosphorus and nitrogen concentrations in the three media were remarkably similar. Elmes and Mosse[61] used rock phosphate or bone meal as the P source, which undoubtedly maintained the available P at very low levels. Howeler et al.[13] obtained VAM colonization of cassava by maintaining P concentrations of 0.0031 ppm and 0.031 ppm, but not at 0.31 or 3.1 ppm. Growth responses due to mycorrhizal fungi were observed only at 0.031 ppm P. Ojala and Jarrell[14] obtained VAM colonization in tomato at initial P concentrations of 0.1 ppm and 0.3 ppm P. Growth responses were observed only at 0.3 ppm P. In this latter study, P concentrations were not kept strictly constant, since P was replenished only twice during the 68-day experiment. Phosphorus was intentionally allowed to diminish in the solution in order to determine the point between the initial concentrations and zero available P which would be compatible with both colonization and growth response.

Nitrogen concentrations which were compatible with VAM colonization under hydroponic conditions ranged from 1 to 25 ppm. Elmes and Mosse[61] reported inhibition of VAM colonization when the N concentration was 11 ppm (95% NO_3 and 5% NH_4). This inhibition disappeared when N concentrations were reduced to 1 ppm. Howeler et al.[13] found no inhibition of VAM colonization with 14 ppm N as NO_3. Ojala and Jarrell[14] obtained root colonization by using 14 ppm N as NO_3 for the first 31 days and thereafter applying N as a 1:1 ratio of NO_3 and NH_4. The ammonia form of N is thought to be more toxic to infection than the nitrate form.[33]

III. CONTROLLING MICROORGANISMS WHICH CONTAMINATE MYCORRHIZAL INOCULUM

Since VAM fungi are obligate symbionts and must be grown on the roots of living hosts, many scientists will not consider commercial greenhouse production of VAM inoculum because of the danger of plant pathogenic organisms contaminating mycorrhizal inoculum. The danger is a real one and must be addressed directly, since VAM inoculum could be responsible for the spread of plant disease. Since VAM inoculum is often used in steamed or fumigated nursery or greenhouse soils, the danger is especially acute. Realistically, however, under controlled conditions and with proper safeguards, pathogen-free VAM inoculum can be produced. Many nursery plants grown under far less scrutiny are shipped freely throughout the world.

The essentials involved in producing pathogen-free VAM inoculum can be seen in Figure 1. Inoculum produced for sale should always be produced in a controlled environment such as a greenhouse or growth chamber so that proper sanitation procedures can be maintained. Sterilized pots and growth media should always be used. The use of cover crops to produce VAM inoculum in the field in fumigated nursery beds may be an ideal way for a nursery to produce its own inoculum,[62] but lack of quality control for such a method makes the sale of such inoculum impractical. Initial inoculum should be produced from surface-disinfested spores on aseptically grown plants in petri plates or test tubes (Figure 1, Chapter 5). Single spore cultures are preferable. Roots from these cultures should be plated on agar to test for possible pathogens, and they should be stained and examined to make sure that unwanted organisms are not present in or on the roots of initial inoculum. These procedures should be repeated with "mother cultures". Most commercial companies repeat the test on inoculum before it is sold, as part of the quality control process. Obligate root parasites such as *Olpidium* are often especially troublesome and care should be taken to eliminate them from mycorrhizal inoculum. Rhizoplane organisms which may or may not affect plant growth should also be eliminated.

Heat treatments may be effective in removing some pathogens and rhizosphere organisms from the roots of plants if they are killed by temperatures which are lower than those which kill VAM fungi.[63] The major tool for producing clean inoculum, however, is the continual alteration of host plants. In this manner, the wide host range of most VAM fungi can be used to advantage. Inoculum should be produced on selected hosts which have no root diseases in common with the crop for which the inoculum is intended; inoculum from one host should never be used to inoculate the same plant species. For instance, inoculum for citrus should be produced on sudangrass, but never on citrus (Figure 1). This prevents build-up of pathogenic or other unwanted rhizosphere or rhizoplane organisms.

As another precaution against propagating pathogens along with mycorrhizal inoculum, commercial inoculum should be routinely drenched with pesticides chosen to eliminate pathogens known to infect the crop for which the inoculum is intended. Mycorrhizal inoculum intended for citrus should be drenched with a nematicide to control the citrus nematode and fungicides to control *Phytophthora* and *Rhizoctonia*.[63] Pesticides which can be used on mycorrhizae with little or no damage to VAM symbiont have been summarized in a recent review.[58]

Further precautions which serve to reduce or eliminate unwanted contaminants from VAM inoculum include standard sanitary greenhouse techniques such as eliminating insects and animals, treating benches with copper napthanate, raising pots off the bench to prevent root contact with the bench, watering from below or with a drip system, and even filter sterilizing water and covering pots with plastic lids.

It is proposed that clean, pathogen-free VAM inoculum can be produced in the greenhouse by observing six rules: (1) sterilize pots and growth media, (2) begin with clean inoculum, (3) test for unwanted organisms and discard questionable cultures, (4) alternate selected host species, (5) drench with selected pesticides, and (6) practice sanitary greenhouse procedures. VAM inoculum produced with these methods will not be a microbial monoculture but it will be as clean as many planting mixes, peat moss, and organic fertilizers. Microorganisms which frequently colonize sterile soil such as *Peziza ostracoderma, Pyronema, Trichoderma,* and many bacterial species are common in mycorrhizal inoculum. These organisms do not appear to affect mycorrhizal symbiosis or plant growth. Microorganisms which may be plant pathogens under certain conditions, such as *Fusarium, Pythium, Alternaria,* and *Papulospora* can often be isolated from the cleanest of mycorrhizal cultures. In carefully produced VAM cultures, these organisms are usually saprophytes, but they provide serious problems for the commercial producer of VAM inoculum. If populations of these organisms are high, the best policy is to discard the inoculum.

IV. PARASITES OF VAM FUNGI

Parasites of VAM fungi have been frequently observed in VAM inoculum from pot cultures. Chytridiaceous fungi such as *Phlyctochytrium* sp. and *Rhizidiomycopsis stomatosa* have been observed producing sporangia on a variety of VAM spores, including those of *Glomus, Gigaspora,* and *Acaulospora*.[64-67] An unidentified, "*Pythium*-like" fungus is found in both spores and hyphae of *G. macrocarpum* var. *geosporum* inside and outside soybean roots.[65] *Humicola fuscoatra,* and *Anguillospora pseudolongissima* have been found parasitizing chlamydospores of *Glomus* spp.[64]

Other parasites or predators of VAM fungi may include mycophagous nematodes, fungus flies, mites, spring tails, a variety of insects, amoeba, and bacteria. Ross[68] concluded that unknown biotic factors from nonsterile soils could severely retard spore production by VAM fungi.

It has not yet been proven conclusively that all parasites or predators can reduce inoculum potential of VAM fungi or reduce plant growth responses to VAM fungi. However, it is advisable to remove potential parasites or predators of VAM fungi from inoculum wherever possible. Fluctuations in VAM inoculum potentials or erratic results from inoculations with VAM may well be ascribed to activities of these organisms. Normal greenhouse sanitation and the steps cited above for preventing contamination by parasites of VAM inoculum will normally prevent contamination by parasites of VAM fungi as well. Starting inoculum from single spores should eliminate much of the problem and selective fungicides can be used to prevent parasites from becoming established. Ethazole was particularly effective at preventing parasitism by *Phlyctochytrium*.[64] Application of selected pesticides can result in the control of insects, nematodes, and other predators or parasites of VAM fungi. The most effective deterrent for "diseased inoculum" is frequent and careful examination of the inoculum. Recognition of these problems is often more difficult than their control.

V. STORAGE OF INOCULUM

Crude soil inoculum is normally air dried to a point where there is no free water (1 to 10% moisture content depending upon the growth medium). After drying, cultures are packed in plastic bags, sealed, and stored at 5°C. Cultures have been successfully stored in this manner for up to 4 years.[69]

While freeze-drying is feasible, it has not been a highly successful method for storing VAM inoculum.[70,71] Tommerup and Kidby,[72] however, have developed an L-drying method where inoculum is first dried in air and then over silica gel. This technique has been used successfully to preserve both soil inoculum and VAM spores which have been extracted from soil. *Glomus epigaeum* sporocarps apparently store very well fresh, and individual spores of this species can be stored in water-saturated bentonite clay at 5°C.[16] A variety of other storage methods were unsuccessful. Spores of *Gigaspora margarita* retained viability after storage in a solution of 200 ppm streptomycin and 100 ppm gentomycin for 1 year.[73]

VI. PRINCIPLES OF SYMBIOSIS ARE PART OF INOCULUM PRODUCTION

A. Growth of Plant and the Nutrition of VAM Fungi

Since VAM fungi are obligate symbionts, they obtain much of their nutrition from the host plant. Since host membranes surround the arbuscule and the rest of the fungus is usually outside the cell walls of the cortical cell, it appears that any nutrition must be derived from root cortical cell exudates which pass through root membranes. Since root exudation is an accepted phenomenon even without mycorrhizal fungi and since the presence of mycorrhizal fungi reduces root exudation of carbohydrates,[31,51] then it is fair to assume that VAM fungi are utilizing at least some of the exuded carbohydrate as a nutrient source for growth. Carbon transfer from host to mycorrhizal hyphae has been documented.[74] If we accept the fact that VAM fungi are obtaining nutrition from root cortical cells, then we can draw some general conclusions about the production of mycorrhizal inoculum. First, any factor which decreases the nutrient supply to the roots and, indirectly, root exudation is likely to decrease VAM inoculum production. In general, factors which increase photosynthesis, such as high light intensity, optimum temperatures, and long days are likely to result in increased mycorrhizal inoculum.[30,48,51] Conversely, low light, low temperatures, short days, pruning, leaf damage by heat, chemicals, air pollutants, or pathogens will reduce the quality and quantity of mycorrhizal inoculum.[21,30,41,47,49-51,53,75,76] Similarly, plants that are extremely nutrient-deficient or are growing so poorly that photosynthesis is inhibited usually produce poor inoculum. Less obvious factors which reduce translocation of photosynthesis to roots, such as girdling, flowering, or fruiting may also reduce inoculum production.[77] In soil, mycorrhizae appear to be inhibited by water-saturated conditions,[39,40,42] but will normally produce good inoculum under conditions in which aeration and water potentials are adequate for root growth and plant health. Lastly, competition for root nutrients by root pathogens or even other microorganisms normally results in poor mycorrhizal inoculum.[76] It has been postulated that phosphorus may reduce root cortical cell exudates by reducing the permeability of cell membranes, thus interrupting the nutrient flow to VAM fungi.[31]

Ideally, VAM inoculum should be produced on healthy plants, at high light intensities, under long daylength periods, under conditions which encourage root growth, and at slightly less than optimum phosphorus levels.

B. VAM Growth Dynamics

Analysis of growth dynamics of VAM fungi indicate that mycorrhizal infections progress in a sigmoid manner as do most other biological populations.[6,78,79] There is a lag phase of variable time period, an abrupt period of exponential growth, and a culminating plateau phase or maximum amount of colonization (Figure 2). Environment and soil type appear to regulate the maximum amount of root colonization achievable in a particular VAM association.[80,81] In pot cultures, maximum percent VAM colonization of host roots normally occurs 10 to 14 weeks after inoculation, although maximum intensity of colonization may occur somewhat later. Spore production by VAM fungi usually commences about 6 weeks after infection and often reaches a maximum shortly after cessation of root growth.[6,20] In single species VAM pot cultures, the numbers of spores produced are often proportional to the percent colonization of the roots. However, some VAM fungi sporulate very little, if at all, and under field conditions or in mixed cultures, spore numbers have little relation to the amount of colonized root present.

Assuming that suitable symbionts are chosen to produce maximum amounts of VAM colonization, inoculum efficiency as well as type of inoculum will depend strongly upon the stage of VAM colonization at harvest.[82] Abbott and Robson[82] have concluded that inoculum infectivity normally increases with the amount of host root infected, except for some fungi such as *Acaulospora laevis* which may produce spores at the expense of hyphae inside roots. Therefore, root and hyphal inoculum with maximum infectivity may be obtained in 10 to 14 weeks, but the production of inoculum with a heavy, mature spore population often requires an entire growing season. The idea that increasing periods of time will result in superior inoculum is incorrect. Once the VAM fungus has completed its life cycle, the inoculum potential of pot cultures

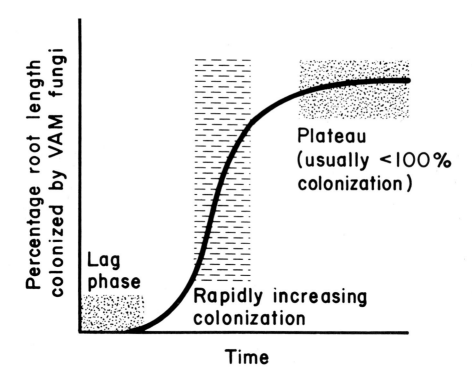

FIGURE 2. Illustrations of stages of root colonization by VAM fungi. Infectivity of VAM inoculum is usually highest after the root colonization plateau has been reached.

rarely improves with age. Mosse[6] believes that spore production declines in pot cultures after the first year and virtually ceases after 3 years. Ross and Ruttencutter[65] describe a similar phenomenon which inhibits VAM spore production in nonsterile soil. Since older pot cultures are easily contaminated with undesirable organisms, it is usually best to harvest pot cultures and store them after 1 year.

C. Inoculum Potential

The concept of inoculum potential is an elusive one, but it is the key to producing high quality VAM inoculum. There are many definitions, of varying practicality, for the term "inoculum potential". The one which is most useful is that of Dimond and Horsfall,[83] which defines inoculum potential as the potential of a specific amount of inoculum to cause root infection under a standard set of conditions, i.e., the infectivity of inoculum. According to this definition, the VAM fungus, the plant, and the environment all interact to produce a given infectivity. The usefulness of such a definition is that it can be measured. Porter[84] has adapted the "most probable number" method to measure the inoculum potential of VAM fungi.

Inoculum potential of VAM inoculum is critical because it is the only factor which can be consistently related to inoculum quality and efficiency.[51,82,85-87] Despite these persistent warnings, researchers continue to regard VAM inoculum from pot cultures as constants. Variability in VAM culture inoculum is no doubt responsible for innumerable experimental failures and also fallacious beliefs about mycorrhizal efficiency or survival. Apparent mycorrhizal dependency of host plants can vary by as much as 700% depending on the inoculum potential of VAM inoculum.[21] Carling et al.[88] found that colonization of soybeans by VAM fungi was clearly dependent upon inoculum potential. Numbers of spores in VAM inoculum means little if it is not known whether they are alive, dead, or dormant. Some mycorrhizal fungi produce a few very infective spores, others appear perfectly able to complete their life cycle without sporulating at all. Hence, spore density is usually not related to inoculum potential. Comparison between species of VAM fungi becomes very difficult because each species possesses a different inoculum potential per unit of inoculum[89] and these must be adjusted to similarity before intrinsic efficiency of VAM species can be compared.[90,91] The effectiveness of the species examined by Abbott and Robson[85] and Hall[86] were largely governed by the infectivity of the inoculum, not the VAM species themselves.

There has been considerable debate about whether infected roots, spores, or hyphae of VAM fungi are best for inoculum. It would appear that experimental conditions determine which type of inoculum is best for a given situation.[82,86,87,92-94] It is the inoculum potential of the various inoculum types which is being compared. It seems reasonable to conclude from these papers that hyphae and root pieces, which quickly produce hyphae that can colonize roots, may have the ability to produce rapid infections; i.e., they have a high initial inoculum potential. However, if conditions become stressful, either during storage or in the soil/plant environment, the hyphae may die and the spores, which can survive unfavorable conditions, will have the greater inoculum potential. It therefore appears desirable to utilize a mixture of spores, hyphae, and root pieces in VAM inoculum to assure maximum inoculum potential under a variety of environmental conditions. This is probably why soil inoculum containing a mixture of inoculum structures is usually superior to spores or root pieces alone.[11,21,86]

VII. CONCLUSIONS

In conclusion, it could be said that the principles of VAM inoculum production are (1) reduction in the ability of the host plant to produce or translocate photosynthate to its roots will reduce VAM inoculum production; (2) factors which result in poor root growth will reduce the substrate for VAM fungi and therefore reduce VAM inoculum production; (3) organisms which compete with VAM fungi for root substrate will reduce the quality and quantity of VAM inoculum; (4) luxury amounts of phosphorus will reduce the amount of mycorrhizal inoculum; (5) mycorrhizal inoculum increases in a sigmoid fashion and percent VAM colonization eventually reaches a maximum plateau value which may be well below 100%. Spore production as well as maximum biomass of VAM fungi are normally produced at the end of the growth phase; allowing the fungus to progress through further life cycles rarely improves either quantity or quality of VAM inoculum; and (6) the only factor which can consistently be related to inoculum quality and efficiency is inoculum potential; the key to producing high quality VAM inoculum is to produce inoculum with persistently high inoculum potential.

REFERENCES

1. Mosse, B. and Hayman, D. S., Mycorrhizae in agricultural plants, in *Tropical Mycorrhizal Research*, Mikola, P., Ed., Oxford University Press, Oxford, 1980, 213.
2. Menge, J. A., Lembright, H., and Johnson, E. L. V., Utilization of mycorrhizal fungi in citrus nurseries, *Proc. Int. Soc. Citric.*, 1, 129, 1977.
3. Menge, J. A. and Johnson, E. L. V., Commercial production of mycorrhizal inoculum may benefit citrus growers, *Citrograph*, 63, 139, 1978.
4. Hayman, D. S., Mycorrhiza and its significance in horticulture, *Plantsman*, 2, 214, 1981.
5. Hayman, D. S., Practical aspects of vesicular-arbuscular mycorrhiza, in *Advances in Agricultural Microbiology*, Subba Rao, N. S., Ed., Oxford V. J. B. H. Publishing, New Delhi, 1981, 325.

6. Mosse, B., Vesicular-arbuscular mycorrhiza research for tropical agriculture, *Res. Bull. Hawaii Agric. Exp. Stn.*, 194, 1981.
7. Gerdemann, J. W., Relation of a large soil-borne spore to phycomycetous mycorrhizal infections, *Mycologia*, 47, 619, 1955.
8. Byrd, D. W., Barker, R. K., Ferris, H., Nusbaum, C. J., Griffin, W. E., Small, R. H., and Stone, C. A., Two semi-automatic elutriators for extracting nematodes and certain fungi from soil, *J. Nematol.*, 8, 206, 1976.
9. Daniels, B. A. and Skipper, H. D., Methods for the recovery and quantitative estimation of propagules from soil, in *Methods and Principles of Mycorrhizal Research*, Schenck, N. C., Ed., American Phytopathology Society, St. Paul, 1982, 29.
10. Tommerup, I. C. and Kidby, D. K., Production of aseptic spores of vesicular-arbuscular endophytes and their viability after chemical and physical stress, *Appl. Environ. Microbiol.*, 39, 1111, 1980.
11. Menge, J. A. and Timmer, L. W., Procedures for inoculation of plants with vesicular-arbuscular mycorrhizae in the laboratory, greenhouse and field, in *Methods and Principles of Mycorrhizal Research*, Schenck, N. C., Ed., American Phytopathology Society, St. Paul, 1982, 59.
12. Hepper, C. M., Techniques for studying the infection of plants by vesicular-arbuscular fungi under axenic conditions, *New Phytol.*, 88, 641, 1981.
13. Howeler, R. H., Edwards, D. G., and Asher, C. J., Application of the flowing solution culture techniques to studies involving mycorrhizas, *Plant Soil*, 59, 179, 1981.
14. Ojala, J. C. and Jarrell, W. M., Hydroponic sand culture systems for mycorrhizal research, *Plant Soil*, 57, 297, 1980.
15. Hoagland, D. R. and Arnon, D. I., The water culture method for growing plants without soil, *Calif. Agric. Exp. Stn. Circ.*, 347, 1938, 39.
16. Daniels, B. A. and Menge, J. A., Evaluation of the commercial potential of the vesicular-arbuscular mycorrhizal fungus, *Glomus epigaeus, New Phytol.*, 87, 345, 1981.
17. Daniels, B. A. and Trappe, J. M., *Glomus epigaeus* sp. nov., a useful fungus for vesicular-arbuscular mycorrhizal research, *Can. J. Bot.*, 57, 539, 1979.
18. Johnson, C. R. and Menge, J. A., Mycorrhizae may save fertilizer dollars, *Am. Nurseryman*, 156, 79, 1982.
19. Bagyaraj, D. J. and Manjunath, A., Selection of a suitable host for mass production of VA mycorrhizal inoculum, *Plant Soil*, 55, 495, 1980.
20. Nemec, S., Menge, J. A., Platt, R. G., and Johnson, E. L. V., Vesicular-arbuscular mycorrhizal fungi associated with citrus in Florida and California and notes on their distribution and ecology, *Mycologia*, 73, 112, 1981.
21. Ferguson, J. J., Inoculum Production and Field Application of Vesicular-Arbuscular Mycorrhizal Fungi, Ph.D. thesis, University of California, Riverside, 1981.
22. Holevas, C. D., The effect of a vesicular-arbuscular mycorrhiza on the uptake of soil phosphorus by strawberry (*Fragaria* sp., var. Cambridge Favourite), *J. Hortic. Sci.*, 41, 57, 1966.
23. Jasper, D. A, Robson, A. D., and Abbott, L. K., Phosphorus and the formation of vesicular-arbuscular mycorrhizas, *Soil Biol. Biochem.*, 11, 501, 1979.
24. Johnson, P. N., Effects of soil phosphate level and shade on plant growth and mycorrhizas, *N.Z. J. Bot.*, 14, 333, 1976.
25. Menge, J. A., Labanauskas, C. K., Johnson, E. L. V., and Platt, R. G., Partial substitution of mycorrhizal fungi for phosphorus fertilization in the greenhouse culture of citrus, *Soil Sci Soc. Am. Proc.*, 42, 926, 1978.
26. Mosse, B., Plant growth responses to vesicular-arbuscular mycorrhiza. IV. In soil given additional phosphate, *New Phytol.*, 72, 127, 1973.
27. Pairunan, A. K., Robson, A. D., and Abbott, L. K., The effectiveness of vesicular-arbuscular mycorrhizas in increasing growth and phosphorus uptake of subterranean clover from phosphorus sources of different solubilities, *New Phytol.*, 84, 327, 1980.
28. Pugh, L. M., Roncadori, R. W., and Hussey, R. S., Factors affecting vesicular-arbuscular mycorrhizal development and growth of cotton, *Mycologia*, 73, 869, 1981.
29. Sanders, F. E., The effect of foliar-applied phosphate on mycorrhizal infections of onion roots, in *Endomycorrhizas*, Sanders, F. E., Mosse, B., and Tinker, P. B., Eds., Academic Press, London, 1975, 261.
30. Graham, J. H., Leonard, R. T., and Menge, J. A., Interaction of light intensity and soil temperature with phosphorus inhibition of vesicular-arbuscular mycorrhiza formation, *New Phytol.*, 91, 683, 1982.
31. Graham, J. H., Leonard, R. T., and Menge, J. A., Membrane-mediated decrease in root exudation responsible for phosphorus inhibition of vesicular-arbuscular mycorrhiza formation, *Plant Physiol.*, 68, 548, 1981.
32. Hayman, D. S., Influence of soils and fertility on activity and survival of vesicular-arbuscular mycorrhizal fungi, *Phytopathology*, 72, 1119, 1982.

33. Chambers, C. A., Smith, S. E., and Smith, F. A., Effects of ammonium and nitrate ions on mycorrhizal infection, nodulation and growth of *Trifolium subterraneum*, *New Phytol.*, 85, 47, 1980.
34. Johnson, C. R., Joiner, J. N., and Crews, C. E., Effects of N, K and Mg on growth and leaf nutrient composition of three container grown woody ornamentals inoculated with mycorrhizae, *J. Am. Soc. Hortic. Sci.*, 105, 286, 1980.
35. Hepper, C. M., Germination and growth of *Glomus caledonius* spores: the effects of inhibitors and nutrients, *Soil Biol. Biochem.*, 11, 269, 1979.
36. McIlveen W. D. and Cole, H., Jr., Influence of zinc on development of the endomycorrhizal fungus *Glomus mosseae* and its mediation of phosphorus uptake by *Glycine max* cultivar 'Amsoy-71', *Agric. Environ.*, 4, 245, 1979.
37. McIlveen, W. D., Spotts, R. A., and Davis, D. D., The influence of soil zinc on nodulation, mycorrhizae and ozone sensitivity of Pinto beans, *Phytopathology*, 65, 647, 1975.
38. Mosse, B., Stribley, D. P., and LeTacon, F., Ecology of mycorrhizae and mycorrhizal fungi, *Microb. Ecol.*, 5, 137, 1981.
39. Reid, C. P. P. and Bowen, G. D., Effects of soil moisture on VA mycorrhiza formation and root development in *Medicago*, in *The Soil-Root Interface*, Harley, J. L. and Russell, R. S., Eds., Academic Press, London, 1979, 211.
40. Redhead, J. F., *Endogone* and endotrophic mycorrhizae in Nigeria, *Proc. XV IUFRO Congr. Sec.*, 24, 1971.
41. Baylis, G. T. S., Host treatment and spore production by *Endogone*, *N.Z. J. Bot.*, 7, 173, 1969.
42. Saif, S. R., The influence of soil aeration on the efficiency of vesicular-arbuscular mycorrhizae. I. Effect of soil oxygen on the growth and mineral uptake of *Eupatorium oderatum* L. inoculated with *Glomus macrocarpus*, *New Phytol.*, 88, 649, 1981.
43. Lambert, D. H. and Cole, H., Jr., Effects of mycorrhizae on establishment and performance of forage species in mine spoil, *Agron. J.*, 72, 257, 1980.
44. Abbott, L. K. and Robson, A. D., The distribution and abundance of vesicular arbuscular endophytes in some Western Ausralian soils, *Aust. J. Bot.*, 25, 515, 1977.
45. Hayman, D. S. and Mosse, B., Plant growth responses to vesicular-arbuscular mycorrhiza. I. Growth of *Endogone*-inoculated plants in phosphate-deficient soils, *New Phytol.*, 70, 19, 1971.
46. Mosse, B., The influence of soil type and *Endogone* strain on the growth of mycorrhizal plants in phosphate deficient soils, *Rev. Ecol. Biol. Sol.*, 9, 529, 1972.
47. Daft, M. J. and El-Giahmi, A. A., Effect of arbuscular mycorrhiza on plant growth. VIII. Effects of defoliation and light on selected hosts, *New Phytol.*, 80, 365, 1978.
48. Furlan, V. and Fortin, J. A., Effect of light intensity on the formation of vesicular-arbuscular endomycorrhizas on *Allium cepa* by *Gigaspora calospora*, *New Phytol.*, 79, 335, 1977.
49. Hayman, D. S., Plant growth responses to vesicular-abuscular mycorrhiza. IV. Effect of light and temperature, *New Phytol.*, 73, 71, 1974.
50. Ferguson, J. J. and Menge, J. A., The influence of light intensity and artificially extended photoperiod upon infection and sporulation of *Glomus fasciculatus* on Sudan grass and on root exudation by Sudan grass, *New Phytol.*, 92, 183, 1982.
51. Johnson, C. R., Menge, J. A., Schwab, S., and Ting, I. P., Interaction of photoperiod and vesiculararbuscular mycorrhizae on growth and metabolism of sweet orange, *New Phytol.*, 90, 665, 1982.
52. Baylis, G. T. S., Experiments on the ecological significance of phycomycetous mycorrhizas, *New Phytol.*, 66, 231, 1967.
53. Furlan, V. and Fortin, J. A., Formation of endomycorrhizae by *Endogone calospora* on *Allium cepa* under three temperature regimes, *Nat. Can. (Quebec)*, 100, 467, 1973.
54. Walker, J. J., One degree increments in soil temperature affect maize seedlings behavior, *Soil Sci. Soc. Am. Proc.*, 33, 729, 1969.
55. Schenck, N. C. and Schroder, V. N., Temperature response of *Endogone* mycorrhiza on soybean roots, *Mycologia*, 66, 1600, 1974.
56. Schenck, N. C., Graham, S. O., and Green, N. E., Temperature and light effect on contamination and spore germination of vesicular-arbuscular mycorrhizal fungi, *Mycologia*, 67, 1189, 1975.
57. Smith, S. E. and Bowen, G. D., Soil temperature, mycorrhizal infection and nodulation of *Medicago trunculata* and *Trifolium subterraneum*, *Soil Biol. Biochem.*, 11, 469, 1979.
58. Menge, J. A., Effect of soil fumigants and fungicides on vesicular-arbuscular fungi, *Phytopathology*, 72, 1125, 1982.
59. Pope, P. E. and Holt, H. A., Paraquat influences development and efficacy of the mycorrhizal fungus *Glomus fasciculatus*, *Can. J. Bot.*, 59, 518, 1981.
60. Schwab, S. M., Johnson, E. L. V., and Menge, J. A., Influence of simazine in formation of vesiculararbuscular mycorrhizae in *Chenopodium quinona* Willd., *Plant Soil*, 64, 283, 1982.
61. Elmes, R. and Mosse, B., Vesicular-arbuscular mycorrhiza: nutrient film technique, *Rothamsted Exp. Stn. Annu. Rep.*, 1, 188, 1980.

62. Kormanik, P. P., Bryan, W. C., and Schultz, R. C., Increasing endomycorrhizal fungus inoculum in forest nursery soil with cover crops, *South. J. Appl. For.*, 4, 151, 1980.
63. Menge, J. A., Johnson, E. L. V., and Minassian, V., Effect of heat treatment and three pesticides upon the growth and reproduction of the mycorrhizal fungus *Glomus fasciculatus*, *New Phytol.*, 82, 473, 1979.
64. Daniels, B. A. and Menge, J. A., Hyperparasitization of vesicular-arbuscular mycorrhizal fungi, *Phytopathology*, 70, 584, 1980.
65. Ross, J. P. and Ruttencutter, R., Population dynamics of two vesicular-arbuscular endomycorrhizal fungi and the role of hyperparasitic fungi, *Phytopathology*, 67, 490, 1977.
66. Schenck, N. C. and Nicolson, T. H., A zoosporic fungus occurring on species of *Gigaspora margarita* and other vesicular-arbuscular mycorrhizal fungi, *Mycologia*, 69, 1049, 1977.
67. Sparrow, F. K., A Rhizidiomycopsis on azygospores of *Gigaspora margarita*, *Mycologia*, 69, 1053, 1977.
68. Ross, J. P., Effect of nontreated field soil on sporulation of vesicular-arbuscular mycorrhizal fungi associated with soybean, *Phytopathology*, 70, 1200, 1980.
69. Ferguson, J. J. and Woodhead, S. H., Production of endomycorrhozal inoculum. A. Increase and maintenance of vesicular-arbuscular mycorrhizal fungi, in *Methods and Principles of Mycorrhizal Research*, Schenck, N. C., Ed., American Phytopathology Society, St. Paul, 1982, 47.
70. Crush, J. R. and Pattison, A. C., Preliminary results on the production of vesicular-arbuscular mycorrhizal inoculum by freeze drying, in *Endomycorrhizas*, Sanders, F. E., Mosse, B., and Tinker, P. B., Eds., Academic Press, London, 1975, 485.
71. Jackson, N. E., Miller, R. H., and Franklin, F. E., The influence of vesicular-arbuscular mycorrhizae on uptake of ^{90}Sr from soil by soybeans, *Soil Biol. Biochem.*, 5, 205, 1973.
72. Tommerup, I. C. and Kidby, D. K., Preservation of spores of vesicular-arbuscular endophytes by L-drying, *Appl. Environ. Microbiol.*, 37, 831, 1979.
73. Mertz, S. M., Heithaus, J. J., and Bush, R. L., Mass production of axenic spores of the endomycorrhizal fungus *Gigaspora margarita*, *Trans. Br. Mycol. Soc.*, 72, 167, 1979.
74. Ho, I. and Trappe, J. M., Translocation of ^{14}C from *Festuca* plants to their endomycorrhizal fungi, *Nature (London)*, 244, 30, 1973.
75. McCool, P. M., Menge, J. A., and Taylor, O. C., Effects of ozone and HCl gas on the development of the mycorrhizal fungus *Glomus fasciculatus* and growth of 'Troyer' citrange, *J. Am. Hortic. Sci.*, 104, 151, 1979.
76. Schenck, N. C. and Kellam, M. K., The influence of vesicular-arbuscular mycorrhizae on disease development, *Fla. Agric. Exp. Stn. Tech. Bull.*, 798, 1978.
77. Johnson, C. R., Graham, J. H., Leonard, R. T., and Menge, J. A., Effect of flower bud development in chrysanthemum on vesicular-arbuscular mycorrhiza formation, *New Phytol.*, 90, 671, 1982.
78. Saif, S. R., The influence of host development on vesicular-arbuscular mycorrhizae and endogonaceous spore population in field-grown vegetable crops. I. Summer-grown crops, *New Phytol.*, 79, 341, 1977.
79. Sutton, J. C., Development of vesicular-abuscular mycorrhizae in crop plants, *Can. J. Bot.*, 51, 2487, 1973.
80. Buwalda, J. G., Ross, G. J. S., Stribley, D. P., and Tinker, P. B., The development of endomycorrhizal root systems. III. The mathematical representation of the spread of vesicular-arbuscular mycorrhizal infection in root systems, *New Phytol.*, 91, 669, 1982.
81. Warner, A., Spread of Vesicular-Arbuscular Mycorrhizal Fungi in Soil, Ph.D. thesis, University of London, 1980.
82. Abbott, L. K. and Robson, A. D., Infectivity and effectiveness of vesicular arbuscular mycorrhizal fungi: effect of inoculum type, *Aust. J. Agric. Res.*, 32, 631, 1981.
83. Dimond, A. E. and Horsfall, J. G., Inoculum and the diseased population, in *Plant Pathology an Advanced Treatise*, Vol. 3., Horsfall, J. G. and Dimond, A. E., Eds., Academic Press, New York, 1960, 1.
84. Porter, W. M., The "most probable number" method for enumerating infective propagules of vesicular arbusclar mycorrhizal fungi in soil, *Aust. J. Soil Res.*, 17, 515, 1979.
85. Abbott, L. K. and Robson, A. D., Infectivity and effectiveness of five endomycorrhizal fungi: competition with indigenous fungi in field soils, *Aust. J. Agric. Res.*, 32, 621, 1981.
86. Hall, I. R., Response of *Coprosma robusta* to different forms of endomycorrhizal inoculum, *Trans. Br. Mycol. Soc.*, 67, 409, 1976.
87. Smith, F. A. and Smith, S. E., Mycorrhizal infection and growth of *Trifolium subterraneum*: comparison of natural and artificial inocula, *New Phytol.*, 88, 311, 1981.
88. Carling, D. E., Brown, M. F., and Brown, R. A., Colonization rates and growth responses of soybean plants infected by vesicular-arbuscular mycorrhizal fungi, *Can. J. Bot.*, 57, 1769, 1979.
89. Daniels, B. A., McCool, P. M., and Menge, J. A., Comparative inoculum potential of spores of six vesicular-arbuscular mycorrhizal fungi, *New Phytol.*, 89, 385, 1981.

90. Menge, J. A., Utilization of vesicular-arbuscular mycorrhizal fungi in agriculture, *Can. J. Bot.*, 60, in press, 1982.
91. Abbott, L. K. and Robson, A. D., The role of vesicular arbuscular mycorrhizal fungi in agriculture and the selection of fungi for inoculation, *Aust. J. Agric. Res.*, 33, 389, 1982.
92. Johnson, P. N., Mycorrhizal Endogonaceae in a New Zealand forest, *New Phytol.*, 78, 161, 1977.
93. Powell, C. L., Development of mycorrhizal infections from *Endogone* spores and infected root segments, *Trans. Br. Mycol. Soc.*, 66, 439, 1976.
94. Manjunath, A. and Bagyaraj, D. J., Components of VA mycorrhizal inoculum and their effects on growth of onion, *New Phytol.*, 87, 355, 1981.

Chapter 10

FIELD INOCULATION WITH VA MYCORRHIZAL FUNGI

Conway L. Powell

TABLE OF CONTENTS

I.	Introduction	206
II.	Inoculum	206
	A. Fungal Selection	206
	B. Inoculum Type	207
	C. Inoculum Formulation	207
	D. Inoculum Placement	208
	1. Plants Raised in the Nursery	208
	2. Plants Seed Sown in the Field	209
III.	Growth Responses in the Field	209
	A. Crops	209
	1. Cereals	209
	2. Legumes	210
	3. Other Field Crops	213
	B. Horticulture	214
	1. Fruit Trees	214
	2. Timber Trees	214
	3. Ornamentals	215
	C. Pasture	215
IV.	Appropriate Technology	216
	A. Objectives	216
	B. Tree Crops	216
	C. Transplanted Field Crops	217
	D. Field Sown Crops and Mechanization	217
	1. Developing Countries	217
	2. Industrialized Countries	218
	E. Permanent Pasture	218
V.	Conclusions	219
References		219

I. INTRODUCTION

It became obvious during the 1960s and early 1970s that infection by vesicular-arbuscular mycorrhizal (VAM) fungi could increase plant growth in many sterilized and unsterilized soils, because of better phosphorus uptake from soil by plants. This general principle had been established over a wide variety of soil types, host species, and fungal endophytes by researchers all over the world.[1-11] These early results have since been confirmed and reported on ad nauseam in sterilized soils.

In contrast to this, there have been relatively few published reports of successful field inoculation with VAM. One might have expected a spate of experiments to extend and confirm the initial and exciting promise of manipulating VAM in the field, but from 1970 to 1975 we have only the reports of Ross,[12,13] Khan,[14,15] Rich and Bird,[16] Gerdemann and others,[17,18] and Schenck.[19,20] By the late 1970s, the important role of VAM in phosphorus cycling in soils had been firmly established and optimistic projections made for responses to VAM inoculation in the field.[15,21] However, there had been no clear indication that growth responses from VAM inoculation in the field were possible, predictable, or long-lived. Accordingly, soil scientists and agronomists have recently begun to question the potential for VAM inoculation in agriculture:[22] "Oh yes, we know that the indigenous mycorrhizal fungi are already present in soil and that they increase plant growth but show us field trials where mycorrhizal inoculation really has stimulated plant yield". It has been well established, however, that many tree crops which are raised in sterilized media in nursery beds greatly respond to VAM inoculation with increased shoot growth, stem diameter, and percentage of saleable plants ready for outplanting.[23]

There have been suggestions that we should not move too rapidly into the unknown area of VAM inoculation of annual and perennial field crops until more is known of the autecology and physiology of VAM endophytes,[24,25] and indeed most field trials to date have been very empirical in the choice of fungal inoculants, soil types, and inoculation procedures. Without definite field evidence of the usefulness of VAM inoculation quite soon, however, the current interest, expectations, and funding of all areas of VAM research will probably (and justifiably) fall away.[22] Nevertheless, there has been an upsurge in reported field trials since 1975 and the results of these and the earlier experiments are discussed in this chapter, with an emphasis on their deficiencies, reliability, and applicability to current agricultural practices.

II. INOCULUM

A. Fungal Selection

Ideally, the inoculant fungi used for any field trial should already have been chosen for their efficiency in the soil type from the trial site and with the host plant being used.[19,26-29] Several pot trials have shown that there are very significant effects of plant host (species and/or cultivar), soil type, and mycorrhizal inoculant on the size of the mycorrhizal growth response.[24,28,30-34] In many soils under permanent vegetation (such as permanent pasture in temperate climates) nonsporing VAM fungi are common,[35] and may be the most suitable inoculants. Otherwise a spore-producing fungus should be chosen as the fungal inoculant, and prior to any field trial, pot trials should be carried out to determine the best inoculants to try in the field. This is a tedious and impossible procedure if the soil from all paddocks in a farm were to be individually tested for the best inoculant fungus. One always hopes to find or select one fungus which is highly effective on a wide range of hosts over many soil types, although this is probably unlikely to occur.

Before laying down VAM field trials with onions,[36] we determined in a pot experiment that a mixture of *Glomus* species was the best inoculant for the onion cultivar

(PLK, Pukekohe Long Keeper),[34] when grown on Patumahoe clay loam (a typic haplohumult) from the field site. The *Glomus* spp. inoculum was then shown to be the best inoculant with PLK onions in a field trial on the Patumahoe soil, amply vindicating the effort of prior fungal selection.

There have been few attempts to select for efficient VAM fungi on a rational basis (such as ability to stimulate plant growth),[24,29,37] and most fungi are selected on arbitrary characters such as the presence (and numbers) of spores in soil. Very few workers have cited good reasons for using the mycorrhizal endophytes they have chosen. There are a few "favorite" spore types for field inoculation including *Glomus fasciculatum*, *G. mosseae*, *G. etunicatum*, *G. tenue*, and *Gigaspora margarita*. It is most unlikely that the strains of these fungi are universally the most effective, rather, they have been available (convenient for the researcher) and have produced recognizable spores in soil. This is not to criticize their use, but perhaps we should expect even better field responses when fungal inoculants are selected specifically for the intended crop and field soil types.

B. Inoculum Type

One of the greatest impediments to VAM inoculation of field crops is the bulk and weight of inoculum required. Pure cultured inoculum of VAM fungi is not yet available, thus, in many field trials, plants have been inoculated by placing pads of mycorrhiza-infested soil below each seed. Inoculum rates have varied from 2 to 50 g per seed,[36,38] with 10 to 20 g rates common.[27,39-41] Very high soil inoculum rates of 100 and 167 t/ha have been used as layers of inoculum soil (10- and 16-mm deep) to establish mycorrhizal infection in cotton-growing soil and coal mine spoil.[16,42]

Where seedlings or cuttings can be preinoculated before transplanting to the field, the bulkiness of mycorrhizal soil inoculum is of no problem since hundreds of field plants can be inoculated with a few hundred spores,[14,15,43] a few liters of inoculum soil,[20,44,45] or a layer of chopped mycorrhizal roots.[46,47] Inoculum bulk is a problem for crops grown in the field from seed. Rates used have varied from 0.8 to 167 t/ha,[16,27] with rates of 20 to 30 t/ha commonly used.[48,49] Most of these rates would be unacceptably high in practical agriculture.

Hayman et al.[40] managed to concentrate their soil inoculum sevenfold by wet sieving through a 100-μm diameter mesh sieve and removing the clay and sand particles. This inoculum was then in a suitable form for fluid drilling,[50] and rapidly established high levels of mycorrhizal infection in inoculated red clover in the field. Chopped, mycorrhiza-infested roots produced on host plants grown under nutrient film technique (NFT)[51] have been used as an inoculum source, but they rapidly lose viability during drying and storage.[40] Inoculum produced by NFT does not really seem to be a better alternative to "ordinary" soil inoculum or concentrated inoculum. Since ordinary soil inoculum and concentrated inoculum were equally capable of infecting plant roots,[40] the decision on which inoculum type to use will depend on the inoculum formulation required (see Section II. C.) and on whether the cost of concentrating the raw inoculum is offset by lower transport and spreading costs.

C. Inoculum Formulation

In experiments where seedlings were preinoculated before outplanting, mycorrhizal inoculum soil was freshly harvested (and the spores extracted and used if necessary) and the inoculum immediately mixed or layered in the seed bed.[47,52,53] Similarly, in most trials with seed sown crops, the inoculum soil has been freshly harvested before use. In practice we may have to use inoculum which has been processed either by wet sieve concentration (ready for fluid drilling)[40] or by drying and reconstituting as a nonseed granule or a multiseeded pellet.[36,40,54-56] Freshly harvested inoculum can be spread by hand satisfactorily, but is very unsuitable for most forms of machine placement as it rapidly blocks pipes and milled rollers.

Pelleting is a successful inoculum formulation when the seedling radicle passing down through the inoculum becomes primarily infected with the fungi in the pellet before it comes into contact with the indigenous mycorrhizal fungi in the field soil.[56] The inoculant fungi have an immediate spatial advantage over the indigenous mycorrhizal fungi. Nonseeded, granulated inoculum has the advantage that it can already be made commercially, stored for several months at 4 to 20°C with minimal loss of viability and is suitable for machine drilling in the field or use in a nursery.

The granulation process can depress inoculum viability, and in multiseeded pellets, germination of the seeds can be impaired probably through competition with the soil pellet for moisture.[40] Hattingh and Gerdemann[18] avoided these problems by pelleting citrus seed with fungal spores rather than with unconcentrated inoculum soil. This method worked well for large citrus seed and avoided the bulkiness of most pelleted seed but is probably not practical for seed sown field crops because the cost of spore separation from inoculum soil and coating onto the seeds would undoubtedly be too high. The spore inoculum would probably also have a shorter shelf life than whole soil inoculum. If and when VAM fungi can be grown in artificial culture it should be possible to coat seeds with fungal mycelium if the inoculum on the seed can be protected from desiccation. All seed pelleting techniques, however, suffer from one major disadvantage: they restrict the positioning of mycorrhizal inoculum to the seed surface which may or may not be the best place for maximum infection of the initial seedling root system (see Section II.D.).

Fresh mycorrhizal roots or spores have occasionally been lyophilized as a means of indefinite storage of inoculum.[6,57] Viability was generally good but production costs and difficulties in rehydration would probably prevent use of the method on a large scale.

In the absence of pure cultures on artificial media, inoculum for field crops will probably be produced in low-cost sterilized soil conditions (one could sterilize 1 ha of land and convert it to mycorrhizal inoculum production by the use of a suitable cover crop), with the inoculum bulk dried and granulated or concentrated before commercial use.

For tree crops such as avocado, citrus, and liquidambar,[28,46,58] and field crops such as chili, finger millet, and asparagus which are routinely raised in nurseries (and could be inoculated before outplanting to the field),[43,45] simple unprocessed VAM inoculum soil could be used. Inoculum could be produced and a supply maintained in the nursery by use of convenient cover crops.[59]

D. Inoculum Placement
1. Plants Raised in the Nursery

Seeds have been pelleted with inoculum before planting,[18,54] but most commonly, mycorrhizal inoculum has been incorporated into the potting mix[52,59] at rates as high as one part inoculum to eight parts of potting mix.[44] Inoculum has also been banded just below the seed for inoculation of transplanted field crops,[15,17] vegetables, and tree crops.[20,28,43] La Rue et al.[60] successfully inoculated each P seedling by burying the inoculum 10 to 15 cm below the seed. Actual inoculum rates used for transplanted tree crops are often not mentioned in published papers, but have been in the range of 1 to 10 g inoculum soil per seed or cutting. It should be relatively easy and involve little expense to produce VAM inoculum for use with transplanted tree crops at these rates. In addition, inoculum placement is not crucial, as the nursery potting mix can be thoroughly infested with VAM inoculum, and there are (usually) no indigenous mycorrhizal fungi to compete with the inoculant fungi for root space.

2. Plants Seed Sown in the Field

Choice of inoculum rate and placement is very crucial, however, for crops which can be inoculated only at seeding in normal unsterilized field soils.[36] There are two main problems: (1) the introduced fungi must compete successfully with the indigenous mycorrhizal fungi for root space; and (2) field crops, such as onions, maize, and soybeans, are sown at rates of 50,000 to 500,000 plants per hectare, while tree crops or shrubs are outplanted at field densities of 500 to 3000 plants per hectare therefore, requiring only 500 to 3000 inoculation points.

A 10-g inoculum pad for trees to be planted out at 3000 plants per hectare will require only 30-kg inoculum per hectare, and the inoculum can probably be reused once the tree crop is lifted from the nursery. A 10-g inoculum for onions sown at 5×10^5 plants per hectare requires 5 t inoculum per hectare and the inoculum soil cannot be reused. These and other problems prompted Maronek et al.[23] to question the chances for successful inoculation of field crops on a wide scale.

There have been only two reported field trials in which the effects of different inoculum placements in the seed bed have been studied. Hayman et al.[40] tried four inoculum placements (all at rates equivalent to 3 t/ha): (1) loose inoculum broadcast on soil surface; (2) pelleted seed broadcast on the soil surface; (3) loose inoculum in the seed furrow; and (4) wet sieved inoculum fluid drilled in the furrow.

Powell and Bagyaraj[36] drilled loose or granulated inoculum with the onion seed or 30 mm below the seed at 1.9 t/ha, and in both experiments, inoculation in the seed furrow lead to highest mycorrhizal infection levels and best growth responses. This may not necessarily be the best inoculum placement for other crop species, although inoculum has been placed in the seed furrow in most other recorded field trials,[13,41,48,61] probably because it is the easiest way of distributing inoculum by hand.

Mycorrhizal inoculum soil was placed at 30 to 80 mm below the seed for inoculation of onions,[49] barley,[39,49] lucerne,[49] and cowpea,[27] although no reasons were given for the chosen depths. A few other placement methods have been used. Ross[12] and Rich and Bird[16] rototilled inoculum throughout the topsoil before seed sowing, and Powell[62] broadcast inoculum soil over the soil surface and still recorded large growth responses in barley. Multiseeded inoculum pellets have been used successfully to introduce VAM fungi into permanent pasture.[56,63,64] Yost and Fox[65] and Van der Zaag et al.[66] examined the effect of phosphorus fertilizer and the indigenous mycorrhizal fungi on crop growth in the field by sterilizing the control plots with methyl bromide.

While there are at least two reported field trials in which different VAM inoculum placements have been used, there are no reports at all on the effect of different inoculum rates on mycorrhizal growth responses in the field, and no attempts to quantify minimum required rates. Mycorrhizal inoculation in the field is still at the stage that researchers are happy to record and report any mycorrhiza effects at all. Almost all of the inoculum rates used to date (in the 20 to 30 t/ha range) would be quite impractical on a field scale and trials must now be carried out to determine minimum inoculum requirements and optimum placements.

III. GROWTH RESPONSES IN THE FIELD

In this section, the actual field responses reported for a wide range of cropping, horticultural, and pastoral plant species will be examined. The progress we make in the future largely depends on what we learn from a critical analysis of past results.

A. Crops
1. Cereals

Most work has been done on barley. In the first reported experiment, Saif and Khan[53] reported a 290% response in grain yield to VAM inoculation with a similarly large response in vegetative plant growth, although they gave no statistical analyses. Their impressive results are not really realistic, however, as the mycorrhizal plants were inoculated as seedlings in the glasshouse and transplanted to the field at 4 weeks old and approximately 85% of their root systems were already mycorrhizal. Although the field soil contained 40 *Glomus mosseae* spores per 50 g soil, buildup of mycorrhizal infection in the control plants was very slow with only 20% of the root system mycorrhizal with the indigenous fungi after 3 months. It must be concluded that a large part of the apparent response to inoculation of the barley crop was due to the preinoculation of the seedlings. This would be a useful result in a crop which is normally transplanted but has no relevance in crops normally grown from seed in the field.

In the other reported field trials, barley seed was inoculated with mycorrhizal soil pads placed under the seed by hand. Owusu-Bennoah and Mosse[49] planted barley seeds in 1.8-m long rows above 10- and 20-g inoculum pads and found vegetative responses up to 167%. Mycorrhizal infection was measured only at the end of the experiment but showed that inoculation had greatly increased mycorrhizal infection levels in inoculated plants. Similarly Powell et al.[67] sowed barley seed above 15-g inoculum pads and found a 92% response to *Gigaspora margarita* at 98 days after sowing, with a 25% response in vegetative yield at 143 days when the grain at the soft dough stage was lost to bird damage. In this trial, plots were very small (400 × 400 mm) and contained only 80 plants which were subject to considerable edge effects. As before however,[49] mycorrhizal infection levels were significantly higher in inoculated than uninoculated plots even though the indigenous mycorrhizal fungi were quite infective. In a later trial,[62] barley was sown in 1.0 × 1.0 m plots with mycorrhizal inoculum broadcast and raked in the soil at the rate of 2 kg/m². There were responses to inoculation of 27% increased seed yield and 35% increased seed phosphorus uptake. Clarke and Mosse[39] (at Rothamsted) found very large grain yield responses to VAM inoculation in unfertilized soil (Olsen P 10 ppm), but no response in plots given phosphorus fertilizer at 82.6 kg P/ha. In this and several other trials,[14,15,48,53] mycorrhizal responses have been recorded in unfertilized plots, but not in highly fertilized plots, and no intermediate fertilizer rates were used.

These results highlight the need to describe the mycorrhizal growth response (in the field as well as in pot trials) against a growth response curve obtained from a range of phosphorus fertilizer rates.[12,68,69,75] Mycorrhizal inoculation will only be useful in agriculture if it enables the farmer to reach his optimum yield level (e.g., 70, 80, or 90% of maximum as required) with a significantly lower level of phosphorus (or other) fertilizer. The savings from the reduced fertilizer requirement (Figure 1) will be dependent on the shape of the phosphorus response curves (Figure 1) and can then be compared with the costs of supply and introduction of the mycorrhizal inoculant.

Khan[14,15] has carried out the only published field trials on VAM inoculation of maize and wheat and found very large growth responses in shoot dry matter (DM) and grain yield to mycorrhizal inoculation. As in his field trial with barley,[53] however, mycorrhizal seedlings were preinoculated before transplant and the field soils were deliberately chosen to have low populations of indigenous mycorrhizal fungi.

Some cereal crops are transplanted from nursery beds to the farmer's field, and Govinda Rao et al.[45] found that mycorrhizal inoculation of finger millet with a selected VAM strain (M6) gave 18% better grain yield than uninoculated plants over three fertilizer rates. More importantly, the inoculated finger millet reached optimum grain yield at 19 kg P ha^{-1}, while uninoculated plants needed 38 kg P ha^{-1} to reach the same yield.

2. Legumes

Soybeans are the most researched legume crop in VAM field trials. In some early

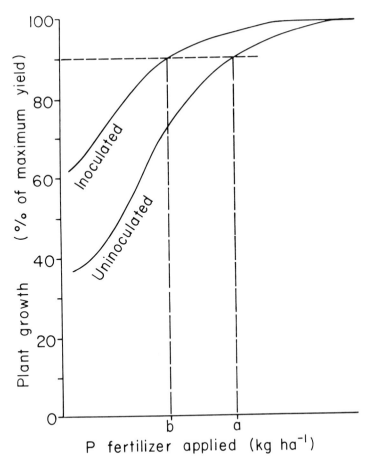

FIGURE 1. Theoretical P response curves of mycorrhiza inoculated and uninoculated plants. At an optimum yield level of 90% of maximum, mycorrhizal inoculation results in a reduction in P fertilizer requirement from a to b kg ha^{-1}.

trials reported in 1970, Ross and Harper[13] inoculated soybeans in sterilized and unsterilized soil in Mississippi and found a 29% response in grain yield to VAM inoculation in one sterilized soil. There was an 8% (nonsignificant) response to inoculation in an unsterilized soil. Ross[12] repeated the experiment in bins of sterilized soil (1 × 1.5 × 0.9 m) set in the field, which were given phosphorus fertilizer at rates of 0, 44, and 176 kg P ha^{-1}. Although these results were from sterilized soil, they showed conclusively that the indigenous mycorrhizal fungi were already increasing growth and phosphorus uptake from soil. Despite the early lead of Ross and Harper[13] in attempting VAM inoculation of field crops in "real" unsterilized soil, most VAM researchers in the U.S. have failed to move away from work in sterilized soil, usually under glasshouse conditions.

Schenck and Hinson[19] transplanted soybean seedlings with or without *Gigaspora gigantea* inoculation into a sterilized field soil and found a 53% response in grain yield of a nodulating isoline but no response to inoculation of a non-nodulating isoline. Despite the use of sterilized soil, this trial did give early evidence for the indirect role of VAM in stimulating nodulation and N fixation in legumes via enhanced phosphorus uptake. In a field trial with several crop species grown on a phosphorus deficient oxisol, Yost and Fox[65] examined mycorrhizal responses over a wide range of phosphorus fertilizer levels by sterilizing half the plots with methyl bromide and comparing the yields of the nonmycorrhizal plants with those from unsterilized (and mycorrhizal) plots. They showed that soybean yield was very responsive to mycorrhizal infection, but only over the soil solution P range of 0.003 to 0.025 µg P ml^{-1}. Since the optimum solution P level for maximum soybean yield was 0.02 to 0.2 µg P ml^{-1}, they concluded that VAM inoculation would only be useful at stimulating yield in underfertilized soils.

There is evidence, however, for increased seed yield and DM yield of soybeans after VAM inoculation in some unsterilized field soils.[41,61] Bagyaraj et al.[61] inoculated soybeans (40 plants in each 1.5 m² plot) in a P-deficient soil with *Glomus fasciculatum* and/or *Rhizobium*. The *Glomus* and *Rhizobium* treatment gave a 137% increase in shoot DM over control plots at harvest although neither inoculum (alone) had a significant effect. Mycorrhizal inoculation led to increases in grain yield of 26% and 19% with and without *Rhizobium* inoculation, although these responses were not significant at $p \leq 0.05$.

Kuo and Huang[41] planted soybean seeds above 15-g pads of a mixed *Glomus* inoculum in the stubble of recently harvested paddy rice and found a highly significant (21%) increase in grain yield. This was also significant since application of 60 kg P ha^{-1} resulted in only a 14% grain yield increase. Mycorrhizal inoculation may be very important for VAM crops such as soybean, which follow paddy rice in a multiple cropping cycle where populations of the indigenous mycorrhizal fungi have been depleted under the anaerobic paddy conditions. Kuo and Huang[41] also carried out prior selection of suitable endophytes in a glasshouse pot trial, an infrequently reported but advisable procedure.

There have been two VAM field trials reported with cowpea.[27,65] Yost and Fox[65] found that cowpea responded to mycorrhizal infection at solution P levels as high as 0.1 µg P ml^{-1}, suggesting a much greater mycorrhizal dependence in cowpea than in soybeans. Islam and Ayanaba[27] carried out a pot trial with cowpea in sterilized soil and chose an isolate of *Glomus mosseae* as the most efficient inoculant for their cowpea cultivar. A major part of their field experiment was in comparing preinoculation (and subsequent seedling transplant) with seed inoculation directly into the field. Both inoculation techniques increased plant shoot DM and nodule DM equally over control plants. Grain yield data were supplied for only the cultivar VITA-4, in which VAM inoculation in the field seed bed increased grain yield by 50% (significant at $p \leq 0.10$), while preinoculation of transplanted seedlings gave a 26% (nonsignificant) increase. They emphasized the vital importance of preselection of the most suitable VAM inoculants because a prior field trial with *Glomus fasciculatum* in the same field had no effect on cowpea yields. Their results (in a soil with a high level of indigenous mycorrhizal fungi) provide one of the few indications that seedling preinoculation may be a valid test of VAM inoculation of a field soil. It is not, of course, a valid experimental method for inoculating a crop which is normally seed sown.

Field inoculation of lucerne was reported by Azcon et al.[21] who transplanted preinoculated lucerne seedlings into nonrandomized field plots; after 12 weeks growth, they recorded a 49% response to *G. mosseae* inoculation and a 111% response to dual inoculation with *G. mosseae* and *Rhizobium meliloti*. These results cannot be accepted as an indication of likely VAM responses with lucerne, even though the soil was a normal, high fertility lucerne field soil since (1) the soil had been fallowed for 6 months prior to the experiment; more importantly (2) transplanted preinoculated seedlings were used (the use of transplanted seedlings could at least be improved by preinoculating the control plants with the indigenous mycorrhizal fungi rather than raising them as nonmycorrhizal plants); (3) the sole harvest at 12 weeks gave no indication of longevity of the mycorrhizal response in this perennial crop.

These criticisms were largely overcome in a subsequent experiment[70] in which lucerne seeds were inoculated *in situ* in the field and two herbage cuts of the 1.0 m² plots were taken at 25 and 35 weeks after seed sowing. There were mycorrhizal growth responses of 59% and 52% to *G. mosseae* inoculation at harvests 1 and 2, giving evidence of persistence of the mycorrhizal effect. This experiment was still basically unreplicated, however, as all "replicates" of each of the four treatments were placed together in a paddock divided into four main (treatment) plots.

Owusu-Bennoah and Mosse[49] placed lucerne seeds above 10- or 20-g inoculum pads in a field experiment and recorded enormous responses in shoot DM yield of 5- to 11-fold in plots not treated with formalin and generally smaller responses in formalin-drenched plots. Mycorrhizal infection levels were already high (64%) in control plants at harvest and not much higher (70 to 77%) in inoculated plants. These very large growth responses were obtained in small single row plots (1.8-m long) confounded by a marked soil phosphorus gradient across the trial. One would not expect VAM responses to persist at 5- to 11-fold, but this was not assessed.

3. Other Field Crops

The role of indigenous mycorrhizal fungi in increasing plant growth rate, yield, and ion uptake in cassava has been demonstrated in field trials by sterilizing the control (nonmycorrhizal) plots with methyl bromide.[65,66] Mycorrhizal cassava plants were able to produce high yields at very low solution P levels of 0.003 to 0.025 ppm P and were generally nonresponsive to phosphorus fertilizer, while nonmycorrhizal cassava plants grown in sand or solution culture are always very phosphorus responsive. There have been no reports of VAM inoculation of cassava in unsterilized field soils but Kang et al.[71] reported significant growth depressions to mycorrhizal inoculation in unsterile soil in pots.

Growth and phosphorus uptake of onions was much lower in nonmycorrhizal than mycorrhizal field plots at solution P levels up to 0.1 μg P mℓ^{-1},[65] and Owusu-Bennoah and Mosse[49] demonstrated large (12- to 13-fold) responses to field inoculation of onions over a short (13 week) experiment. Limitations of this experiment have already been discussed. Powell and Bagyaraj[36] found agronomically realistic increases in onion bulb production (18% increase, equivalent to an additional 8.4 t ha^{-1}) from large field plots on fertile soil (80 μg P mℓ^{-1}, Olsen P). This trial is being repeated under a variety of different conditions before farmer recommendations for inoculation can be issued.

Black and Tinker[48] obtained a 20% increase in potato tuber yield to VAM inoculation, but the P-deficient soil had been fallowed previously for 2 years and had unusually low levels of indigenous mycorrhizal fungi.[48] Most farmers planting potatoes will be growing them under continuous rotations under which low populations of VAM fungi are unlikely to occur. It is possible that the potato crop would merely act as a convenient host plant to build up and revitalize the indigenous VAM populations,[59] and that mycorrhizal responses in subsequent potato crops might not occur.

The early report of increased cotton growth in unsterilized soil after VAM inoculation has little practical significance in view of the huge inoculum rate (167 ha^{-1})[16] that was used. Iqbal and Qureshi[38] reported a mycorrhizal response of 85% in the height of sunflower plants in a field soil, but these results have limited value since the seedlings were transplanted after prior inoculation and controls were not inoculated with the indigenous mycorrhizal fungi. There is little indication of experimental design, replication, or statistical analyses.

Chili plants are normally raised in nursery beds before transplanting to the field in Southern India. Bagyaraj and Sreeramulu[43] raised chili plants in inoculated and uninoculated nursery beds on a farm property and transplanted the seedlings at 6 weeks (as normally done by the farmer) into field plots at 3 P fertilizer levels. At harvest, the most efficient fungal inoculant increased shoot weight and fruit yield by 54% and 55%, respectively, equivalent to an application of 37.5 kg P ha^{-1}. Mycorrhizal infection levels were very high in inoculated plants; up to 100% of the root system was mycorrhizal and even the noninoculated plants had 73% infection although this was measured at the end of the experiment with no earlier infection assessments.

B. Horticulture
1. Fruit Trees

Kleinschmidt and Gerdemann[17] first showed that VAM inoculation could increase fruit tree (citrus) growth in sterilized nursery soils and could overcome the stunting effects of sterilization on plant vigor which had been frequently reported.[72,73] They placed mycorrhizal inoculum soil in a 4-cm deep layer directly below seeds of rough lemon, mandarin, citrange, and sour orange and obtained increases in plant growth ranging from 10 to 142%.[17] These results were corroborated by nursery trials in California,[18,58] Texas,[74] and Florida.[20] In particular, Menge et al.[58] emphasized the differences between cultivars in their mycorrhizal dependence. The benefits of VAM inoculation in nursery-grown citrus are now well recognized and inoculation is now standard commercial practice in many parts of the U.S.

Schenck and Tucker[20] attempted to see whether VAM preinoculation of nursery stock actually led to greater plant growth once the inoculated plants were transferred to unsterilized field soils. They found that sour orange root stocks preinoculated with VAM had 34% more shoot length in disked unfumigated soil than in equivalent control plants, at 100 days after transplanting. In other published papers on mycorrhizal inoculation there is an almost complete lack of information on the effects (long-lasting or not) of VAM inoculation on the growth of plants in the field once they leave the nursery. It is usually possible to substitute P and/or Zn fertilizer for mycorrhizal inoculation in nursery culture at relatively little additional cost although fertilizer response curves for mycorrhizal and nonmycorrhizal plants have not been determined for many species.[75] If mycorrhizal inoculation in the nursery offers only a fleeting response in unsterilized field soils before the indigenous mycorrhizal fungi again dominate, inoculation may not be economically worthwhile. Field trials should be carried out to monitor the size and persistence of mycorrhizal growth responses in the field as well as in the nursery. There is no doubt of the value of mycorrhizal transplants for eroded and disturbed soils in which there are generally low levels of mycorrhizal fungi, but most orchard soils probably have or develop reasonable levels of indigenous VAM.

Fruit trees other than citrus have been investigated. Menge et al.[28] increased the growth of avocado seedlings by 49 to 254% in sterilized loamy sand by inoculation with two isolates of *Glomus fasciculatum*. Menge et al.[76] observed decreased transplant shock in mycorrhizal rather than in nonmycorrhizal avocado plants. Similarly, La Rue et al.[60] observed a 79% increase in the height of peach seedlings when inoculated wth *G. fasciculatum* in sterilized nursery soil. Mycorrhizal inoculation was more effective in stimulating yield than combined Zn and P fertilizers applied together at 1.3 and 50 kg ha^{-1}, respectively in the Zn-deficient nursery soil.

Plenchette et al.[47] presented novel and useful data on the longevity of mycorrhizal responses in an unsterilized field soil. They grew apple seedlings with or without VAM inoculum in pots of sterilized potting mix and transplanted them to an unsterilized field soil after 8 weeks and before any mycorrhizal growth responses were apparent. A mycorrhizal growth response (142% increase) in shoot length then developed in the field soil over the next 3 months despite a high population of indigenous VAM in the field soil. Hopefully we will see more field trials of this nature carried out in unsterilized soil with mycorrhizal responses followed through to fruit production.

2. Timber Trees

Many of the reported experiments on VAM inoculation of timber tree seedlings have been carried out in the U.S. at the Institute for Mycorrhiza Research and Development,

Athens, Ga. Sweetgum, an important hardwood species, has proven extremely responsive to VAM inoculation in sterilized nursery mix and mycorrhizal growth responses of 1- to 80-fold have been reported.[26,46,77] Sweetgum appears to be almost obligatorily mycorrhizal for good growth as nonmycorrhizal seedlings failed to thrive even when given as much as 1120 kg ha^{-1} of a 10-10-10 fertilizer.[46] Mycorrhizal inoculation improved survival rate in the nursery and the proportion of seedlings suitable for outplanting.

Kormanik et al.[78] also demonstrated mycorrhizal growth responses of 43 to 167% in different clonal lines of yellow poplar propagated in sterilized sand and soil mix. VAM inoculation increased root growth and decreased the time required for plants to reach a saleable size.[78] Despite this extensive work (especially on VAM inoculation of sweetgum), I have been unable to find any published reports of experiments where the initial growth responses to VAM inoculation in the nursery were followed up in unsterile field soils.

3. Ornamentals

Maronek et al.[44] showed that VAM inoculation of southern magnolia in sterile potting mix increased growth by 100 and 117% at low and high (the recommended) fertilizer rates. Similarly, Johnson et al.[79] found that inoculation of semihardwood cuttings of three ornamental species with a mixed *Glomus* inoculum greatly increased root and shoot growth in sterilized and highly fertile peat plus vermiculite medium. We[80] have also found that VAM inoculation increased root strike and plant growth of *Coprosma* and *Lophomyrtus* in sterilized peat plus pumice mix. Johnson and Crews[81] have gone one step further and transplanted *Azalea* with or without prior VAM inoculation into an unsterilized field site and found much better survival and growth of the inoculated plants at 4 months after transplant. I emphasize again that this is the most important demonstration of the benefit of VAM: long-term responses to inoculation in unsterilized field soils.

C. Pasture

The lack of papers on field inoculation of VAM into permanent pasture probably reflects the level of difficulties involved. Most permanent pasture soils have a high level of indigenous VAM fungi, and there is an established aerial sward with abundant roots already highly colonized by the indigenous VAM. Powell[82] transplanted white clover plants preinoculated with the indigenous VAM or an inoculant strain *(G. fasciculatum)* into hill country pasture sites in New Zealand. *Glomus* inoculation increased clover shoot growth at one site, but this was significant at only $p \leq 0.10$ and the results are completely unrealistic because of the use of transplanted seedlings. White clover seed was also sown above pads of inoculum soil (3 g) on hill country pasture sites, with increases in clover DM yield after VAM inoculation of up to 50%. These were significant at only $p \leq 0.10$, however, and it is just not physically or economically feasible to inoculate clover seed with such a high rate (3 g/seed) on steep hill country pasture.

In a later experiment, multiseed inoculum pellets were sown onto a hill country pasture site and there were significant DM responses to inoculation of up to 79%.[56] These were only short-term responses, however, on individually measured clover plants in a vigorous sward and the introduced fungi which established in the inoculated seedlings[56] may or may not have spread to neighboring plants with time. In recent field trials with pelleted clover seed,[83] we have found significant increases in clover shoot DM of up to 30% over a range of field sites (sheep tracks and intertrack zones) for 14 months after seed sowing. This shows that efficient VAM can be established in permanent pasture and give recurrent growth responses, but there are still severe practical difficulties in oversowing the inoculated seed onto steep hill country pasture. Hall[55] has found responses in white clover in field plots of up to 28% to his most effective inoculants, with responses persisting for more than 2 years. Hall's sites were on rolling hill country in New Zealand, accessible to tractor and on which mycorrhizal fungi could be established into a prepared seed bed. This has the advantage of removing the original pasture root mat, and probably decreasing the infectivity of the indigenous VAM fungi with which the introduced VAM fungi have to compete. It should be possible to develop suitable methods for mechanical inoculum placement for these soils if the potential for VAM inoculation can be demonstrated on this class of rolling country.

Hayman and Mosse[52] transplanted VAM inoculated white clover plants into hill country grassland sites in Wales and found that inoculation greatly increased shoot DM at two sites with largest responses occurring at the highest rates of P fertilizer applied (90 kg P ha^{-1}). Mycorrhizal growth responses continued for the 2 years that they were monitored at one unimproved site. In this sort of experiment, however, the apparent mycorrhizal response may be suspect because the inoculant fungi *Glomus mosseae* and *G. fasciculatum* were well established in the roots of the inoculated plants while the control plants were not even inoculated with the indigenous mycorrhizal fungi. It could be argued, however, that the original use of transplanted seedlings could not still be the primary cause of the very large (1- to 12-fold) growth responses to VAM inoculation recorded in the second year at one of their hill country sites. Nevertheless, for practical VAM manipulation in hill country grasslands, it must be shown that clover can be inoculated in the field from seed coated with some form of mycorrhizal soil inoculum.

There is no doubt that VAM inoculation greatly enhances pasture establishment on disturbed soils with very low populations of indigenous mycorrhizal fungi. Lambert and Cole[42] inoculated 0.25 m^2 plots of coal mine spoil with 10-mm deep layers of field soil as a VAM inoculum source and recorded mycorrhizal growth responses of 7- to 18-fold in flat pea, crownvetch, and lotus. This is a very heavy inoculum rate (100 t ha^{-1}), but one which could easily be spread by machine during mine spoil reclamation. Hall[64] found a fourfold response to inoculation of lotus by VAM soil pellet when oversown onto an eroded mountain grassland site in Canterbury, New Zealand.[64] It should be quite possible to aerially oversow large areas of such eroded soils with mycorrhiza-treated lotus seed to enhance the survival and reestablishment of plant cover.[84]

IV. APPROPRIATE TECHNOLOGY

A. Objectives

I suggest that there are now three main objectives for VAM field research: (1) to enhance plant yield in the field with reduced fertilizer requirement; (2) to enable farmers and growers to inoculate their crops effectively with existing and appropriate machinery and labor resources; and (3) to show that VAM inoculation is economically viable.

To a large extent the field trials already reported have at least demonstrated the potential for increased crop yield and decreased phosphorus fertilizer requirement in the field. We now have to concentrate on objectives (2) and (3): taking the demonstrated potential of VAM inoculation to the farmer and showing him that it will save him money.[85,86]

B. Tree Crops

Progress has been most rapid in the commercial use of VAM inoculation in the nursery production of citrus and other tree crops.[58] Most of the research and development effort in this field has occurred in the U.S. where tree crops are routinely raised in nursery beds of sterilized media to reduce competition from weeds and infection from soil-borne pathogens such as *Phytophthora* spp., e.g., under the Premium Quality Nursery Tree Programme.[58] Starter cultures of VAM and VAM soil inocula are produced commercially for sale to growers, and the technology needed for inoculation is relatively minimal and available. Growers only need incorporate inoculum into their seedling or cutting beds at the appropriate rate by hand, or by mixing in a concrete mixer or with a rotary hoe to ensure successful inoculation. Keeping the VAM soil inoculum free from pests and diseases has become one of the main problems. VAM inoculation of tree crops should become standard practice in all nurseries as it has been for ectomycorrhizal tree species for many decades.[35] Inoculum is already being exported for this purpose from California to some Middle East countries.

C. Transplanted Field Crops

There is also immediate potential for farmer use of VAM inoculation in many Third World countries, in particular, where seedlings of some vegetables and crops are raised in nursery beds and planted out by hand in the field.[85,86] VAM inoculum of suitably selected fungal strains can be bulked up and distributed for seedling inoculation of crops such as finger millet and chili in the farmers' own nursery beds.[29,43,46] As the seedlings are always hand planted, there is no additional labor or machinery cost to the farmer at planting. Mycorrhiza technology is again simple but appropriate to the local farming methods, and the selection and distribution of suitable VAM starter cultures will probably be the limiting factors.

D. Field Sown Crops and Mechanization

1. Developing Countries

Development of appropriate technology for mycorrhizal inoculation is much more difficult for crops sown from seed in the field. In most developing countries, primary tillage of small peasant farms is usually achieved using human- or animal-drawn ploughs, and seed sowing and fertilizer placement is done by hand directly into the plough furrow.[85] Under these conditions, VAM inoculum could also be hand-placed in the seed furrow by use of another person in the "family planting team".[85] This simple method suffers from several problems: (1) inoculum cannot easily be placed well below the seed if that is the required (optimum) placement; (2) accurate metering of inoculum will be difficult and require additional labor; and (3) large amounts of inoculum must be produced by the farmer (probably in the order of 1 t inoculum per hectare) in advance of normal crop sowing time. This may be possible where intensive multiple cropping is carried out in the humid tropics.[85]

As with other crops the best inoculant fungi must be chosen, mass produced, and made available to the farmer as a starter inoculant. He must also be educated to use the inoculum service. Lower inoculum rates may be acceptable in soils with very few indigenous mycorrhizal fungi, e.g., after a long fallow, or following a very hot scrub or bush fire under shifting cultivation. VAM inoculation by hand may be most feasible for root crops such as yam, potato, and cassava which are normally sown at relatively low field densities and are clearly responsive to VAM inoculation.[48,66,87]

A great advance in appropriate technology for small farming in the semiarid tropics is the research development and farmer acceptance of animal-drawn wheeled tool carriers,[86] which can be used for primary tillage and also for seed drilling and fertilizer banding. It should be possible to mount a VAM inoculum bin on the tool bar with appropriate colters to allow optimum metering and placement of inoculum in the seed bed.

The increased yields from mycorrhizal inoculation under these conditions should be very worthwhile because of the high cost of phosphorus fertilizer (relative to the value of the crop yield), and its infrequent use.

2. Industrialized Countries

In agriculturally advanced countries where crops are sown by tractor-drawn seed drill, mycorrhizal inoculation (if it is used) must also be carried out by machine at seeding. The many demonstrations of potential field responses to VAM inoculation from hand-placed inoculum in Western countries have certainly not satisfied objectives (2) and (3) (see Section IV.D.1.) in agricultural situations where labor costs are high. Adaptation of suitable farm machinery for sowing mycorrhizal inoculum is possible and has several advantages: (1) optimum placement of inoculum for any crop above, below, or to the side of the seed which may lead to a consequent reduction in inoculum requirements; (2) large areas of land can be inoculated at low labor cost.

If and when VAM responses are shown to be long-lived, and a single inoculation is sufficient to introduce an effective VAM inoculant, widespread field inoculation should be possible.

At Ruakura Soil & Plant Research Station, we have successfully used a self-propelled cone seeder (model Oyjord) and a tractor-drawn seed drill (model Nodet) to handle fresh or granulated VAM inoculum for onion and maize crops.[36] The cone seeder did not have to be modified to dispense VAM inoculum, and the revolving partitions under the cone easily coped with the maximum inoculum rate of 57 g per row meter that we used. Granulated inoculum flowed easily through the delivery pipes to the colters and inoculum could be sown at any depth in the seed bed by altering the depth setting on the tool bar, as usual. We were unable to sow onion seed through the machine and these were sown after the inoculum placement using a tractor-drawn Stanhay drill. Mycorrhizal inoculation at 30 mm depth increased onion bulb yield by 17% (significant at $p \leq 0.05$). A cone seeder is not used commercially to sow onions; our trial merely demonstrated that VAM inoculum can be mechanically placed in soil at reasonable rates and stimulate crop yield.[36] Suitable engineering modifications could be made to a standard onion drill to allow mycorrhizal inoculation concurrent with seeding.

The Nodet seed drill had some simple modifications so that inoculum, fertilizer, and maize seed could all be drilled into the soil concurrently, viz., the drive shaft to 2 of the 4 fertilizer bins was speeded up (8-fold speed increase) so that mycorrhizal inoculum from those 2 fertilizer bins could be drilled at any required depth in soil at rates ranging from 400 to 2000 kg inoculum per hectare. Mycorrhizal inoculation (at the rate of 400 kg ha^{-1}) increased grain yield by 15% (significant at $p \leq 0.05$) in one field trial.

These field trials show that existing farm machinery can be modified to accept mycorrhizal inoculum and give biologically and statistically significant increases in crop yield in the field. The economic practicability of these ventures will depend on (1) the longevity and value of the field responses; (2) the rate and cost of inoculum required; and (3) the cost of machine modifications.

So, even for VAM inoculation of seed sown field crops, appropriate technology is already available and should be developed further. The main factors limiting widespread use of VAM inoculum will be, as before, selection and production of inoculum. A commercial company in New Zealand has already suggested that a 1 ha paddock when sterilized, inoculated (with pelleted seed), and turned over to inoculum production with maize or clover as a host plant would produce about 10^3 t inoculum soil if mined of its topsoil to a depth of 100 mm. The inoculum soil from 1 ha would then be enough to inoculate 500 to 2000 ha of land, depending on inoculum rate, and the inoculum paddocks could be situated where needed to reduce cartage costs.

E. Permanent Pasture

Permanent hill country pastures on hillsides which are too steep for tractor-drawn implements are the most difficult situation in which to manipulate VAM fungi populations. On New Zealand hill country pastures mycorrhizal inoculum could only be applied economically by aerial topdressing of the loose or granulated material as is done for fertilizers. The aerially delivered inoculum landing on the steep pasture surface must survive surface moisture deficits and become incorporated into the soil and compete with the indigenous, although less efficient, mycorrhizal fungi already present in the roots of the existing vegetation, if inoculation is to succeed in stimulating pasture growth. Nevertheless, there have been some successful field inoculation trials using hand-placed multiseeded inoculum pellets on a range of hill country sites.[31,56,63,83]

V. CONCLUSIONS

There are many indications, already published, of worthwhile field responses to VAM inoculation in cropping, horticulture, and pastoral agriculture. Nevertheless, a high percentage of past trials have suffered (at least in hindsight) from:

Use of very small plots
Lack of replication
Inappropriate use of preinoculated seedlings
Short growth period (harvested before the responses disappeared)
Excessive rates of randomly placed inoculum
Incompatibility with agricultural technology and economics
Lack of correlation of mycorrhizal responses to growth responses obtainable from P fertilizer alone

In the horticultural field we need to be far more adventurous and actually move away from the comforting security of working only in sterilized potting mixes and see what worthwhile and long-term responses can be achieved by transplanting VAM inoculated plants into unsterile field soils. In field crops grown in agriculturally mechanized countries, inoculum must be machine drilled at reasonably low rates and at optimum placement relative to the seed. For permanent pastures we must devise means of pelleting or coating seed or use nonseed inoculum granules which will result in seedling infection by the inoculant fungi, despite the overwhelming presence of the indigenous mycorrhizal fungi in the soil.

There is great potential for increasing the field growth of plants by VAM inoculation even at our present level of understanding of the physiology and autecology of VAM fungi, but we need to demonstrate this potential in properly designed, agronomically valid trials. We also need research workers with great imagination and ideas but their feet thoroughly planted in agricultural reality to carry out these field trials. Ultimately though, the extent of our manipulation of VAM in agriculture will probably depend on whether we can culture the organisms on artificial media and on how much we can learn of the physiology and ecology of VAM fungi and their host plants.

REFERENCES

1. **Baylis, G. T. S.**, Effect of vesicular-arbuscular mycorrhizas on growth of *Griselinia littoralis* (Cornaceae), *New Phytol.*, 58, 274, 1959.
2. **Baylis, G. T. S.**, Experiments on the ecological significance of phycomycetous mycorrhizas, *New Phytol.*, 66, 231, 1967.
3. **Baylis, G. T. S.**, Root hairs and phycomycetous mycorrhizas in phosphorus-deficient soil, *Plant Soil*, 33, 713, 1970.
4. **Daft, M. J. and Nicolson, T. H.**, Effect of *Endogone* mycorrhiza on plant growth, *New Phytol.*, 65, 343, 1966.
5. **Hayman, D. S. and Mosse, B.**, Plant growth responses to vesicular-arbuscular mycorrhiza. I. Growth of *Endogone* inoculated plants in phosphate deficient soils, *New Phytol.*, 70, 19, 1971.

6. Jackson, N. E., Franklin, R. E., and Miller, R. H., Effects of vesicular-arbuscular mycorrhizae on growth and phosphorus content of three agronomic crops, *Soil Sci. Soc. Am. Proc.*, 36, 64, 1972.
7. Mosse, B., Effects of different *Endogone* strains on the growth of *Paspalum notatum, Nature (London)*, 239, 221, 1972.
8. Mosse, B. and Hayman, D. S., Plant growth responses to vesicular-arbuscular mycorrhizas. II. In unsterilized field soils, *New Phytol.*, 70, 29, 1971.
9. Gerdemann, J. W., Vesicular-arbuscular mycorrhizae formed on maize and tuliptree by *Endogone fasciculata, Mycologia*, 57, 562, 1965.
10. Mosse, B., Hayman, D. S., and Ide, G. J., Growth responses of plants in unsterilized soil to inoculation with vesicular-arbuscular mycorrhiza, *Nature (London)*, 224, 1031, 1969.
11. Mosse, B., Growth and chemical composition of mycorrhizal and nonmycorrhizal apples, *Nature (London)*, 179, 222, 1957.
12. Ross, J. P., Effect of phosphate fertilization on yield of mycorrhizal and nonmycorrhizal soybeans, *Phytopathology*, 61, 1400, 1971.
13. Ross, J. P. and Harper, J. A., Effect of *Endogone* mycorrhiza on soybean yields, *Phytopathology*, 60, 1552, 1970.
14. Khan, A. G., The effect of vesicular-arbuscular mycorrhizal associations on growth of cereals. I. Effects on maize growth, *New Phytol.*, 71, 613, 1972.
15. Khan, A. G., The effect of vesicular-arbuscular mycorrhizal associations on growth of cereals. II. Effects on wheat growth, *Ann. Appl. Biol.*, 80, 27, 1975.
16. Rich, J. R. and Bird, G. W., Association of early season vesicular-arbuscular mycorrhizae with increased growth and development of cotton, *Phytopathology*, 64, 1421, 1974.
17. Kleinschmidt, G. D. and Gerdemann, J. W., Stunting of citrus seedlings in fumigated nursery soils related to the absence of endomycorrhizae, *Phytopathology*, 62, 1447, 1972.
18. Hattingh, M. J. and Gerdemann, J. W., Inoculation of Brazilian sour orange seed with an endomycorrhizal fungus, *Phytopathology*, 65, 1013, 1975.
19. Schenck, N. C. and Hinson, K., Response of nodulating and nonnodulating soybeans to a species of *Endogone* mycorrhiza, *Agron. J.*, 65, 849, 1973.
20. Schenck, N. C. and Tucker, D. P. H., Endomycorrhizal fungi and the development of citrus seedlings in Florida fumigated soils, *J. Am. Soc. Hort. Sci.*, 99, 284, 1974.
21. Azcon-G. de Aguilar, C., Azcon, R., and Barea, J. M., Endomycorrhizal fungi and *Rhizobium* as biological fertilizers for *Medicago sativa* in normal cultivation, *Nature (London)*, 279, 325, 1979.
22. Tinker, P. B., Mycorrhizas: the present position, in *Whither Soil Research,* Trans. 12th Int. Congr. Soil Sci., New Delhi, 1982.
23. Maronek, D. M., Hendrix, J. W., and Kiernan, J., Mycorrhizal fungi and their importance in horticultural crop production, *Hortic. Rev.*, 3, 1972, 1981.
24. Carling, D. E. and Brown, M. F., Relative effect of vesicular-arbuscular mycorrhizal fungi on the growth and yield of soybeans, *Soil Sci. Soc. Am. J.*, 44, 528, 1980.
25. Tinker, P. B., Personal communication, 1980.
26. Bryan, W. C. and Ruehle, J. L., Growth stimulation of sweetgum seedlings induced by the endomycorrhizal fungus *Glomus mosseae, Tree Pl. Notes,* 27, 9, 1976.
27. Islam, R. and Ayanaba, A., Effect of seed inoculation and preinfecting cowpea *(Vigna unguiculata)* with *Glomus mosseae* on growth and seed yield of the plants under field conditions, *Plant Soil,* 61, 341, 1981.
28. Menge, J. A., La Rue, J., Labanauskas, C. K., and Johnson, E. L. V., The effect of two mycorrhizal fungi upon growth and nutrition of avocado seedlings grown with six fertilizer treatments, *J. Am. Soc. Hortic. Sci.*, 105, 400, 1980.
29. Govinda Rao, Y. S., Bagyaraj, D. J., and Rai, P. V., Selection of an efficient VAM mycorrhizal fungus for finger millet. I. Glasshouse screening, *Zentralbl. Microbiol.*, 138, 409, 1983.
30. Azcon, R. and Ocampo, J. A., Factors affecting the vesicular-arbuscular infection and mycorrhizal dependency of 13 wheat cultivars, *New Phytol.*, 87, 677, 1981.
31. Hall, I. R., Effect of vesicular-arbuscular mycorrhizas on two varieties of maize and one of sweetcorn, *N.Z. J. Agric. Res.*, 21, 517, 1978.
32. Johnson, P. N., Mycorrhizal endogonaceae in a New Zealand forest, *New Phytol.*, 78, 161, 1977.
33. Ramirez, B. N., Mitchell, D. J., and Schenck, N. C., Establishment and growth effects of three vesicular-arbuscular mycorrhizal fungi on papaya, *Mycologia*, 67, 1039, 1975.
34. Powell, C. L., Clark, G. E., and Verberne, N. J., Growth response of four onion cultivars to several isolates of VA mycorrhizal fungi, *N.Z. J. Agric. Res.*, 25, 465, 1982.
35. Powell, C. L., Mycorrhiza, in *Experimental Microbial Ecology,* Burns, R. G. and Slater, J. H., Eds., Blackwell Scientific, Oxford, 1982.
36. Powell, C. L. and Bagyaraj, D. J., VA mycorrhizal inoculation of field crops, *Proc. N.Z. Agron. Soc.*, 12, 85, 1982.
37. Powell, C. L., Selection of efficient VA mycorrhizal fungi, *Plant Soil,* 68, 3, 1982.

38. Iqbal, S. H. and Qureshi, K. S., The effect of vesicular-arbuscular mycorrhizal associations on growth of sunflower (*Helianthus annuus* L.) under field conditions, *Biologia (Lahore),* 23, 189, 1977.
39. Clarke, C. and Mosse, B., Plant growth responses to vesicular-arbuscular mycorrhiza. XII. Field inoculation responses of barley at two soil P levels, *New Phytol.,* 87, 695, 1981.
40. Hayman, D. S., Morris, E. J., and Page, R. J., Methods for inoculating field crops with mycorrhizal fungi, *Ann. Appl. Biol.,* 99, 247, 1981.
41. Kuo, C. G. and Huang, R. S., Effect of vesicular-arbuscular mycorrhizae on the growth and yield of rice stubble cultured soybeans, *Plant Soil,* 64, 325, 1982.
42. Lambert, D. H. and Cole, H., Jr., Effects of mycorrhizae on establishment and performance of forage species in mine spoil, *Agron. J.,* 72, 257, 1980.
43. Bagyaraj, D. J. and Sreeramulu, K. R., Preinoculation with VA mycorrhiza improves growth and yield of chilli transplanted in the field and saves phosphatic fertilizer, *Plant Soil,* 69, 375, 1982.
44. Maronek, D. M., Hendrix, J. W., and Kiernan, J., Differential growth response to the mycorrhizal fungus *Glomus fasciculatus* of southern magnolia and "Bar Harbor" juniper grown in containers in composted hardwood bark shale, *J. Am. Soc. Hortic. Sci.,* 105, 206, 1980.
45. Govinda Rao, Y. S., Bagyaraj, D. J., and Rai, P. V., Selection of an efficient VA mycorrhizal fungus for finger millet. II. Screening under field conditions, *Zentralbl. Microbiol.,* 138, 415, 1983.
46. Kormanik, P. P., Bryan, W. C., and Schultz, R. C., Influence of endomycorrhizae on growth of sweetgum seedlings from eight mother trees, *For. Sci.,* 23, 500, 1977.
47. Plenchette, C., Furlan, V., and Fortin, J. A., Growth stimulation of apple trees in unsterilized soil under field conditions with VA mycorrhizal inoculation, *Can. J. Bot.,* 59, 2003, 1981.
48. Black, R. L. B. and Tinker, P. B., Interaction between effects of vesicular-arbuscular mycorrhiza and fertilizer phosphorus on yields of potatoes in the field, *Nature (London),* 267, 510, 1977.
49. Owusu-Bennoah, E. and Mosse, B., Plant growth responses to vesicular-arbuscular mycorrhiza. XI. Field inoculation responses in barley, lucerne and onion, *New Phytol.,* 83, 671, 1979.
50. Salter, P. J., Fluid drilling — a new approach to crop establishment, in *Proc. Sect. M Agric. 139th Ann. Meet. Br. Assoc. Adv. Sci.,* Hayes, W. A., Ed., University of Aston, Birmingham, England, 1978, 16.
51. Elmes, R. and Mosse, B., Vesicular-arbuscular mycorrhiza: nutrient film technique, in *Rothamsted Exp. Stn. Annu. Rep.,* 1, 188, 1980.
52. Hayman, D. S. and Mosse, B., Improved growth of white clover in hill grasslands by mycorrhizal inoculation, *Ann. Appl. Biol.,* 93, 141, 1979.
53. Saif, S. R. and Khan, A. G., The effect of vesicular-arbuscular mycorrhizal associations on growth of cereals. III. Effects on barley growth, *Plant Soil,* 47, 17, 1977.
54. Gaunt, R. E., Inoculation of vesicular-arbuscular mycorrhizal fungus on onion and tomato seeds, *N.Z. J. Bot.,* 16, 69, 1978.
55. Hall, I. R., Personal communication, 1982.
56. Powell, C. L., Inoculation of white clover and ryegrass seed with mycorrhizal fungi, *New Phytol.,* 83, 81, 1979.
57. Crush, J. R. and Pattison, A. C., Prelimiary results on the production of vesicular-arbuscular mycorrhizal inoculum by freeze drying, in *Endomycorrhizas,* Sanders, F. E., Mosse, B., and Tinker, P. B., Eds., Academic Press, London, 1975, 485.
58. Menge, J. A., Lembright, H., and Johnson, E. L. V., Utilization of mycorrhizal fungi in citrus nurseries, in *Proc. Int. Soc. Citric.,* 1, 129, 1977.
59. Kormanik, P. P., Bryan, W. C., and Schultz, R. C., Increasing endomycorrhizal fungus inoculum in forest nursery soil with cover crops, *South. J. Appl. For.,* 4, 151, 1980.
60. La Rue, J. H., McClellan, W. D., and Peacock, W. L., Mycorrhizal fungi and peach nursery nutrition, *Calif. Agric.,* 29, 6, 1975.
61. Bagyaraj, D. J., Manjunath, A., and Patil, R. B., Interaction between a vesicular-arbuscular mycorrhiza and *Rhizobium* and their effects on soybean in the field, *New Phytol.,* 82, 141, 1979.
62. Powell, C. L., Inoculation of barley with efficient mycorrhizal fungi stimulates seed yield, *Plant Soil,* 59, 487, 1981.
63. Hall, I. R., Soil pellets to introduce vesicular-arbuscular mycorrhizal fungi into soil, *Soil Biol. Biochem.,* 11, 85, 1979.
64. Hall, I. R., Growth of *Lotus pedunculatus* Cav. in an eroded soil containing soil pellets infested with endomycorrhizal fungi, *N.Z. J. Agric. Res.,* 23, 103, 1980.
65. Yost, R. S. and Fox, R. L., Contribution of mycorrhizae to P nutrition of crops growing on an oxisol, *Agron. J.,* 71, 903, 1979.
66. Van der Zaag, P., Fox, R. L., de la Pena, R. S., and Yost, R. S., P nutrition of cassava, including mycorrhizal effects on P, K, S, Zn and Ca uptake, *Field Crops Res.,* 2, 253, 1979.
67. Powell, C. L., Groters, M., and Metcalfe, D., Mycorrhizal inoculation of a barley crop in the field, *N.Z. J. Agric. Res.,* 23, 107, 1980.
68. Abbott, L. K. and Robson, A. D., Growth stimulation of subterranean cover with vesicular-arbuscular mycorrhiza, *Aust. J. Agric. Res.,* 28, 639, 1977.

69. Abbott, L. K. and Robson, A. D., Growth of subterranean clover in relation to the formation of endomycorrhizas by introduced and indigenous fungi in a field soil, *New Phytol.*, 81, 575, 1978.
70. Azcon-G. de Aguilar, C. and Barea, J. M., Field inoculation of *Medicago* with V-A mycorrhiza and *Rhizobium* in phosphate fixing agricultural soil, *Soil Biol. Biochem.*, 13, 19, 1981.
71. Kang, B. T., Islam, R., Sanders, F. E., and Ayanaba, A., Effect of phosphate fertilization and inoculation with VA mycorrhizal fungi on performance of cassava (*Manihot esculenta* Cranz) grown on an alfisol, *Field Crops Res.*, 3, 83, 1980.
72. Martin, J. P., Effect of fumigation, fertilization and various other soil treatments on growth of orange seedlings in old citrus soils, *Soil Sci.*, 66, 273, 1948.
73. Martin, J. P., Aldrich, D. G., Murphy, W. S., and Bradford, G.R., Effect of soil fumigation on growth and chemical composition of citrus plants, *Soil Sci.*, 75, 137, 1953.
74. Timmer, L. W. and Leyden, R. F., Stunting of citrus seedlings in fumigated soils in Texas and its correction by phosphorus fertilization and inoculation with mycorrhizal fungi, *J. Am. Soc. Hortic. Sci.*, 103, 533, 1978.
75. Menge, J. A., Labanauskas, C. K., Johnson, E. L. V., and Platt, R. G., Partial substitution of mycorrhizal fungi for phosphorus fertilization in the greenhouse culture of citrus, *Soil Sci. Soc. Am. Proc.*, 42, 926, 1978.
76. Menge, J. A., Davis, R. M., Johnson, E. L. V., and Zentmyer, G. A., Mycorrhizal fungi increase growth and reduce transplant injury in avocado, *Calif. Agric.*, 32, 6, 1978.
77. Bryan, W. C. and Kormanik, P. P., Mycorrhizae benefit survival and growth of sweetgum seedlings in the nursery, *South. J. Appl. For.*, 1, 21, 1977.
78. Kormanik, P. P., Bryan, W. C., and Schultz, R. C., Endomycorrhizal inoculation during transplanting improves growth of vegetatively propagated yellow poplar, *Plant Prop.*, 23, 4, 1977.
79. Johnson, C. R., Joiner, J. N., and Crews, C. E., Effects of N, K, and Mg on growth and leaf nutrient composition of three container grown woody ornamentals inoculated with mycorrhizae, *J. Am. Soc. Hortic. Sci.*, 105, 286, 1980.
80. Powell, C. L., Bagyaraj, D. J., and Clark, G. E., Unpublished data, 1982.
81. Johnson, C. R. and Crews, C. E., Survival of mycorrhizal plants in the landscape, *Am. Nurseryman*, 150, 15, 1979.
82. Powell, C. L., Mycorrhizas in hill country soils. III. Effect of inoculation on clover growth in unsterile soils, *N.Z. J. Agric. Res.*, 20, 343, 1977.
83. Powell, C. L., Bagyaraj, D. J., and Clark, G. E., Unpublished data, 1982.
84. Powell, C. L., The need for mycorrhizas for cocksfoot growth at high altitude: a note, *N.Z. J. Agric. Res.*, 18, 95, 1975.
85. Harwood, R. R., *Small Farm Development — Understanding and Improving Farming Systems in the Humid Tropics,* Westview Press, Boulder, Colorado, 1979, 160.
86. ICRISAT, *Proc. Int. Symp. Dev. Transfer Tech. for Rain-Fed Agric. and the SAT Farmer,* International Crops Research Institute for the Semiarid Tropics, Patancheru, India, 1980, 324.
87. Van der Zaag, P., Fox, R. L., Kwakye, P. K., and Obigbesan, G. O., The phosphorus requirements of yams (*Dioscorea* spp.), *Trop. Agric. (Trinidad)*, 57, 97, 1980.

INDEX

A

Acaulospora
  bireticulata, 67
  laevis, 60—61, 68
  sp., 1, 67—68
  spinosa, 67
Acetic acid, 105
Acetylene reduction, 134
Acid phosphatase, 22, 160
Actinomycetes, 136—137
Actinomycin D, 103
Actinorhizal nodules, 137
Active dissemination, 36—37
Aerial topdressing, 218—219
Afforestation programs, 135
Agricultural plant yield, 36
Agricultural practices, 206
Agricultural systems, role of legumes in, 134
Agriculture, potential for VAM inoculation in, 206
Alkaline phosphatase, 17, 22—23, 162, 166
Amino acid metabolism, 164—165
Amino acids, 105, 143, 169, 179
Ammonium and nitrogen, assimilation of, 164
Ammonium-N, 164
Ammophyla arenaria, 8
Amoebae, 49
Amyloplasts, 168
Anaerobic paddy conditions, 212
Anatomy of VA mycorrhizae, 5—33
  angiosperms, 6—23
    extramatrical phase, 8—9
    host-fungus relationships, 20—23
    host response to fungal colonization, 19—20
    intraradical phase, 9—19
  bryophytes, 23—24
  gymnosperms, 26
  pteridophytes, 24—26
Angiosperm roots, 6
Angiosperms, see also specific topics
  arbuscules, 8, 14—17
  external vesicles, 7
  extramatrical hyphae, 7—9
  extramatrical phase, 7—9
  fungal infection in, 6—23, 28
    extramatrical phase, 8—9
    host-fungus relationships, 20—23
    host response to fungal colonization, 19—20
    intraradical phase, 9—19
  host-fungus relationships, 20—23
  host response to fungal colonization, 19—20
  intercellular hyphae, 8, 11—14
  intracellular hyphae, 8—11
  intraradical phase, 7—19
  morphology of roots, 7
  root hairs, 7
  root penetration, 9—10
  size of roots, 7
  spores, 7
  vesicles, 8, 17—19
Anguillospora pseudolongissima, 145
Animal-drawn ploughs, 217
Aphelenchoides sp., 146
Appressorium, 7, 105
  formation, 101
  Glomus fasciculatum, 8, 10
Arabis mosaic virus, 139
Arbuscular clumps, 17, 19
Arbuscular hyphae, 16
Arbuscular mycorrhizae, 66
Arbuscular walls, 17—18, 169
Arbuscules
  bacterium-like organelles, 144
  defined, 14
  development of, 173
  external mycelium, 108
  fungal infection
    in angiosperms, 8, 14—17
    in bryophytes, 23
  fungal structures, 1
  glycogen and lipid in, 168
  glycolysis, 169
  haustorial organ, 162
  host plasmalemma similarity to normal plasmalemma, 22
  intracellular structures, 6—7
  lipid droplets, 167
  nutrient exchange, 162, 178
  phosphorus concentrations, 161
  physiological properties, 156
  position in normal infection, 108
  septum formation, 17, 19
  site of P transfer between fungus and host, 168
  subapical septum, 17
Artificial media, 219
Aspetate phycomycetous hyphae, 26
Aspergillus niger, 135—136
Assessing infection, 119—120
ATP, 162, 168
ATPase activities, 17, 23, 162, 168
  localization along host and fungal plasmalemmas, 22—23
Autecology of VAM endophytes, 206
Auxins, 173
Axenic mycorrhizal plants, 98—103
  methods of preparation, 98—100
  use, 100—103
Azospirillum lipoferum, 135
Azotobacter
  chroococcum, 135
  paspali, 135
  sp., 135
Azygospores, 62, 64, 67
  defined, 89

B

Bacillus circulans, 136

Bacteria, 48—49, 139
Bacterial associates, 144
Bacterial fertilizers, 136
Bacterium-like organelles (BLOs), 14, 144
Basidiomycetes, 162
Beijerinkia mobilis, 135
Beneficial soil organisms, 132—137
Biological interactions with VA mycorrhizal fungi, 131—153
  actinomycetes, 136—137
  bacteria, 139
  beneficial soil organisms, 132—137
  components of rhizosphere, 132
  foliar pathogens, 139
  free-living nitrogen-fixing bacteria, 135
  fungi, 137—139
  legume bacteria, 132—135
  mechanisms of suppression of pathogens, 142—143
  microorganisms intimately associated with VAM fungi, 143—144
  nematodes, 139—142
  parasitic flora, 144—145
  phosphate-solubilizing organisms, 135—136
  plant pathogenic organisms, 137—143
  predatory microflora, 145—146
  root pathogenic fungi, 137—139
  viruses, 139
Biological nitrogen fixation, 1
Biological significance, 2
Biomass of VAM fungi, 122, 161
Biotrophic fungi, 103—104, 106, 108—109, 114
Bone meal, 100, 103
Boron nutrition, 165—166
Bouteloua gracilis, 100
Bryophytes
  description of colonization process, 6
  fungal infection in, 23—24, 28
Bulbous subtending hypha, 89

C

Calcium, 17, 24
Calcium nutrition, 165—166
Carbohydrate metabolism, 161, 166
Carbohydrate transport, 159
Carbon compounds, see Carbon physiology and biochemistry
Carbon cost of the symbiosis, 170—173
Carbon demand of fungal symbiont, 170
Carbon metabolism, 168
Carbon physiology and biochemistry, 166—173
  factors affecting mycorrhizal development, 166—167
  lipid biochemistry, 169—170
  symbiosis cost, 170—173
  transfer from fungus to host, 169
  transfer from host to fungus, 167—169
Carbon transfer, 167—169
Caryophyllaceae, 1
Casein, 105
Cell-to-cell passage, 11, 13
Cellulose, 105
Centrifugation, 60
Centrosema pubescens, 133
Cereals, 209—210
Chemotactic substances, 37
Chenopodiaceae, 1
Chitin, 8, 106
  polymerization, 17
Chitinase activity, 178
Chitosan, 106
Chlamydospores, 61—63, 66—67, 70—71
  defined 89
  Glomus, 68
p-Chloronitrobenzene (PCNB), 175
Chlorophyll, 100
Chytridiomycetes, 58
Circulating hydroponic culture, 193—194
Citric acid, 105
Clavate spore, 89
Clumps, 6—7
  fungal infection in bryophytes, 23—24
Coal mine spoil, 216
Coarse endophyte, *Athyrium* roots, 24
Codinella fertilis, 139
Coenocytic fungi, 176
Coiled hyphae, 7
Coil formation, 9—12
Collection of sporocarps and spores, 59—62
Collembola, 49, 145
Colonization, see Fungal colonization; Root colonization
Colonization processes, 6
Companion fungi, 143—144, 146
Competition in root colonization and sporulation, 43—45
Complexipes
  moniliformis, 61, 67
  sp., 58, 69
Concolorous, 89
Constricted subtending hypha, 89
Continental drift, 36
Copper, 121, 134
Copper nutrition, 165—166
Cortical parenchyma, 11—13
Cover crops, 208
Cropping plant species, 209—214
Cropping schedule, 47—48
Crown gall pathogen, 132
Cruciferae, 1
Cultures, see specific types
Current agricultural practices, 206
Cyanide-insensitive respiratory pathway, 169
Cycloheximide, 103
Cyperaceae, 1
Cylindrocarpon destructans, 138
Cystine, 105

Cytobiotic state, 14
Cytochemical tests, 17
Cytokinin activity, 173, 176
Cytokinin levels, 173, 176
Cytokinins, 173, 178
Cytological compatibility, 28—29
Cytological organization, 28

D

Decanting, 59—60
Defoliation, 167
Dehydrogenases, 169
Developing countries, 217
Dialysate, 105, 108
Dinitrogen fixation, 165
Disbudding of plants, 173
Disease, freedom from, 217
Dispersal of VAM fungi, 36—39
Disturbed soils, 137
DNA, 103
Dormancy factor in spore germination, 39
Double infection, 6
Drought resistance, 1
Drought tolerance, 176
Dual inoculation with VAM and *Rhizobium*, 134

E

Echinulated spores, 66, 90
Ecology of VA mycorrhizal fungi, 35—55, 219
    colonization, 42—46
    dispersal, 36—39
    spore germination, 39—42
    sporulation, 42—46
    survival, 46—50
Economic usefulness, 2
Ectendomycorrhizas, 69
Ectocarpic spores, 90
Ectomycorrhiza, 6, 58, 63
Ectomycorrhizal tree species, 217
Ectoparasites, 140
Effectiveness of species in plant growth, 123—124
Efficiency of inoculant fungi, 206
Electron microscopy, 69, 178
Electron transport to oxygen, 169
Elutrition, 59
Embedding spores, 62
Endogonaceae, 57—94
    Acaulospora, 67—68
    collection of, 59—62
    Complexipes, 69
    describing, 69—71
    dissemination of, 37—38
    Endogone, 63
    Entrophospora, 68
    extraction of sporocarps and spores, 59—62
    features of the genera, 62—69
    Gigaspora, 63—66

    Glaziella, 68
    Glomus, 68
    identifying, 69—71
    incidence of, 58—59
    informal descriptions, 69
    keys to, 72—89
    Modicella, 68—69
    Sclerocystis, 68
    sectioning spores,
    storage of, 62
    VA mycorrhizae formed by, 1, 6
Endogone
    flammicorona, 63, 65
    incrassata, 60—61
    pisiformis, 58, 63
    sp., 58, 63
Endomycorrhizae, 6, 66, 68
Endophytes
    autecology, 206
    host plant associations, 1
    improvement of water relations of plants, 177
    physiology, 206
    stimulation of rooting, 173
Endopolyphosphatase, 162
Enhanced nutrient uptake, 120—121
Entrophospora
    infrequens, 60, 67, 68
    sp., 58, 68, 69
Environmental conditions in spore germination, 40—42
Environmental factors in root colonization and sporulation, 44—46
Epigeous sporocarps, 58, 59, 68
Eroded soils, 216
Ethidium bromide, 103
Evapotranspiration rates, 176
Evolutionary basis, 29
Exopolyphosphatase, 162
External hyphae, 123, 124
External mycelium, 157, 170
    absorbing sites, 159
    arbuscule-like structures on, 108
    carbon transfer, 169
    development of, 157—158
    inoculum for axenic mycorrhizal plants, 99—100
    physiological properties, 156
    rate of movement of phosphorus, 100
    translocation of phosphorus fed to, 158
External vesicles, 7
Extraction of sporocarps and spores, 59—62
Extramatrical hyphae, 6—9
Extramatrical mycelium production, 156
Extramatrical phase, 7—9

F

Fallow, 217
Fallow year, 47
Fatty acids, 170
Fern evolution, 24

Ferns, see Pteridophytes
Fertilization of soil culture, 190—192
Fibrous root systems, 121
Field inoculation with VA mycorrhizal fungi, 205—222
  aerial topdressing, 218—219
  anaerobic paddy conditions, 212
  animal-drawn ploughs, 217
  appropriate machinery and labor resources, 216
  appropriate technology, 216—219
  cereals, 209—210
  crops, 209—214
  economic viability, 216
  field plots, 216
  field sown crops and mechanization, 217—218
  fruit trees, 214
  growth depressions, 213
  growth response curve, 210
  growth responses in the field, 209—216
  horticulture, 214—215
  impediments to, 207
  inoculum, 206—209, 216, 218, 219
  inoculum rates, 207
  legumes, 210—213
  longevity of mycorrhizal response, 212
  long-term responses to inoculation in unsterilized field soils, 215, 219
  multiple cropping, 217
  nursery beds, 213, 217
  objectives, 216
  optimum yield, 210
  ornamental trees, 215
  other field crops, 213—214
  pasture, 215—216
  pasture establishment, 216
  peasant farms, 217
  permanent pasture, 215, 218—219
  preinoculation, 207, 210, 212, 214
  reduced fertilizer requirement, 216
  root growth, 215
  seed inoculation, 212
  self-propelled cone seeder, 218
  starter inoculant, 217
  sterilized nursery mix, 215
  survival rate, 215
  timber trees, 214—215
  tractor-drawn seed drill, 218
  transfer to unsterilized field soils, 214
  transplanted field crops, 217
  transplanted preinoculated seedlings, 212
  transplanted seedlings, 216
  transplanting, 207
  tree crops, 216—217
  trial deficiencies, 219
  wheeled tool carriers, 217
Field plots, 216
Field sown crops and mechanization, 217—218
Field trials, 206
Fine endophyte
  Athyrium roots, 24
  Glomus tenue, 70

isolation, 59
  VAM fungi from soil, 96
Finger millet, 137
Fixing spores, 62
Flaking spore, 70
Flammenkrone, 90
Flotation-adhesion technique, 97
Fluid drilling, 207, 209
Fluorescein diacetate, 157
4-Fluorouracil, 103
Foliar pathogens, 139
Folsomia candida, 145
Forestry applications, 135
Fossil representatives, 58
Frankia sp., 137
Free-living nitrogen-fixing bacteria, 135
Fruit trees, 214
Funding of VAM research, 206
Fungal colonization
  host response to, 19—20
  relationship with fern evolution, 24
Fungal digestion, 22
Fungal efficiencies, 2
Fungal genome, 28
Fungal infection
  angiosperms, see also Angiosperms, 6—23, 28
  bryophytes, 23—24, 28
  gymnosperms, 26, 28
  pteridophytes, 24—26, 28
Fungal respiration, 170
Fungal selection for field inoculation, 206—207
Fungal species, root colonization in, 43—44
Fungal spores, see also Sporocarps and spores, 208
Fungal structures, presence of, 28
Fungal translocation, 161
Fungal vacuoles, 161
Fungus-host relationship, fungal infection in pteridophytes, 24—25
Fungal wall, 21
Funnel-shaped subtending hypha, 90
Fusarium oxysporum f. sp. *lycopersici*, 138

G

Gaeumannomyces graminis var. *tritici*, 139
Gametangia, 61, 63
Gametangium, 90
Gas-liquid chromatography (GLC), 58
Genera, key to, 88—89
Germination compartments, 65, 68, 71
  defined, 90
Germination warts, 90
Giant cells, 140, 142
Gibberellins, 173
Gigaspora
  aurigloba, 65, 68, 71
  coralloidea, 70
  decipiens, 60, 66, 69, 71
  margarita
    field inoculation, 207, 210

spore germination, 65, 71
spore variability, 69
sp., 1, 63—66
Glaziella
aurantiaca, 59, 68
sp., 58, 68
Gleba, 90
Glomus
caledonicum, 62, 66
convolutum, 58, 65
etunicatum, 207
fasciculatum
field inoculation, 207, 212, 214—216
interaction with free-living nitrogen-fixing bacteria, 135
taxonomy, 69
Vitis vinifera roots infected by, 8, 10—12
fuegianum, 64, 68
gerdemanii, 137
infrequens, 69
invermaium, 60
macrocarpum, 59, 70
macrocarpum var. *macrocarpum*, 60, 69
merredum, 69
monosporum, 66
mosseae, 69, 72
features of, 60
field inoculation, 207, 210, 212—213, 216
spore variability, 69
sporocarp, 65
multicaule, 67, 68
pallidum
chlamydospores, 66, 71
isolation of, 59
sporocarp damage, 64
pulvinatum, 60, 70
reticulatum, 72
sp., 1
features of, 68
field inoculation, 206—207, 212, 215
spore germination, 68
tenue, 59, 69—70, 207
Glossary, see also specific terms, 89—92
Glucosamine, 106
Glucose, 105, 167
Glycogen, 167—168
Glycolsis, 169
Glysine, 105
Gondwanaland, 36
Grain yield, 209—210
Granulated inoculum, 209, 218
Granulation process, 208
Granule, 207
Growth response curve, 210
Growth sinks, 168
Gymnosperms, 6, 26, 28

H

Harvested inoculum, 207

Heavy metals, 100
Hemp seeds, 104, 105
Histochemical studies, 178
Histopathology of nematode galls, 142
Historical aspects, 6
Hormonal effects of mycorrhizal fungi, 170—171, 173—174, 177
Hormone levels, differences in, 123
Hormone production by mycorrhizal fungi, 1, 173
Horticultural plant species, 209, 214—215
Host
choice to produce VAM inoculum, see also Inoculum production; Soil culture, 190
field inoculation with VAM, 213
metabolically very active, 29
morphogenesis regulated by, 28
root colonization and sporulation, 45—46
spore germination, 42
transfer of phosphorus to, 162
Host-fungus cytological interactions, 29
Host-fungus relationships
fungal infection in angiosperms, 20—23
fungal infection in gymnosperms, 26
interface, 20—21
morphological description of contact patterns, 20
VAM interactions, 22
wall-to-wall contact, 21
Host growth enhancement, 158
Host plasmalemma, 9—10
arbuscular hyphae surrounded by, 16
ATPase activity, 162, 168
continuous, 11
fungal infection in gymnosperms, 26
fungal infection in pteridophytes, 25
interface formation, 2
invagination by cell-to-cell passage of fungus, 11, 13
proliferation, 20
separation from fungal walls by osmiophilic fibrilla layer of matrix material, 10, 12
similarity to normal plasmalemma, 22
Host response to fungal colonization, 19—20
Host root cell, 29
Host walls, 17—18
Humicola fuscoatra, 145
Hyaline, 90
Hydraulic conductivity, 174—176
Hydroxyapatite, 100
Hyperparasites, 47
Hyperparasitism
Endogonaceae mixtures, 63
Entrophospora infrequens, 68
spores, 48
surface-sterilization of resting spores, 98
survival of VAM fungi, 49—50
Hyperparasitization, 145
Hyphae, see also specific types, 104, 176
Hyphal constriction, 11—12
Hyphal growth, 104—106
Hyphal inflow, 157—158, 162
Hyphal translocation, 158, 161—162, 164, 166

Hyphal wall, *Glomus fasciculatum*, 8, 10
Hypogeous, 90
Hypogeous fungi, dispersal of, 37
Hypogeous sporocarps, 59, 68

I

Identity of symbiotic fungus, 27
Impure inocula, 114
Indigenous mycorrhizal fungi, 208—211, 213, 216
Indigenous VAM species, root colonization, 43
Industrialized countries, 218
Infection pattern, 9
Infection unit, 8
Infectivity, differences in, 124
Inoculum contaminants, 115
Inoculum density, root colonization and sporulation, 43
Inoculant fungi, see Field inoculation with VA mycorrhizal fungi
Inoculum for field inoculation, see Field inoculation with VA mycorrhizal fungi
Inoculum pad, 209—210
Inoculum potential, 47, 198—199
Inoculum production, 187—203
    circulating hydroponic culture, 193—194
    controlling microorganisms which contaminate, 194—195
    inoculum potential, 198—199
    methods, 188—194
    nutrient film culture, 193—194
    nutrition of VAM fungi, 196—197
    parasites of VAM fungi, 196
    plant growth, 196—197
    principles of, 199
    soil culture, see also Soil culture, 188—193
    storage, 196
    symbiosis principles, 196—199
    VAM growth dynamics, 197—198
Inoculum rates, 217
Inoculum viability, 208
Inorganic nutrients in soil, 41
Inorganic sulfur-containing compounds, 110
Inositol, 103, 167
Insoluble inorganic P, 157
Intercalary spore, 90
Intercellular hyphae
    bacterium-like organelles, 144
    fungal infection in angiosperms, 8, 11—14
    osmiophilic walls, 14—15
    regeneration of hyphae, 104
Interface
    host-fungus relationships, 20—21
    polysaccharides, 22
Intracellular coils, 9, 12
Intracellular hyphae, 6—7
    amount influenced by host, 9
    behavior influenced by host, 9
    fungal infection in angiosperms, 8—11
    looped arrangement, 9
    outer cortical layers of root, 9—11
Intraradical phase, fungal infection in angiosperms, 7—19
Ion uptake, 157, 213
Iron nutrition, 165—166
Isolation of VA mycorrhizae from soil, 96—97
Isotopic dilution kinetics of soils, 160

K

Keys
    Endogonaceae, 72—89
    genera, 88—89
    nonsporocarpic species
        subtending hyphae not observable, 86—88
        subtending hyphae observable, 79—86
    sporocarpic species, 72—79

L

Lactophenol, 90
Laminated spore wall, 90
Lateral projection, 90
Leaf water potentials, 174, 176
Leaves
    Glomus tenue, 24
    mycorrhizal infection, 19—20
Legume bacteria, 132—135
Legumes, 134—135, 210—213
Leucaena leucocephala, 133
Light intensity, 167
    plant growth, 122
    root colonization and sporulation, 44
    soil culture, 192
Light microscopy, 7
Lipid biochemistry, 169—170
Lipids, 14, 167—168
    reserves, 10
    synthesis of, 104
Liverwort, see Bryophytes
Longevity of mycorrhizal response, 212
Lyophilization, 208
Lysine, 105

M

Machine drilled inoculum, 219
Magnesium nutrition, 165—166
Maltose, 105
Manganese, 100
Manganese nutrition, 165—166
Mannitol, 105, 167
Mantle, 90
Mass spectrometry, 58
Mature spore, 90
Mechanical placement of inoculum, 216, 218
Medicago sativa, 133
Meloidogyne

arenaria, 140
incognita, 140
Melzer's reagent, 90
Membrane, 90
Membrane permeability, 176
Metabolically active fungal structures, 157
Metabolic capabilities of VAM fungi, 103—104
Metabolic pathways in carbon utilization, 169
Metabolism
  amino acid, 164—165
  carbohydrates, 161, 166
  carbon, 168
  nitrogen, 164—165
  phenol, 178
Methyl bromide, 2, 211, 213
Michaelis-Menten parameters, 158
Micromonospora sp., 144
Micronutrients
  deficiencies, 165
  soil culture, 191—192
Microorganisms contaminating mycorrhizal inoculum, 194—195
Microorganisms intimately associated with VAM fungi, 143—144, 146
Migratory endoparasites, 140—142
Mine spoils, 137, 216
Mineral nutrition, 157—166, 177
  boron, 165—166
  calcium, 165—166
  copper, 165—166
  iron, 165—166
  magnesium, 165—166
  manganese, 165—166
  nitrogen, see also Nitrogen, 164—165
  phosphorus, 162—164
    utilization of, see also Phosphorus, utilization of, 157—162
  potassium, 165—166
  sodium, 165—166
  sulfur, 165—166
  uptake of other nutrients, 165—166
  zinc, 165—166
"Miracle" mounting fluid, 90
Mitscherlich equation, 116—117
Mobility of ions in soil, 157
Modicella
  malleola, 60—61
  sp., 58, 68—69
Morphogenesis of VAM fungi, 28
Morphological homogeneity, 28—29
Morphology of VA mycorrhizae, 5—33
  angiosperms, 6—23
    extramatrical phase, 8—9
    host-fungus relationships, 20—23
    host response to fungal colonization, 19—20
    intraradical phase, 9—19
  bryophytes, 23—24
  gymnosperms, 26
  pteridophytes, 24—26
Mortierella, 63
Mortierellaceae, 58, 69

Mosses, see Bryophytes
Mother hypha, 61, 90
Mother spore, 90
Mother vesicle, 91
Mucoraceae, 58
Mucoraceous mycoparasites, 109
Mucorales, 58
Multiple cropping, 217
Mycoparasitism, 49
Mycophagous nematodes, 146
Mycophagy, 37—38
Mycorrhizae
  anatomy of, 71
  defined, 1
  role, 1
Mycorrhizal development, factors affecting, 166—167
Mycorrhizal fungi
  efficient or superior strains, 123
  hormone production by, 173
Mycorrhizal growth responses, see Field inoculation with VA mycorrhizal fungi
Mycorrhizal infection, see also Fungal infections; Physiology of VA mycorrhizal associations, 162—164
Mycorrhizal inoculation, see Field inoculation with VA mycorrhizal fungi
Mycorrhizal root respiration, 170—171
Mycorrhizal soil inoculum, 2
Mycorrhizal symbiosis, 14, 51
Mycorrhizal technology, 217
Mycorrhizal transplants, 214
Mycothalli, 23

N

Nematode galls, 140—142
Nematodes, 178
  ectoparasites, 140
  interaction with VAM fungi, 139—142
  migratory endoparasite, 140—142
  mycophagous, 146
  root-knot, 140—141
  sedentary endoparasites, 140—141
Nitrate-N, 164
Nitrate reductase, 103
Nitrification inhibitors, 134
Nitrogen, 164—165, 191
Nitrogenase activity, 133—134
Nitrogen fixation, 119, 132, 211
  enhancement by *Rhizobium*, 133
  legume bacteria, 132—135
  legumes, 132—135
Nodulation, 119, 211
  legumes, 132—135
Nonhost plants, 42, 47
Nonindigenous VAM species, 43
Non-nutritional effects of VAM fungi, see also Plant growth, 122—123, 126
Nonseptate phycomycetous fungi, 1

Nonsporing VAM fungi, 206
Nonsporocarpic species, keys to
    subtending hyphae not observable, 86—88
    subtending hyphae observable, 79—86
Nurse plants, 45
Nursery beds, 213, 217
Nutrient content of soil, 40—42
Nutrient film culture, 193—194
Nutrient film technique (NFT), 207
Nutritional effects on infection, 122
Nutrition of plants, 114

O

Oomycetes, 58
Open pot cultures, 114
Optimum yields, 210
Organic matter, 104, 109
Organic phosphates, 160—161
Organic phosphorus, 157
Ornamental trees, 215
Osmiophilic fibrillar layer of matrix material, 10, 12
Osmiophilic walls, 14—15
Osmotic shock, 97
Oxaloacetic acid, 105
Oxygen, 108, 169

P

Parasitic microflora, 144—145, 196
Parent hypha, 68, 91
Parent spore, 106
Passive dissemination, 37—39
Pastoral plant species, 209, 215—216
Pasture establishment, 216
Pathogen-free VAM inoculum, production of, see Inoculum production
Pattern of branching, SEM analysis of, 16
Peasant farms, 217
Peg, 91
Pelleting, 208
Penicillum funiculosum, 136
Peridium, 91
Permanent pasture, 215, 218—219
Pests, freedom from, 217
Peuraria phaseoloides, 133
pH
    soil culture, 192
    spore germination, 41
    survival of VAM fungi, 48
Phenol metabolism, 178
Phenols, 143
Phosphatase, 160, 163
    root surface, 100
Phosphate, enhanced uptake of, 120—121
Phosphate-fixing soils, 160
Phosphate-solubilizing bacteria, 135—136, 161
Phosphobacteria, 135
Phospholipids, 169

Phosphorus, 103
    cycling, 206
    deficiency, 163
    depletion zones, 157—158
    effect on mycorrhizal infection, 162—164
    flux rates in fungal hyphae, 161
    flux values, 162
    fungal infection in bryophytes, 24
    granules of mycorrhizae, 17
    inflow, 162, 176
    nutrition, 156—157, 163, 165, 172, 174, 176—177
    root colonization and sporulation, 45
    soil culture, 190—191
    specific activity of, 120
    survival of VAM fungi, 48
    toxicity, 123
    transfer, 166—168
    transport mechanism, 22, 159
    uptake, see Phosphorus, utilization of
Phosphorus, utilization of, 157—162
    inhibition of mycorrhizal infection, 163
    storage, 161
    transfer to host, 162
    translocation, 161
    uptake, 173—175, 206, 211—213
        carrier-mediated active transport mechanism, 162
        insoluble P sources, 159—161
        kinetics, 159
        mechanisms, 158—161
        rates, 158
        soil solution, 157—159
Phosphorus-tolerant fungi, 164
Photoperiod, 167
Photosynthate, 134
Phylyctochytrium sp., 145
Physiological relationships, 28
Physiology of VAM, 219
Physiology of VAM endophytes, 206
Physiology of VA mycorrhizal associations, 155—186
    carbon, see also Carbon physiology and biochemistry, 166—173
    hormonal effects, 173—174
    mineral nutrition, see also Mineral nutrition, 157—166
    problems causing difficulty, 156—157
    water relations, see also Water relations; Water transport, 174—177
Phytates, 100, 161
Phythium ultimum, 138
Phytohormone levels, 100
Phytophthora
    megasperma var. *sojae*, 138
    palmivora, 138
    parasitica, 138
    sp., 217
Plant growth
    assessment of infection, 119—120
    biomass of VAM fungi, 122

comparison of fungi, 121
control treatments, 114—116
depression, 2, 122—123, 126, 171—173, 177—178, 213
effectiveness, 123—124
effect of different fungal species and strains on, 123—125
effect of VA mycorrhizae on, 113—130
enhanced nutrient uptake, 120—121
hormone levels, 123
impure inocula, 114
increase in, 206
ineffectivity, differences in, 124
inoculum contaminants, 115
inoculum production, 196—197
light intensity, 122
non-nutritional effects, 122—123, 126
open pot cultures, 114
other nutrients, 121
partially sterilized soils, 122
phosphate, 120—121
phosphorus toxicity, 123
propagule number, 124
response curves, 114, 116—119
rhizosphere effect, 132
root temperature, 122
sequential harvesting, 114, 119—120
soil sterilization, 114—115
soluble carbohydrates, 122
stimulation, 123, 126
suitable controls, 114—116
water uptake, 123
Plant growth responses
field inoculation with VA mycorrhizal fungi, see also Field inoculation with VA mycorrhizal fungi, 209—216
inoculation with VAM fungi leading to, 2
measurement and prediction of, 1
Plant hormones, 134
Plant/mycorrhizal fungus interaction, complexity of, 7
Plant nutrients, rhizosphere effect, 132
Plant pathogenic organisms, 137—143
Plant viruses, 139
Plasmodesmata, 20
Plastid morphology, 20
Plug, 70, 91
Polyols, 167
Polyphosphatase, 17, 162
Polyphosphate, 158, 161, 162, 168
Polyphosphate kinase, 161—162, 168
Polysaccharides
alkali-insoluble, 8, 10, 17
cytochemical tests revealing, 17
hyphal wall formation, 8, 10
interfacial material, 22
VAM endophytes, 14—15
Polyvinyl lactic acid, 91
Polyvinyl lactophenol, 91
Pore, 91
Potassium metabisulfite, 110

Potassium nutrition, 165—166
Potassium sulfite, 110
Pot culture, parasitized, 145
Potting mix, 208
Predatory microflora, 145—146, 196
Preferential association, root colonization and sporulation, 46
Preinoculation, 207, 210, 212, 214
Proflavine hemisulfate, 103
Propagule number, 124
Propagules, 46
Proteins
alkali-insoluble, 8, 10
cytochemical tests revealing, 17
hyphal wall formation, 8, 10
VAM endophytes, 14
Protein synthesis, 103
Protoplasm, 20
Pseudomonas
solanacearum, 139
syringae, 139
Pteridophytes, 6, 24—26, 28
Pure culture, 2, 173, 207
Pure culture of VA mycorrhizae, 96, 103—109
failure of growth, reasons for, 106—108
growth from resting spores, 104—106
metabolic capabilities of VAM fungi, 103—104
other possible approaches, 108—109
progress toward, 104—109
requirement for, 103
Pyruvic acid, 105

R

Radophilus similis, 142
Rarely sporing fungi, 59
Reducing sugars, 163, 166
Respiration, 170—171
Response curves, 114, 116—119
Resting spores, 67—68
bacterium-like organelles, 144
pure culture of VAM fungi from, 104—106
surface-sterilization, 97—98
Reticulate spore, 91
Rhizobium
meliloti, 212
solani, 139
sp., 133, 134, 212
Rhizosphere, 146
increase in free-living nitrogen-fixing bacteria, 135
plant growth, 132
plant hormone production, 134
plant nutrients, 132
VAM fungi as components of, 132
Rhizosphere microflora, 132
Rhizoplane bacteria, 136
RNA, 103—104
Rock phosphate, 120, 133, 160
Rodent mycophagy, 37—38

Role of VAM in nutrition of plants, see also Plant growth, 114, 121
Root, 28
Root colonization, 42—46
Root exudates
   interaction of VAM fungi and plant pathogenic organisms, 142—143
   light intensity, 163
   promotion of growth from spores, 105
   reducing sugar content, 166
   root colonization and sporulation, 45
   spore germination, 39, 42
   stimulation of growth of VAM fungi, 100
   survival of VAM fungi, 48
Root growth, 215
Root hair depletion zones, 158
Root hairs, 7, 121, 175
Rootings of cuttings, 173
Root infection, 156
Root-knot nematodes, 140—141
Root morphology, 176
Root organ cultures, 100
Root pathogenic fungi, 1, 143, 178
   interaction with VAM fungi, 137—139
   suppression by VAM mechanisms, 142—143
Root penetraton, 9—10
Root rot of citrus, 138
Root space, 209
Root temperature, 122
Rust fungi, 103, 109

S

Salinity stress, 177
Saprophytes, 104
Scanning electron microscopy (SEM)
   extramatrical fungal hyphae observed in, 8, 10
   identification of Endogonaceae, 69
   pattern of branching observed, 16
Sclerocystis
   coremioides, 64
   dussii, 58, 64, 68
   indicus, 72
   sinuosa, 65
   sp., 162, 68
Sclerotium rolfsii, 139
Season, survival of VAM fungi, 50
Sectioning spores, 62
Sedentary endoparasites, 140—141
Seed furrow, 209
Seed inoculation, 212
Self-propelled cone seeder, 218
Septum, 17, 19, 91
Sequential harvesting, 114, 119—120
Simple subtending hypha, 91
Size of roots, 7
Sodium nitrate, 105
Sodium nutrition, 165—166
Soil animals, 48
Soil-borne vesicles, 66, 70—71

Soil condition, 48
Soil culture, 188—193
Soil fertility, 44—45, 48
Soil microflora, 39—40
Soil organisms, 48—50
Soil phosphorus, 157—162
Soil-root interface, 175
Soil sieving, 59
Soil solution P, 212
Soil sterilization, 114—115
Soil water potential, 40
Soluble carbohydrates, 122, 163, 166—168
Soluble inorganic P, 157
Solution P levels, 213
Sorbitol, 167
Spontaneous germination, 50
Sporangia, 62—63
Sporangioles, 14
Sporangiospores, 68, 91
Sporangium, 91
Spore attachment, 91
Spore damage, 62
Spore death, 50
Spore formation, see also specific genera, 59
Spore germination, see also Sporocarps and spores, 39—42
Spore production, see also Root colonization, 42—46
Spores, see Sporocarps and spores
Sporocarpic species, keys to, 72—79
Sporocarps and spores, see also specific types
   anatomy, 63
   axenically maintained cultures, 103
   collection and extraction, 59—62
   definitions, 91
   embedding spores, 62
   fixing spores, 62
   fungal infection in angiosperms, 7
   germination of spores, see also Spore germination, 63—65, 68—69, 71, 104
   isolation of VAM fungi from soil, 96
   sectioning spores, 62
   storage, 62, 97
   variability in spores, 69
Sporulation, see Root colonization
Springtails, 49
Starch, 167
Starter inoculant, 217
Stems
   Glomus tenue, 24
   mycorrhizal infection, 19
Sterilant, 97—98
Sterilization, stunting effects of, 214
Sterilized inoculum soil, 208, 218
   experiments in pot trials, 1
Sterilized media, 206
Sterols, 170, 176
Storage
   carbon compounds, 167—168
   phosphorus, 161
   sporocarps and spores, 62, 97

VAM inoculum, 196
Storage sinks, 168
Streptomyces cinnamomeus, 137
Striated spore wall, 91
Stunting effects of sterilization, 214
Stylosanthes guyanensis, 133
Subapical septum, 17—18
Subtending hypha, 63—64, 68—71
    defined, 91
    keys, 79—88
Sucrose, 105, 167—168
Sugar transport systems, 168
Sulfur, 100, 121
Superphosphate, 133
Surface-sterilization of resting spores, 97—103
Survival of VAM fungi, 46—50
Survival rate, 215
Symbiosis principles of VAM inoculum production, 196—199
Symbiotic theory, 29
Synergistic interaction, 135—136

T

Take-all of wheat, 138, 142
Tartaric acid, 105
Taxonomic classification of VA mycorrhizae, 96, 100
Taxonomy of VA mycorrhizal fungi, see alo Endogonaceae, 57—94
    Acaulospora, 67—68
    collection, 58—62
    Complexipes, 69
    describing, 69—72
    Endogone, 63
    Entrophosphora, 68
    Gigaspora, 63—66
    Glaziella, 68
    Glomus, 68
    glossary, 89—92
    identifying, 69—72
    incidence, 58—62
    keys
        genera, 88—89
        nonsporocarpic species, subtending hyphae not observable, 86—88
        nonsporocarpic species, subtending hyphae observable, 79—86
        sporocarpic species, 72—79
    Modicella, 68—69
    Sclerocystis, 68
    sectioning, 62
    storage, 58—62
TEM, see Transmission electron microscopy
Temperature
    root colonization and sporulation, 44
    soil culture, 192—193
    spore germination, 42
Thiamin, 105
Thick-walled hyphae, 8

Thielaviopsis basicola, 138
Thin-walled hyphae, 8
Third World countries, 217
Timber trees, 214—215
Tissue culture, 109
Tobacco mosaic virus, 139
Tomato mosaic virus, 139
Trace elements, 134
Tractor-drawn seed drill, 218
Transfer of phosphorus from fungus to host, 162
Translocation
    control, 173
    fungal, 161
    hyphal, 161—162, 164, 166
    phosphorus, 161
Transmission electron microscopy (TEM)
    arbuscular hyphae, 16
    Glomus caledonicum, 145
    host nuclei in close association with hyphae, 9, 12
    identification of Endogonaceae, 69
    intercellular hyphae, 12, 14—16
    sectioning spores, 62
Transplanted field crops, 208
Transplanted preinoculated seedlings, 212
Transplanted seedlings, 216
Transplanting, preinoculation before, 207
Transport, see also Water transport
    carbon compounds, 167—168
Tree crops, 216—217
Trehalose, 105, 167
Tricalcium phosphate, 120, 160
Trifolium
    parviflorum, 100—103
    repens, 133
    sp., 98—99
Triglycerides, 14
Tripartite associations, 137
Tropical legumes, 133
Trumpet-shaped hyphae, 92
Two membered cultures, 59

U

Ultrathin frozen sections, 17—18
Unsterilized field soils, 209, 214—215
UV microscopy, 157

V

VA endophytes, see Endophytes
VA mycorrhizae, see also VAM topics
    agricultural plant yield, use to increase in, 36
    anatomy, see also Anatomy of VAM, 5—33
    ancient symbiotic event, 29
    biological interactions with, see also Biological interactions with VAM fungi, 131—153
    cytological changes in root, 28
    cytological organizations, 28

ecology, 35—55, 219
field inoculation with, see also Field inoculation with VAM fungi, 205—222
functional criteria, 26—28
fungal structures, presence of, 28
historical aspects, 6
host root cell as safe constant habitat for, 29
identification, 27—28
identity of symbiotic fungus, 27
isolation from soil, 96—97
metabolic capabilities of, 103—104
morphogenesis, 28
morphological criteria, 26—28
morphology, see also Morphology of VAM, 5—33
most common association, 1—2
physiological relationships, 28
physiology, 219
plant growth, effect on, see also Plant growth, 113—130
pure culture, 96, 103—109
recognition between plants and, 28
role in plant nutrition, 114
root hairs, 7
size of roots, 7
species and strains, effect on plant growth, 123—125
taxonomic classification, 96, 100
taxonomy, see also Taxonomy of VAM, 57—94
VA mycorrhizal associations, 28—29
  physiology of, see Physiology of VA mycorrhizal associations
Vacuolar system, 14
Vacuolate spore, 92
Vacuoles, 17, 162
VAM, see VA mycorrhizae
VAM cells, 20
VAM fungi, see VA mycorrhizae
VAM inoculation, see Field inoculation with VA mycorrhizal fungi
VAM interactions, 22
VAM research funding, 206
VAM starter cultures, 217
VAM symbiosis, 1—2, 6, 156
VAM systems, 28
Vectors, dispersal by, 38
Verticillium dahliae, 138
Vesicles
  carbohydrate storage organ, 167
  chitinous walls, 169
  defined, 17, 92
  fungal infection in angiosperms, 8, 17—19
  fungal structures, 1
  glycolysis, 169
  lipid droplets, 167
  oval terminal swellings, 6—7
  role of, 19
Vesicular-arbuscular mycorrhizae (VAM), see VA mycorrhizae
Vigna unguiculata, 133
Viruses, 139
Vitamin-producing bacteria, 136
Vitamins, 173
Vitis vinifera roots, 8, 10—12

W

Wall-to-wall contact, host-fungus relationship, 21
Water absorption, 170
Water-aeration in soil culture, 192
Water deficit, 177
Water depletion zones, 176
Water flow rates, 175
Water relations, 174—177
Water stress, 177
Water transport, 174—177
Water uptake, 123
Wet-sieving method, 96, 207, 209
Wheeled tool carriers, 217
Wind, dispersal by, 38—39

X

X-ray analysis of vacuoles, 17

Y

Yeast extract, 105

Z

Zinc, 100, 121, 134
Zinc nutrition, 165—166
Zygomycetes, 58
Zygospores, 62—63, 92

SETON HALL UNIVERSITY
QK604 .V25 1984
VA mycorrhiza /
MAIN

3 3073 00256070 2

DATE DUE

APR 1 5 1994

GAYLORD — PRINTED IN U.S.A.